Roy Wanda 486th

Glenn A. Burshman
486TH FTR. SQDN. 352ND FTR. GRP.,

Bill Straub 487th

Howard Polin - Weather Staff.

Ralph P. McLain 486 F.S.

Bob Landrum
SON of Jack MOOSE LANDRUM 487TH

James E. Woodley 487th SQDN 352 FTR Grp

Duane Palmquist

Richard: learn all you can about your grandfather and follow his example James Bleechner
Armorer 352nd/487th

Sept 2007 Columbus

Bob "Punchy" Powell

BLUENOSER TALES

352ND FIGHTER GROUP WAR STORIES

EDITED BY

ROBERT H. "PUNCHY" POWELL

352ND EDITOR / HISTORIAN / CO-AUTHOR / PILOT

MARC L. HAMEL

352ND ASSOC HISTORIAN & CO-AUTHOR

SAMUEL L. SOX, JR.

352ND PHOTO ARCHIVIST & CO-AUTHOR

Best wishes and
God Bless,
Marc Hamel

FRONT COVER DESIGN BY SUSAN WARD

FEATURING A DETAIL OF THE PAINTING

"SERENADE IN BLUE"

FIRST PUBLISHED IN 2007 AND DISTRIBUTED BY
UNITED WRITERS PRESS, INC.
P.O. BOX 326
TUCKER, GEORGIA 30085
1-866-857-4678 * WWW.UNITEDWRITERSPRESS.COM

THIS PUBLICATION IS AVAILABLE FOR PURCHASE AT A DISCOUNT
IN BULK QUANTITY FOR SALES-PROMOTIONAL USE.
FOR DETAILS, CONTACT ~~THE~~ PUBLISHER AT THE ADDRESS ABOVE.

ISBN-13: 978-1-934216-35-4 (TRADE PAPER)
ISBN: 1-934216-35-6 (TRADE PAPER)

ISBN-13: 978-1-934216-38-5 (CASEBOUND)
ISBN: 1-934216-38-0 (CASEBOUND)

LIBRARY OF CONGRESS CONTROL NUMBER: 2007905176

PRINTED IN CANADA

ILLUSTRATIONS

The more than 400 wartime photos and sketches included in this book are from the 352nd Photo Archives, which now includes more than 8,000 wartime pictures collected from 352nd veterans and their families. These photos are maintained and enhanced by our highly skilled associate, Samuel L. Sox, Jr., with assistance from his friend, Richard Gose, and we are most appreciative to all those who have shared their photos with us, and particularly grateful to Sam Sox for his dedication to their restoration and his personal contribution to perpetuating the history of the 352nd Fighter Group and 1st Service Group.

Our gratitude, also, to USSTRATCOM Public Affairs for permission to reproduce the painting of Gen. John C. Meyer, which hangs in the Pentagon, and to aviation artists Troy White, Nick Trudgian, John Doughty, Charles Taylor, and Robert Bailey for permission to print their wartime action paintings. Limited edition prints of all showcased paintings are available for purchase at fine galleries and via the internet.

PREVIOUSLY PUBLISHED MATERIALS

Our thanks to Michael O'Leary, associate publisher and editor of *Air Classics®* magazine for his cooperation in allowing us to reprint excerpts and articles from past editions of the publication.

Throughout ***Bluenoser Tales***, in the descriptions and interviews with pilots and crew, abbreviations and terms are used, many of which may be unfamiliar to the reader new to aviation during World War II. A listing of aircraft frequently flown by American, British, and German forces, as well as a legend for those abbreviations are provided below to enhance the reader's understanding and enjoyment.

AMERICAN AIRCRAFT	BRITISH AIRCRAFT	GERMAN AIRCRAFT
B-17—Flying Fortress	Hurricane	Ar—Arado
B-24—Liberator	Mosquito	Do—Donier
B-25—Mitchell	Spitfire	Fw—Focke Wulf
P-38—Lightning	Typhoon	He—Heinkel
P-39—Airacobra		Me—Messerschmitt
P-40—Warhawk		
P-47—Thunderbolt		
P-51—Mustang		

AM	Air Medal
DFC	Distinguished Flying Cross
DSC	Distinguished Service Cross
OLC	Oak Leaf Cluster (additional award of medal)
A/C or a/c	Aircraft
A/D or a/d	Airfield or Airdrome
AF	Air Force
API	Armor-piercing incendiary ammo
BC	Bomber Command
C.O.	Commanding Officer
C/C	Crew Chief
E/A or e/a	Enemy Aircraft
ETO	European Theatre of Operations
MTO	Mediteranean Theatre of Operations
GAF	German Air Force (Luftwaffe)
PTO	Pacific Theatre of Operations
FC	Fighter Command
FS	Fighter Squadron
FG	Fighter Group
F.O.	Mission Field Order
F/O	Flight Officer rank
G.I.	A U.S. soldier (Govt. issue)
I/O or S-2	Intelligence Officer
IP	Initial Point of a bomb run
KIA	Killed in action
MIA	Missing in action
M/Y	Railroad Marshaling Yard
OTU	Operational Training Unit
PIF	Pilot Information File
POW	Prisoner of war
PRO	Public Relations Officer
R/T	Radio Telephone
RTU	Replacement Training Unit
R/V	Rendezvous
SOP	Standard Operating Procedure
T/O	Takeoff
Type 16 Control	Radar controlled mission
VHF	Very High Frequency Radio
WIA	Wounded in action
Abort	Return to base (a/c problem)
Bandit	Enemy aircraft
Bogie	Unidentified aircraft
Jackpot	Planned strafing mission
Ramrod	Bomber escort mission
Rhubarb	Strafing sweep
Rodeo	Bomber Baited Mission
Element	Two aircraft (two to a Flight)
Flight	Four aircraft (two to a Section)
Section	Eight a/c (two to a Squadron)
Squadron	Sixteen a/c (three to a Group)
Group	Forty-eight a/c (three squadrons)

Contents

SHORT SNORTERS*

***A "short snorter" was a bond of friendship amongst the crewmembers or comrades in arms. The physical "symbol" was often paper money signed by two or more men and then separated (torn) so that when all were together again they could combine the pieces and buy a drink! (Woe to you, though, if you were caught without yours in hand...you had to pay the whole tab.) It has also been defined as consisting of a roll of bills, each from a different man and/or place, all attached one to the next. Here we use the term to describe short stories and vignettes inserted throughout to demonstrate the humanity of these brave young men and reflect those very bonds the original "short snorters" were intended to celebrate.**

FOREWORD

By Col. Stephen R. Hicks, USAF (Ret.)

As kids, most of us build a galaxy of fantasy heroes to imitate, often super-human figures who perform super-human feats, heroes like Superman and Batman, who don't really exist, and then we move on to a new set of heroes – teachers, sports stars, astronauts or Presidents. I probably wasn't any different from most other youngsters and it was years later before I figured out who my real heroes were.

In the late 90's I took over command of the 192nd Fighter Wing of the Virginia Air National Guard flying F-16 Fighting Falcons. Our military lineage was traced to a squadron of the 352nd Fighter Group, a highly-decorated World War II fighter unit. When we learned that the 352nd was planning its final reunion, we thought it would be fitting to host that reunion in Richmond, Va., where we were based.

During the next two years I would learn more than I could ever imagine about the real-life heroes who wore the insignia of the 352nd FG. I would meet many of them, listen to their stories, and marvel at their deeds. They were just ordinary Americans who were called upon to accomplish a task worthy of any superheroes – saving the world.

When the 352nd was formed in early 1942, the forces of Hitler, Mussolini and Hirohito seemed unstoppable. No one could say for sure that they could be stopped, or at what price. America was desperate for victories.

After completing a brief period of training, the 352nd was dispatched to a small airfield in Bodney, England as part of the Eighth Air Force. From there they would fly and fight over the Continent and later move onto the Continent to engage the much more battle-experienced Luftwaffe. They were flying P-47 Thunderbolts and later P-51 Mustangs, the latter carrying a distinctive blue nose. Legend has it that a German general said, when he first saw the "bluenosed bastards of Bodney" over Berlin, that he knew that the war was lost, giving the 352nd its wartime nickname and a point of pride.

The 352nd recorded some impressive statistics – 420 combat missions, 115,675 sorties, 59,387 combat hours and 776 enemy aircraft destroyed. The Group had 29 aces, and was awarded the Distinguished Unit Citation twice and the French Croix de Guerre with Palm. But statistics don't tell the whole story.

In Bluenoser Tales, *352nd veteran Bob "Punchy" Powell along with associates Marc Hamel and Sam Sox, Jr. use the personal accounts of the "bluenosed bastards" to record a remarkable tale. They have weaved together a book full of new and legendary material and hundreds of wartime photos to take the reader through the training, the expectations, the challenges and the thrill of victory in aerial combat and ground support actions. And, yes, the stories of their friends who never made it back.*

Among these fascinating stories you will find the combat stories of Major George E. Preddy, Jr., commander of the 328th Squadron whose honors and lineage the Virginia Air Guard carries today. He has become one of my personal heroes. On Christmas Day, 1944, Major Preddy, while leading his 328th pilots on a patrol from an advance base in Belgium, downed two enemy fighters and was chasing a third at treetop level when he was hit by friendly groundfire and killed after becoming the leading Mustang ace of WWII. To commemorate his life and achievements, the 192nd Fighter Wing dedicated its operations building in his honor.

To the veterans of the 352nd FG and the others like them we owe a great debt. Because of their commitment, sacrifices and love, we live in a safer world today. As a commander and fighter pilot, I can only hope that the 352nd's legacy of dedication to duty and devotion to each other will live forever in our hearts and souls. By proudly accepting the military honors that the 352nd Fighter Group earned, the men and women of the 192nd Fighter Wing realize this distinction comes with an awesome responsibility – to defend our nation at any cost.

In our external search for heroes, we didn't have to look far. We have met them and talked with them and we vow to do our duty to emulate them – to preserve America's freedoms that they fought for so valiantly and won. We pray that we will be worthy of the sacrifices they made.

Colonel Stephen R. "Steve" Hicks

Star Over the Phillippines
Pursuit Ace I.B. "Jack" Donalson
by Jack Donalson with Marc Hamel

December 9, 1941. The Philippines....

"I continued my patrol until my fuel was low, and headed for Nichols Field in Manila. I switched my radio to the control tower frequency and there was a lot of chatter about the field being attacked by Zeroes. A few minutes later I saw one heading north at a very low altitude. Apparently he was one of those that had strafed the field and had not pulled up from his pass. I did a 180 degree turn, had plenty of altitude, and was in a good position to get him. After a short burst in the cockpit area, he hit the ground. I pulled up to regain my altitude for fear there were other Zeroes in the area, and continued to the vicinity of the airbase. After a few minutes I got clearance to land."

If you had passed I.B. "Jack" Donalson on the street in his later years, you might have guessed he was a retired actor. However, USAAF ace Donalson flew pursuit aircraft in the Pacific, scored three victories on the first two days of the war, fought the Japanese as an infantry leader, and became an ace against appalling odds in the Philippines and Australia. Jack went on to a mastermind the highest scoring squadron to fly in the ETO, and a long and successful career in the USAF. A very modest man, Jack focused here on how he helped others rather than his personal exploits. Here is his story.

Jack Donalson with one of his P-40s named "Mauree" in Australia

Jack was born on July 6, 1915 on his family's cattle ranch near Kyle, Texas. After several moves during his youth, he graduated from Tulsa Central High School in the midst of the Great Depression in 1933, and went to work for the railroad. Playing baseball and basketball for a local oil company–sponsored team led Jack to a job with Texaco. Spotted by a basketball scout, Jack accepted a scholarship to the University of Tulsa in 1937. Here he excelled, and later discovered the Civilian Pilot Training Program; learning to fly in 1940. This brought dreams of flying for the Air Corps, so Jack applied to the flying Cadet Program. Working hard, he was commissioned 2nd Lieutenant I.B. Jack Donalson on August 15th, 1941 at Kelly Field, Class 41-F.

Ordered to Hamilton Field near San Francisco, Jack had pressing business to take care of on his 30 day leave. "I immediately left for Tulsa, Oklahoma. At noon the next day, my college sweetheart and I were married.

"We had a 30 day leave and reported for duty with the 21st Pursuit Squadron on September 15th, 1941. On November 1st, we departed for the Philippines and arrived November 20th. Eighteen days before Japan started the war."

Air War, The Philippines

"During the next two weeks we had to find a place to live, set up an area at Nichols Field for our airplanes, supply, and maintenance. Initially we were given three P-35's for the purpose of pilot familiarization with the other airfields in the area. Our P-40's were still in crates that had been shipped from the U.S.

"On the day the Japs attacked (December 8th, December 7th in Hawaii) I was on alert sitting under the wing of my P-40 waiting for something to happen. Suddenly, a sergeant came running from operations and said the radio just reported the Japs had hit Hawaii. Our C.O. Ed Dyess told Bob Clark and I to take off and patrol over the Manila area.

"As we entered our patrol area over Manila, we heard over the radio that Clark Field was being attacked. The radio was so full of chatter it was impossible to decide what to do, so we headed for Clark Field about 50 miles north of Manila.

"Soon up ahead we saw what looked like a swarm of bees. Fighters, bombers, and dive-bombers over Clark Field, which was covered with smoke and fire. We became very busy looking for airplanes, the mess on the base, and watching for our own tails.

"Suddenly, off to my left and about 200 feet below, I saw a formation of bombers going south. The clouds were broken so I guess they did not see me. I knew I had to act or the opportunity would be lost. I just peeled off and headed for the bombers. There was no use trying to radio anyone as the air was full of excited chatter anyway. I was in the perfect position for the attack. I made a turn onto the tail of the last bomber, closed in, and fired. From the six o'clock position I fired at his tail, which of course meant the bullets went on up into his fuselage and engines too. He lit up with fire and smoke, and went down to the left, burning and coming apart. I continued on to the next one and gave him the treatment…with the same result. He also began to burn and fell off to the side.

"About this time, I got to thinking, 'I wonder where those Zeroes are?' I looked around and here

Colonel Ed Dyess, Jack's Commanding Officer in the Philippines

they are, ready to get me. I stuck the nose down and headed straight down at full throttle for the ground. I pulled away from them, which was easy to do with a P-40. When I leveled off they were not around, so I guess I outran them or they decided to stay with the bombers. I was then low on fuel so I landed at LUBAO Field. It was recently prepared for emergency use only. After refueling, I returned to Nichols Field in Manila.

"With regards to those two planes I shot down, they were definitely bombers. Some authors have gotten it wrong in the past—they just made a mistake. They were Betty bombers. I was

John Landers (later 78th FG CO) and Jack Donalson are at left in this 49th FG photo (Jack Donalson)

too excited to remember what color they were or anything, but they were definitely not Zeroes.

"Later that evening our CO got six of us together and said we were leaving for Clark Field because the 20th Squadron had lost all of their P-40s during the raid that day, and we were to replace them.

"We took off about five p.m. and arrived over Clark Field a few minutes later. Ed Dyess called for landing instructions but got no reply. We circled and finally decided the best way to land and made it with no trouble.

"As we taxied towards Base Operations, we could not believe our eyes. B-17's and P-40's were burned where they stood. Dead bodies on the ground and dead pilots in their airplanes. Bomb craters all over the field and all of the buildings completely destroyed.

"There was no one to guide us or tell us where to park. After parking we all walked towards Ops and finally saw one soldier. We asked him where the people were and he said they all left and he was to guard the place. He said he thought there were some people down "that way," pointing towards the jungle.

"We walked towards the area about a mile and saw another soldier who told us to follow a path down through the jungle and there would be some people there. We found a captain and two enlisted men. The captain said he was in contact by radio with Air Force Operations and we were to take off in the morning before daylight and patrol the area. The Japs would probably arrive early to hit Manila.

"We slept on the ground, got up before daylight, and were going to meet the enemy at 20,000 feet. Ed took off first and made it okay. Sam Grashio waited for the dust to settle and took off second. He hit something, but got into the air. He called and said he was having trouble controlling his plane. Ed said he would escort him to LUBAO for an emergency landing. Then Ben Irvin took off and made it, but Bob Ibold called in and said he had taxied into a bomb crater and was hurt. Then my flight leader Bob Clark started his takeoff. I watched him closely to get the proper compass heading. I could follow his exhaust flames for a few seconds, but then he disappeared into the dust and darkness. Then a huge ball of fire appeared. Bob had hit a parked B-17, and what a terrible sight it was.

"I gulped a few times, got control of myself, turned my heading 20 degrees to the right, and firewalled it. The Old Man above took over and I made it. Ed had gone to LUBAO with Sam, and Ben had gone to go get an ambulance for Ibold. I was left to patrol the area over Clark."

Jack then encountered a Zero and its unfortunate pilot, and that narrative opens this article. With that action, Jack accomplished what no other pilot did in World War II—victories over the enemy in the first two consecutive days of the war.

"After the first two days, a lot of time was spent trying to sort out what was left of our five squadrons. None of the squadrons were able to continue to operate as a unit because we had lost most of our aircraft. So it was decided to pool our aircraft, and the most senior and experienced pilots would continue flying the missions.

"Those of us not included in the pool were assigned to other duties. My duty was to move our squadron to living quarters at LaSalle University, in downtown Manila. At about five p.m. on Christmas Eve I was at LaSalle and a message came over the loudspeaker system. 'Tomorrow Manila will be declared and open city and be subjected to bombing raids. Their ground forces are approaching the city rapidly. All military personnel report to the shipping docks for evacuation from the city.'"

Bataan

"I quickly packed a small bag with personal items, jumped into my car and drove to the dock. Police were directing people to a ship tied up at the dock, and I ended up on what appeared to be a commercial fishing boat. I asked where we were going and someone said Bataan, though I had no idea what or where Bataan was. When loaded the ship headed west across Manila Bay, and the next morning we arrived at Marvielles on Bataan Peninsula. It was a U.S. Navy submarine base. Near the dock there was a one-lane dirt road going east-west, but I had no idea which way to start walking.

"After asking several people which way to go, I gave up and started walking east. Cars, trucks and people filled the road. About noon I was tired and thirsty and hungry, and saw an American soldier standing beside the road. After asking where I could find something to eat, he took me to where two of his buddies were sitting by a fire with a coffeepot on it. They gave me some coffee, two slices of bread and some meat, and this was my Christmas dinner. I consumed it as if it were a turkey dinner.

"The soldiers said they thought there was an airfield about five miles down the road, so I thanked them and continued down the road. After a few minutes I saw a little pine tree decorated with DelMonte tomato can wrappers to make it look like Christmas. A few tears ran down my cheeks as I thought about my wife who did not know where I was, and that I might never see her again. What a miserable Christmas day.

"About three o'clock I saw a field ahead, which turned out to be Cabcaben. It had a dirt strip for a landing area, and a few planes off in the distance. I asked a soldier for directions to the 21st Squadron; he said they were not there, but there was Bataan Field three or four miles down the road. I continued to walk and about five o'clock I saw Bataan Field. I asked for the 21st and a soldier said to follow this little road through the jungle to the end and there would be some pilots there who would know. I walked as fast as I could and came to a clearing with a bamboo shack and a lot of pilots standing around. I saw Ed Dyess and Sam Grashio in the group and they saw me about the same time. We had a happy reunion as they both thought I had been killed. I went with them to their shack, and they invited me to live with them. They had flown P-40s to Bataan and wanted to know how I got there. They also explained we only had nine P-40's, two P-35's, a Beechcraft, and a Bellanca left. The two latter civilian planes would shuttle between Bataan, Cebu, Slo Slo, and Mindanao transporting mail, medicine, food and personnel.

"So, we had five squadrons worth of pilots (about 100), 2000 support personnel, and only nine P-40's to fly. Only a small pool of people would do the flying as well. As they do in the military, the senior people would do the flying and the rest of us would be assigned to the infantry to defend Bataan."

Nightmare in the Jungle

"A few days later the Japs did what we expected and landed a force of about 600 men at Agoloma Bay, about 10 miles west of Marvielles. Records uncovered later stated that they were a battle-hardened unit with extensive experience in the China War. They were well trained. The Japs moved the force into the area at night from warships, which fired cannons, rockets, and machine guns to cover the landing. The Philippines Army unit which was defending the area called for help. Our squadron assembled four units of ten men led by an officer. We were each given a rifle and four hand grenades, loaded on a truck, and driven to the area. We were met by a major who briefed us from a map. He indicated where the Japs were, where our men were, and the four units that needed replacing.

"A Filipino soldier led us down through the jungle to the area we were assigned to cover. This was the beginning of 45 days of misery, fear, sickness, and desperation. I still dream and have nightmares about those days.

"As the soldier led us through the jungle, the smell of dead bodies was awful. We passed wounded men that were waiting to be helped. Two litter men came by with a wounded soldier and asked directions to medical help. I looked at the soldier and he had been hit in the jaw which had been blown away. I pointed in the direction they should go.

"When we arrived at the area, the guide pointed, shook hands, and left. The battle plan was to move forward when we could, clear our area, and keep our line from being penetrated by the Japs. It was only about 300 yards from our line to the ocean. The reason the mission took so long was that the Japs would reinforce their men at night by warship, firing over the heads of their men and shells would fall in our area. We would have to take cover in our foxholes.

"A plan was finally developed where two tanks would push forward into the Jap area. When they finished firing, they sounded their sirens and our line moved forward. I recall a few yards in front of us a Jap popped up from his hole, opened up with a machine gun, and we hit the ground. The grass was tall and I hugged the ground wondering what to do next. A few seconds later a man off to my left saw where the fire was coming from, tossed a grenade into the foxhole, and ended the problem.

"When I saw that happen, I got up and yelled, 'Let's go!' The men on my left and right did not move, and I tried to move them but they were dead. The noise from the firing was awful, but we continued forward. As we proceeded towards the cliffs

"Hey, 'Bataan!'"

Major I. B. "Jack" Donalson was an early war hero in the Pacific Theater along with Major George Preddy and later, as operations officer for the 487th Squadron, the strong right arm of Col. John C. Meyer. In fact, he was responsible for Major Preddy being assigned to the 487th FS and Meyer's keeping him when he had some doubts about that "skinny little guy."

In addition to setting up missions for the 487th, he led the squadron on many of its missions, particularly those in late 1943 and early 1944. His earlier combat experience and quiet leadership was one of the factors responsible for the great success of the 487th.

Stories about 'Jack' are hard to come by, but we managed to fo to the top, his wife, Mauree, to get a few stories about Jack. She found a few of these in his memorabilia.

During his flying days with the 49th Pursuit Squadron in Australia, Jack crashed his P-40 Kittyhawk, named 'Mauree' after his wife, into a tree. It happened on 16 July 1942 a few miles south of Livingstone airfield.

Then Lt. Donalson and Col. John Landers were up on an orientation flight since Jack had only recently joined the 49th after escaping when the Japanese captured the Phillipines. On the way back his engine quit and he was forced to land his plane on a single-lane dirt road. Jack's wing caught a tree as he landed on the narrow road. The plane caught fire and burned, but Donalson was unhurt. Sometime later a truck came by and picked him up and took him back to their base in the jungle. When he got back to his tent he found his pilot friends had eaten all of the cookies he had stored in his tent, thinking he wasn't coming back. Visiting after the war, he and John shared laughs about this incident.

When Jack got out of the Phillipines and had joined the Darwin group, he was nicknamed "Bataan." Whenever the Japs would come over he pooh-poohed everyone hitting their foxholes until one night they hit close to his camp. Who was the first to make it to their foxhole? You guessed it, It was 'Ol Bataan.' After that he was teased unmercifully, "Hey, 'Bataan,' the Japs are coming. Got your foxhole ready!"

over the ocean, we stepped over dead Japs, wounded Japs, and some of our own men. We finally reached the top of the cliffs after about an hour. We looked out at the ocean and saw about 50 more Japs swimming away, to God knows where. It is 500 miles to China. As I said, this force was an elite group, battle trained in China, and was not about to surrender. We quickly solved their dilemma with rifle practice.

"From where we were down to the beach was about two hundred feet. Only two trails led down, and they were steep. We tried going down that way, but heavy fire prevented this approach. They were heavily entrenched along the beach. We had Filipinos who spoke their language. They asked the Japanese to surrender, but the only reply we got was gunfire. They would not surrender!

"I was standing there at the edge of the cliff and saw Ed Dyess coming with a couple of our men. He said, 'Come with me, Jack, *we* are volunteering for a little mission.' I said, 'You are good at that, and now you want me to go with you?' We drove to the supply area and there were 20 men; ten for Ed and ten for me. The men were given rifles and hand grenades; Ed and I were given World War I machine guns with leather straps from the barrel to the stock so we could carry them. We also had asbestos gloves to protect our hands when the barrel got hot. The machine guns were fed from a huge drum on top.

"The next morning we would be towed in boats by the Navy from Marvielles to Agoloma Point and eliminate the Japs entrenched below the cliffs where we could not otherwise get them. We arrived at the base the next morning and the Navy had the boats ready. Our landing boats were towed up to the point about 200 yards offshore. Ed was to land on the east side, and I was to cover the west. The two Navy boats were going to provide heavy machine gun fire so we could cut loose and row in without resistance. As we approached the cut-loose point, I looked up and saw four Zeroes approaching. The Navy cut us loose so they could evade the Japs. Apparently the Jap ground troops radioed for help. We rowed as fast as we could and I was in the front of the boat firing my machine gun wherever I saw something that looked like it needed it! Fortunately, where we hit the beach there was a cave the Japs had been using so we took cover.

"The Zeroes were coming from the east and saw Ed's boat first, so they hit about the time his boat hit the beach. They scrambled for cover, but two of his men were killed and one wounded. When they finished with Ed's boat, they came up to my boat and strafed it, but we had taken cover and none of my men were hurt.

"When the airplanes left, we started east and Ed started west according to plan. As we progressed we found dead Japs every few yards.

Some had been dead for a long time and were in various stages of decomposing. I am sure some had committed suicide rather than surrender. I looked down at one who was lying between several dead ones. He had a large leaf covering his head and I started to pass him by, but noticed his stomach going up and down. I kicked the leaf and he was still alive so I gave him a burst and ended his problem.

"I began to hear Ed's men firing and knew we would soon meet and our 'little job' (as Ed called it) would be over. We met in about another 30 minutes, and followed one of the paths topside where we were greeted by a lot of happy people. That was the ending of 45 days which I wish I could forget, but it will never go away.

"On our way back to Bataan, Ed said, 'Jack, for what you have done, I'm putting you in for the Distinguished Service Cross and you are going to fly again. No more infantry duty.' At the time, I did not know what the Distinguished Service Cross was, and couldn't care less about medals. I was happy to be flying again.

"We went to Bataan Field and Ed's bamboo Shack, and my cot was still there and the place looked better than the Waldorf Astoria Hotel. Especially after 45 days in the jungle."

A Charmed Exodus

"The Agoloma Point job took care of January. During February, March, and through April 8th (the day Bataan fell) all flights were for delivering mail to Cebu, Mindanao, and Slo Slo, or for checking on reports of shipping in the area. The reason for this is that by April 8th, we were down to two P-40's.

"On the afternoon of April 8th, Randal Keator and I were on the alert in the cockpits of our two P-40's. We had been hearing rumors from the front line that the Japs were tearing up the line and may break through at any time.

"Shortly before dark I saw Ed Dyess coming up in his Jeep. He jumped up on the wing of my airplane and said, 'Jack, I think the Japs have broken through the line. Go up and check it out and if they have, come back over the field and wobble your wings. Then go to Cebu or Slo Slo and we will know this is the end and we can prepare for surrender.' I said, 'Ed, you know this is the end. *You* go because all of the higher ranking officers have been leaving since evacuating Manila.' He replied, 'I know it, but I am not leaving my men. Go, Jack, and God bless you.' What a man!! They don't make many like Ed.

"I took off, flew up the road, and the Japs were coming down and it looked like a long trail of men, tanks, and trucks. At the end of the line I turned around and started giving them the .50-caliber machine gun treatment. Then I turned and went back for another run. I saw a bridge over a little creek and hit it with my bombs. Then I went back to our field, made a low pass down the runway and wobbled my wings. In Ed's book, he said when he saw me wobble my wings, they started evacuating the base. It was pretty dark by then, so I got out my map and saw Slo Slo was closer than Cebu. I started navigating for the 400 mile trip. I had never been there before, and there are 7000 islands in the Philippines, some small and some large. It was very dark, so I knew I had better not make any mistakes or I would never make it. It was even more difficult to navigate because it was the time of year the Filipinos burn off their old rice field preparing for a new crop. I also had to stay below an overcast to follow compass headings to each island, and most were small.

"About 100 miles south I was following the coast of an island, and at about two or three hundred feet above the water in order to stay below this overcast. I was startled by bright lights a short distance directly in front of me. Before I knew it, I had flown over this very large ship, from one end to the other. I never told anyone about this until years after I retired from the Air Force because I thought they would never believe me. In 1974, I had a young student from Bob Jones University (Greenville, SC) call saying he was working on a study of the war in the Philippines. He came to visit me and I told him my story and about the ship. Three months later I received a letter saying the ship I flew over certainly was BIG! In fact it was the flagship of the Japanese Philippines theater of war. He had discovered the wartime log of the cruiser Kuma, and the photocopies of the log fit my memories. The ship's log mentioned the time they spotted a single enemy plane on that day. The cruiser carried seven 5.5 inch guns and 3 inch guns as well. It displaced 5000 tons over 500 feet, carried 400 men, and would do 35 knots. I still have this material in my files today.

"After the shock of passing over the ship, I tried to concentrate on navigation, fuel consumption, and time/distance. Finally as I approached the island that I thought was Panay, I continued down the west coast about 100 miles. When I reached the end of it, I turned east and then up the coast for 40 miles, which is where the airfield should be. The area was totally blacked-out, but I thought 'This is where the airfield has to be.' I was going to have to bail out soon as I was almost out of gas. I thought, 'Jazz the throttle, and if somebody is down there, they will turn on some lights.'

"Suddenly, the runway lights came on and I started my approach. The landing gear would not come down! I made one more circle, trying repeatedly to get the gear down, but no luck. So I landed on my belly.

"Captain Benny Putnam came out in his Jeep and I told him about Bataan and not to send the courier up there, which was what they were about to do. They pulled the P-40 off the runway, and we took off in the courier airplane, went to Cebu, and gave them the same message. The next morning a Filipino pilot reported a large number of warships approaching Cebu, so we loaded up again and took off for the base at Mindanao.

"When we got to DelMonte Field, Mindanao, we checked in with the unit there and told them the story. I met 21st Squadron senior pilot John Burns there who had been sent from Manila with a group of pilots headed for Australia to ferry back P-40's to the Philippines. The planes in Australia were never sent to replace ours because the Japs were planning to hit Australia as well. I bunked with John and he

This shot of I.B. Jack Donalson climbing from his P-47 gives some idea of the intensity and integrity he brought to his job as the 487th's Operations Officer. (Jack Donalson)

was scheduled to fly down to Davao, a seaport on the south coast of the island to report on shipping activity. He said, 'We always fly in pairs so we can support each other.' So they scheduled me to fly with him, with takeoff before daylight so we could arrive at Davao about dawn. Like Clark Field, it was again a large grass field with no lights, so we had to takeoff singly when the dust cleared and form up over the field.

"We taxied out, he started his takeoff, and I watched the exhaust flames, checked the compass heading, and just as he disappeared into the darkness there was another huge fireball. Fortunately I had enough sense to taxi back, get a firetruck, and go to the crash site down in a canyon.

"The next day we were told General Royce was coming from Australia with some bombers for the purpose of bombing Jap units in Cebu and Manila. They arrived about three days later. Three B-17's and ten B-25's, which made a couple of raids and had to return to Australia. They took as many pilots as they could, and I was lucky to be one of them.

"My airplane was a B-25, and I traveled in the crawl space between the nose gunner's cabin position and the pilot's cabin. We eventually got to Melbourne, Australia, and I reported to General George who had been with us for a brief time on Bataan.

"He greeted me like a lost son and wanted to know all about Ed Dyess and the surrender. He then told me to check into the hotel, get paid, get some clothes, and in three days I would leave with him for Darwin, Australia. The General's unit would be called the North West Territory Air Force.

"At this point, the only thing I had was the flying suit I was wearing. No clothes, no billfold, no identification. Just me. I was ready to go in three days, and reported to the base at the appointed time. General George was there with Joe Moore (a pilot), Mel Jacoby (a Life magazine correspondent), and an administrative officer. We took off at 7:30 in the C-40, Moore as pilot and I was copilot. We arrived at Darwin about four p.m., after a long flight up through the center of Australia. When we arrived we were instructed to park about 50 yards off of the edge of the runway, and a group of people came out to meet us. As we got out of the airplane, the General walked to the front of the airplane with Jacoby, while Joe and I remained near the door near the tail. I heard two P-40's taking off in formation, and as I looked up, I could see one was headed straight for us. I hit the ground and almost immediately there was this crash that shattered stuff all around us. I then saw the airplane that hit us go crashing down the runway.

"Neither Joe nor I was hurt, and so we went around to the front of the plane. We were shocked. Both engines and the nose were completely gone, and people were lying all over the place. General George and Jacoby were killed. Two people from the greeting party were also killed, and several were injured. An ambulance came out and picked up the dead and injured and took them to the hospital."

Jack Donalson had not only made it out of the Philippines ground and air war alive, he succeeded against the odds, and did so with conspicuous valor.

Not much has been written about the air war over the Philippines, perhaps as it was not a time of glorious victories for the United States. Many ask the question, 'Why didn't we stop the Japanese in the Philippines?' It is a complex question, and the author recommends William H. Bartsch's book *Doomed At The Start* as the most comprehensive book dealing with that issue. From Bartsch's carefully researched vantage point, let us review some of the factors that allowed the Japanese such a stunning victory over our Air Force in the Philippines.

First and foremost, neither the P-40 nor the P-35 were a match for the famed Zero, especially in an interceptor role. While the P-40 was rugged and heavily armed, it simply could not climb fast enough to engage the Japanese, nor maneuver with it once at altitude. The P-35 was at an even greater disadvantage, and was basically obsolete years earlier.

Second, the virtually non-existent early warning did not allow our pilots that time to laboriously climb to altitude and position themselves for an appropriate attack. Though our one radar site worked well enough, it was only in operation the first day and was soon knocked out.

"Ribbons" Donalson

Note: Your editor was one of the witnesses for this next story.

Major Donalson was one of the only early pilots in the 352nd who had a string of "ribbons" on his officer's blouse, having been awarded the DSC and some other medals for his exploits in the Phillipines. It is said that Colonel Meyer might have been a little envious of Jack's awards and there was one incident which might have indicated this.

Late in 1943 several 352nd pilots flew down to Fighter Command Headquarters to be awarded their first Air Medals and some few who were to receive DFCs. Some flew down with Col. J. C. Meyer in the C-45 and others flew down with Major Jack Donalson in another hack ship. Meyer's pilots landed sometime before Donalson arrived and his group were waiting on the tarmac for the second plane to arrive...

Colonel Meyer, looking toward the horizon, said, "Here they come now," having spotted a tiny dot in the sky.

"How do you know that's them, Colonel?" asked someone.

"Easy," Meyer answered. "I can see Donalson's ribbons flashing in the sun."

Colonel Meyer was well known for his sometimes sarcastic sense of humor.

Another factor was the serious shortage of practice in combat tactics and maneuvering, especially in the P-40s. As Jack noted, their planes were not even assembled until after the Squadron arrived two weeks before the attack. Also, the information gleaned about Zeroes in China was not passed down to the actual combat units. If it had been, they would have known that it was not possible to dogfight and maneuver with this enemy aircraft. Similar aircraft training in the P-35 and P-40 did nothing to prepare the young pilots for combat with the nimble Zero. The Zero could be successfully defeated using a P-40 with the right tactics, but dogfighting from an inferior position was not a way to live long and prosper.

Last, and this seems outrageous in retrospect, many aircraft went into combat with non-functioning machine guns. This was due to a shortage of time and supplies for the guns to be installed, sighted, and debugged to insure reliable operation. Consider the plight of a young American pilot, in a disorienting swirling fight against an experienced enemy, getting his sights on a Zero only to discover his guns are not operating. Appalling.

Courage and skill were not lacking in our valiant young pilots defending the Philippines, but the deck was clearly stacked against them. Therefore, Jack Donalson's enviable record of three Japanese downed on the first two days of the war takes on a new and brilliant meaning.

Air War, Southwest Pacific

"A couple of days later I was assigned to the 9th Squadron of the 49th Fighter Group at Darwin. When I arrived I was greeted by several of my flying school classmates and pilots from the Philippines who had been sent over early in the war to ferry back new P-40's.

"I was assigned to Andy Reynold's flight, as he was about a year senior to me, and he assigned me as his element leader. In our flight we had Andy, John Landers as his wingman, myself, and John Sauber as my wingman.

"The Japs were planning to invade Australia and take over the country as they did in China, the Philippines, Malaysia, Indonesia and New Guinea. They were bombing Darwin with aircraft stationed at Timor, an island about 200 miles away.

"The 49th FG consisted of three squadrons with about one hundred P-40's and 80 or 90 pilots. We also had an Australian fighter Squadron with about 25 aircraft and pilots. When the Japs first hit Darwin it was of course a surprise attack with fighters and bombers like at Pearl Harbor and Clark Field. They pretty much destroyed the small town of Darwin, with its airfield, docks, and business area, and it was abandoned after that. Following that first attack, our aircraft, control equipment, and early warning system were prepared and waiting. After that, few of their aircraft ever returned to Timor.

When I joined the Squadron in April of 1942 (five months after the war started) the Japs were making fewer and fewer raids on Darwin, and decided to invade from New Guinea. My first action was on June 14th, and by then their raids consisted of mostly Zeroes with a few bombers. They conducted mainly reconnaissance missions looking for shipping or new construction in the Darwin area. On the 14th our squadron was scrambled at about 6:00 am and was directed to an area north of Darwin, as the usual Jap approach was from Timor over Melville Island. We waited west of the island so we would have the altitude and position to come in behind them to attack. There soon appeared about four bombers and twelve Zeroes. Andy made his attack on a Zero and it burned. I followed and hit the leader, and he did the same. Our wingmen followed us as we hit the deck. We had plenty of help from the rest of our squadron and the other squadrons, so we let them have a share of the pie as we were getting low on fuel.

"My second and last action was on July 30th. The Japs sent a force of about 24 fighters, apparently to attack our fighter bases—which they did! They came in on the deck to avoid our radar warning and their target was the 7th squadron base. However, they were soon detected and we were scrambled. As we arrived at the action it was at low altitude for a change. The Zeroes were strafing and we caught some pulling up off of their targets. Andy Reynolds and John Landers each got one, and I did too. My wingman Clay Tice was a new pilot on his first mission. He was so shocked by the whole show that he got lost and we did not see him until we got back to our base. I will never forget when he came into the briefing. He said, 'What happened? Where did you guys go?' We just looked at each other and shook our heads.

"As I said before, the Japs' visits became few and far between, so it was decided to move our group to New Guinea, leaving the Australians to protect Darwin. We made our move in October. The plan was to move our outfit from Darwin to Townsville on the northeast coast, convert from P-40's to P-38's, then move to New Guinea. We completed the move to Townsville, and the P-38s started to arrive in November. I was told that a tour of duty for combat pilots had been established as one year in combat, and as I had left the States in November of 1941, I could rotate home. I had no argument with that, so I packed my bag, which had never been much more than a pair of socks, a shirt, and shoes. There was a B-24 in the depot in Townsville leaving for San Diego the next day." When Jack arrived back in the States, it was November 1, 1942. One year after he had left for the Philippines.

Meyer's Maulers

"My next duty was at Westover Field, Massachusetts, with a P-47 Replacement Training Unit. Their mission was to check out pilots in the P-47, then form groups to fight in Europe. The first group they formed after I got there was the 352nd Fighter Group. One of the pilots in our class was 1st Lieutenant J. C. Meyer, who had just returned from Iceland. He said the new Group Commander had been with him in Iceland and wanted J.C. to have one of the squadrons. He asked me if I would like to go with him as his Operations Officer. I talked with Mauree, and it was not an easy decision to make. We had only been married two and a half months before I left for the Philippines. Now I was about to do the same again. God love her, she understood the way I felt. Ed Dyess probably saved my life by sending me from Bataan in an airplane he should have taken. The other squadron commanders left, but not Ed. I simply could not live with the fact the Ed and the men were in prison, murdered, starved or tortured, and I was not in combat."

Jack did not know it at the time, but he was joining what would become one of the most successful squadrons ever to fly with the USAAF. Originally instituted as the 34th FS, its name and number would become legendary over Europe as "Meyer's Maulers"; the 487th Fighter Squadron—the highest scoring squadron in the ETO.

Jack's P-40 "Mauree," named for his lovely wife, poses beneath camouflage netting in Australia. (Jack Donalson)

As Jack would find, J.C. Meyer was a flamboyant and highly aggressive squadron commander, who led from the front and tolerated no less than the best from his men, as you can see from the message he sent to the officers and airmen who served beneath him.

But he knew from the start that he needed an experienced cool head to run the organization as his Operations Officer. Jack Donalson would prove to be a perfect choice.

Christmas Greetings
by Major John C. Meyer

With this special Christmas message to the officers and airmen of his 487th Fighter Squadron, then Major John C. Meyer commended his troops for their wartime sacrifices and efforts during their first few months of operations.

SUBJECT: Christmas Greetings, 25 December 1943
TO: All Personnel

It's about time 2:00 AM Christmas morning, and as I am writing this a large number of you men are lying on your backs beneath airplanes. With cold bodies and frozen hands, you are working outside on Christmas Eve so that in case our Squadron is called upon to perform its mission tomorrow, Christmas Day, we will be ready. To you men especially I extend my Christmas Greetings. Also, I extend my Christmas Greetings to you, the pilots of this organization, who have fought the war in the past few weeks and made, in that time, an outstanding record for the 487th. To you, Lt. Grow, who quitted yourself like a man; to you, Lt. Britt, who spends Christmas in a foreign hospital; to you both who have given yourselves for us, and of whom the rest of the 487th are justifiably proud, I extend my heartiest Christmas Greetings. To you others of the 487th, you administrative officers and men who have made it possible for us to do our part in bringing this inferno to an end, I extend my heartfelt wishes for a Merry Christmas.

I offer, also, my sincere thanks for the manner in which you have performed the arduous duties this six months of service in a foreign land has demanded. Neither the rigors of the climate, the efforts of the enemy against you pilots in the air, nor the constant high pressure to which you have all been subjected, have dimmed your enthusiasm or blurred the sparkle of your excellence.

I know that at times it must irk you, those of you who do not fly, to learn about your comrades in arms who are engaged in more spectacular fields of glory, and that you are impatient to join them. I also know that at times you long to be with your loved ones at home.

And I also know, because you have shown me, that yours is the courage and stout-heartedness that can meet any demands cheerfully and willingly without hope of reward other than that which comes with a job well done; and with your determination to apply your every effort to the task at hand, the successful offense against this important objective in which we are making a vital contribution, will lead to the successful conclusion of this war.

I thank you again for your superb performance. It is a great privilege to be able to serve with such a fine body of men.

Merry Christmas to you and your families,
John C. Meyer, Major, Air Corps, Commanding

The Story of WWII Flying Ace George Preddy Jr's Involvement in the Defence of Australia's Top End

By Susan Ward

"Do not go gentle into that good night
Rage, rage against the dying of the light."
– Dylan Thomas

The P-40 aircraft of three 49th Fighter Group aces lined up at Darwin Airfield. The nearest aircraft was assigned to 8th FS pilot Capt. Jim Morehead who became an ace during the big ANZAC Day raid on 25 April by shooting down three Type 97 bombers. Number 36 was flown by Capt. Bill Hennon of the 7th FS who scored his sixth victory in the 25 April air battle. Hennon joined the 352nd FG a year later as a squadron commander but was killed when the BT-14 he was flying disappeared off Long Island. The P-40 with the shark mouth is the mount of 9th FS ace Robert Vaught. He was nearly killed on 16 July 1942 when a three-foot poisonous snake stowed on board his P-40 and bit him in flight. The last aircraft number 57 was flown by nine-victory ace Capt. George Kiser who flew with the 8th FS.

Many people with an interest in WWII aviation know of Major George E. Preddy's exploits with the 8th Air Force in Europe which has seen him famously feted by the late Vice Chief of Staff of the USAF, General John C. Meyer as "the greatest fighter pilot who ever squinted through a gun sight." George also served in the southwest Pacific flying P-40s in the defence of Australia. What were the circumstances in Australia that took Preddy to Darwin with the 49th Fighter Group half a world away?

What was life like for him and his compatriots on that wild frontier? It is fortunate that George was a diarist and his daily musings have been preserved, not only giving us a detailed look at his daily life but also bringing a personal perspective to the political events and preparation for the unfolding world war.

Left to right: Bobby Boaz, Bill Preddy, Bill Teague and George Preddy with his red and white Waco 10 prior to America's entry into WWII. (Sam Sox, 352nd FG Assn.)

George Preddy grew up in leafy Greensboro North Carolina. He was smitten with a love for flying after his first "real airplane ride" in 1938 when he flew to Danville in a '33 Aeronca with his friend Hal Foster. George recorded in his daily journal that "I must learn to fly" and pursued his love of flying ardently. In the late thirties as the Second Sino-Japanese war continued to rage on the other side of the planet he was learning to fly. In 1939 George became a barnstormer pilot flying a Waco, learning all the tricks of the trade in stunt work and aerobatics. It came to him easily as he was a talented natural athlete with excellent spatial awareness. As George took to the skies the world was beginning to realise the danger of Nazism which had taken root and was beginning to threaten Europe. Poland was invaded on September 1st 1939. In response Great Britain, France and the British Commonwealth countries of Australia, New Zealand and Canada all declared war on Germany on the 3rd September. A conservative government was in power in Australia and the Prime Minister Robert "Pig Iron Bob" Menzies made the announcement to Australia thus:

Fellow Australians, it is my melancholy duty to inform you officially that in consequence of a persistence (sic) by Germany and her invasion of Poland, Great Britain has declared war upon her and that, as a result, Australia is also at war.

Despite the initial neutrality of the United States, George decided he wanted to fly with the military. Joe Noah says in his biography of Preddy, *Wings God Gave My Soul,* that the desire to fly bigger more powerful planes was what inspired George. "[H]e wanted more speed, more altitude, more excitement." He flew for the thrill of it.

Described by those who knew him as "aggressive, tenacious and eager," George applied to and was accepted by the USAAF in 1940 after being rejected from the US Navy three times on physical grounds. He spent some time with the Army National Guard, serving with the 252nd Coast Artillery whilst awaiting orders to report for flight training. When he heard of imminent plans to ship the 252nd out to Costa Rica, George applied for transfers about which he heard nothing until on the day his outfit were due to leave. At the eleventh hour George's commanding officer arranged for an internal transfer.

In April 1941 as two thousand American troops were dispatched to the Philippines and Japan signed a pact of neutrality with the USSR, George's orders finally came through. No. 452 Squadron, the first RAAF unit to be formed under the Empire Air Training Scheme was getting underway in England and Rommel was preparing to cut a swathe through North Africa as George tried his hand at flying a Stearman PT-17. Three months later he moved on to the Vultee BT-15 and George's confidence grew along with his skill. He was keen to get into pursuit rather than bomber or observation flying and found his understanding of aerobatics very helpful. He finished basic training in October '41 and a delighted George moved on to Craig Field in Alabama for advanced pursuit pilot training with the AT-6 Texan. He also tried out a Curtiss P-36 during gunnery training at Elgin Field in Florida, glimpsing the power and speed that was to come with the P-40 in the months ahead defending Darwin.

Preddy during pilot training

The capital of the Northern Territory, Darwin is on the northwesterly tip of Australia and was strategically situated for maintaining a supply route and transit base for armed forces heading into the Pacific Theatre. The new Australian Prime Minister John Curtin and President Roosevelt had agreed that a combined Australian/American force was the only hope to halt the steady southward movement of the Japanese. Curtin said at the time: "Without any inhibitions of any kind, I make it quite clear that Australia looks to America, free of any pangs as to our traditional links or kinship with the United Kingdom." When war broke out in the southwest Pacific the majority of the Royal Australian Air Force were in Europe and North Africa. In 1939 when Australia had automatically gone to war in defence of King and country, the RAAF had been consigned to Britain. The RAAF prepared Australian flyers for the Royal Air Force and then shipped them off to fight for the Empire. This not only left the shores of Australia undefended but meant that as the Japanese encroached on other nations in the region there was initially no concerted force to hold them back.

Appreciating the dire situation, the alliance had been formed and a programme undertaken by Australia and the USA to send P-40E Warhawks (known as Kittyhawks to the RAAF) and fighting men to the south west Pacific from the east coast of Australia, dispatching reinforcements to Java and the surrounding area via Darwin. US Far Eastern Air Force Commander General Lewis Brereton had travelled to Australia from the Philippines in mid 1941 to discuss the creation of an aerial supply route to support forces in the Philippines against probable Japanese hostility. With RAAF cooperation a secret network of airfields and supply bases was created which by November of that year became known as the Brereton Route. The route would enable fighter groups such as the 49th to leapfrog across the country, ultimately arriving at one of several bases constructed in and around Darwin.

On the 7th December 1941 Pearl Harbour was attacked. Japan declared on the 8th of December that European domination had ended in Asia as they landed in Malaya with the intention of vanquishing the British Empire forces within Malaya and Singapore. The Australian War Cabinet minutes for that day state: "Note was taken of the following Admiralty message dated 8 December: Commence hostilities against Japan, repetition, Japan at once." Governor General Lord Gowrie and PM John Curtin signed the Proclamation of War with Japan the same day. At this time American forces had already begun steadily arriving in Australia under the command of General Douglas Macarthur. George graduated on December 12, 1941. 2nd Lt. George Preddy Jr. had earned his wings, destined to fly P-40 Warhawks with the 9th Pursuit Squadron of the 49th Fighter Group in Darwin.

On the 16th December 1941 in Australia an edict was issued via Darwin's Newspaper *The Northern Standard* ordering evacuation from Darwin, stating: "Citizens of Darwin, the Federal War Cabinet has decided that women and children must be compulsorily evacuated from Darwin as soon as possible, except women required for essential service." At this time two thousand of the four thousand residents of Darwin departed, soon to be replaced by incoming Allied armed forces. Almost a month later George set sail on a refitted former luxury liner to an "unknown destination…"

January 12th 1942 – Monday

"Left San Francisco Harbour on the Mariposa for unknown destination possibly Australia.

About 4,000 men aboard. Ship is armed with 50 calibre anti-aircraft machine guns and 3 inch guns.

Also loaded with bombs and P-40s. No radios or cameras can be used, blackout is in effect."

On board the USAT Mariposa steaming towards Melbourne the news came through that Rabaul had fallen on the 23rd January and by January 30th Ambon Island had been lost to the encroaching Japanese. By the 31st of January the British forces had entirely withdrawn to Singapore.

The Mariposa arrived in Melbourne on Monday February 2nd 1942. The 49th set up their billet at Camp Darley about 30 miles from Melbourne, near Bacchus Marsh, appropriately named, as according to his diary George "picked up a girl" the same evening. George celebrated his 23rd birthday by getting hammered with friends McGee and Smitty and some local girls on the 5th. He mentions in the same diary entry that he "would surely love to move now and start flying."

Realising the difficulty of invading a continent as vast as Australia the Japanese had formulated a plan whereby they would isolate Australia by occupying various island groups such as Fiji, the New Hebrides, Samoa and the Solomon Islands. By controlling this part of the Pacific, Japan could disrupt American forces trying to get to Australia as well as the severing supply routes. The Japanese also had their eye on the rubber plantations, mineral wealth and oil reserves of the Dutch East Indies.

P-40s had already begun being transferred to the Java campaign via the top end but there were still very few military aircraft permanently based at Darwin in February 1942. At this time Wirraways were the RAAF front line fighter.

A serious Lt. George Preddy poses for a public relations photo at a remote Northern Territory air strip.

There were several "weary" Wirraways and a few Lockheed Hudson Bombers stationed at the RAAF Aerodrome. With no fighters to provide cover for the Hudsons losses had been very heavy. The three squadrons of the 49th FG that would be in Darwin by March were still training in the south. According to a contemporary report by General Brett USAFIA an "excessive and alarming" number of accidents during this period occurred which were almost entirely caused by pilot error as a result of little or no experience flying the P-40Es. Brett insisted "Better trainee pilots must be assigned to this theatre." They actually had plenty of keen talented men but insufficient training aircraft. Although there were no major injuries to the pilots the loss of aircraft due to inexperience was extensive and weakened the force considerably.

Singapore was surrendered to the Japanese on February 15th by General Percival. Some 140,000 people were taken prisoner, including about 15,000 Australians. RAAF Spitfire pilot, Flight Lieutenant Edward "Ted" Sly still remembers the general feeling of helplessness amongst the Australians "I was in the United Kingdom… with the fall of Singapore, Australia could be next in the firing line and I was 13,000 miles from home!" In Melbourne, two Wirraways had finally been provided to train the fifty officers in George's outfit. The first chance George had had to even sit in a 'plane was February 16th only three days before the first Darwin raids. The attacks on Darwin seemed to take everyone by surprise and suddenly make the threat of invasion real.

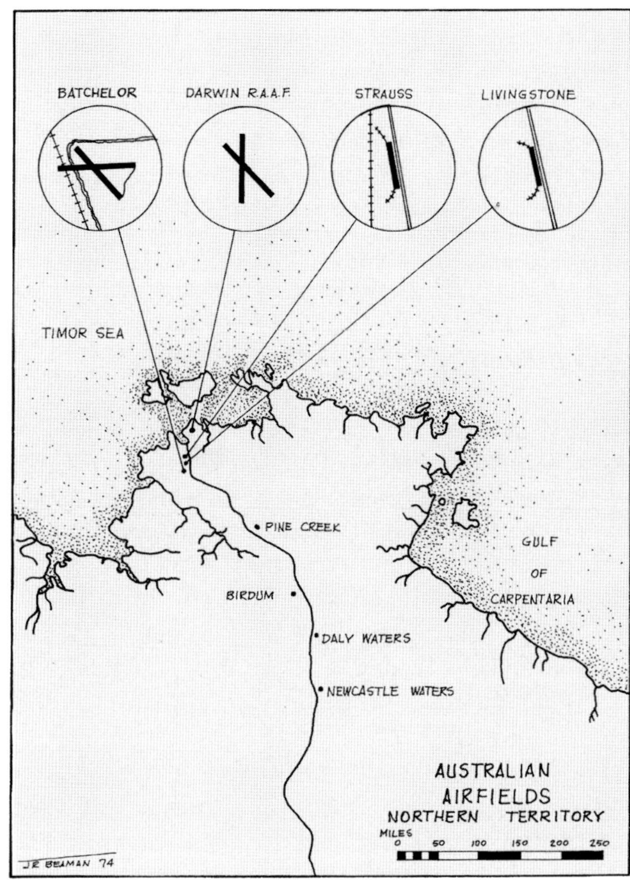

Darwin area airfields

Darwin Raid

On the morning of February 19, 1942 just to the north of Darwin on Bathurst Island, Father John McGrath was alerted by local islanders of a large force of aircraft heading towards Darwin. A volunteer with the Royal Australian Navy coast-watching scheme, he radioed in on the top secret frequency "X" or 6840 Khz, the frequency which all Coast Watcher's radios were crystal locked onto.

The Australian Navy never listened to frequency X. Father McGrath's message was received in Darwin by Lou Curnock at the Coastal Radio Station VID at 0935 hrs. Curnock then contacted RAAF Operations. In a turn of events remarkably similar to what had transpired at Pearl Harbour only two months earlier, the RAAF incorrectly assumed that the approaching Japanese aircraft were P-40s of the 33rd Pursuit Squadron USAAF. The P-40s were returning to Darwin after bad weather had forced them abort their flight to Timor en route to Java. Realising that no alarm had been raised by the RAAF Curnock called the Office of Naval Intelligence and spoke with Lt. Cdr. JCB McManus as the bombs began falling.

Eyewitness Gunner Jack Mulholland said "the enemy planes looked like a neat cemetery advancing across a blue field." The attack was so unexpected that at one of the "ack-ack" stations the predictor crew were actually training on the first flight of approaching Japanese bombers. One of the gunners looking through the telescope yelled, "Hell! They've got red spots on their wings!"

The attack on Darwin was executed by the same force that had visited destruction on Pearl Harbour. From the aircraft carriers Akagi, Kaga, Soryu and Hiryu, commanded by Admiral Chuichi Nagumo came 188 aircraft including 71 D3A "Val" dive bombers, 81 B5N "Kate" high level bombers and 36 escorting A6M2 "Zero" fighters in the morning raid. Flying from newly acquired land bases at Kendari in the Celebes and Ambon, for the noon raid the Japanese launched a force of 27 G4M1 "Betty" and 27 G3M1 "Nell" twin engine bombers from the 1st and Kanoya Kus of the 22nd Koku Sentai.

Despite heavy anti-aircraft fire from the ships in Darwin Harbour, great damage was inflicted during the two raids that day with 23 ships sunk or badly damaged. This accounted for half the ships in port and the two ships Don Isidro and Florence D which were underway at the time of the attack. The USS Peary was lost with 91 lives, almost half of the official death toll from the raid. The US Navy had been in Darwin preparing to transport troops to support the allied and US forces in Java. Of the 10 P-40Es of the 33rd PS (Provisional) which were returning from the aborted mission to Java, five had been on patrol when the Japanese struck and only one survived the day. In all, 26 aeroplanes were destroyed including 23 allied aircraft.

The Post Office which housed the Cable Office, the Police Station and Telegraph Station was completely levelled, killing the postmaster and his entire family. The city was in ruins and all services including the railway line were cut. Mrs Abbott, the Darwin Administrator's wife said, "When we looked out on our so lovely harbour no picture of an inferno could equal what we saw. Black smoke rolled in heavy clouds; flames rose in

Preddy and his crew chief Cpl. Yates stand beside their P-40. (Robert Alford Collection)

great tongues to the sky...the water was covered with black oil. In this water small navy boats were moving, rescuing blackened figures and men were struggling towards our garden where the black water was licking the shore."

There were 252 deaths and 350 people injured or missing. The attacking forces dropped almost 80% of the number of bombs dropped on Pearl Harbour on the 19th February, "using a sledgehammer to crack an egg," as Commander Mitsuo Fuchida reportedly said. Darwin endured 46 raids during the war with a total of 64 in the top end.

George wrote on the 19th, "Flew the Wirraway again today, sure need a lot of practice to get the touch may be leaving here soon because the Japs bombed Darwin today, am anxious to get a crack at them but hope to be familiar with the P-40 first" at this time with the threat of invasion more pressing than ever, practice with the P-40 was still seen as a luxury by those in power.

On the 20th the Japanese invaded Timor. In Australia the Darwin raids were played down in the press. The official death toll for the raid was only 17. A Sydney newspaper ran a photograph on the front page of a huge column of black smoke and the caption," Large grass fires burning in the Darwin area." The Brisbane Courier Mail stated on Feb 20th that "[t]here were some casualties and some damage was done to service installations." The fall of Singapore four days previously was still hot news and there was an article on the front page of the same paper sensationalising Japanese atrocities saying that on hospital ships evacuating Singapore: "...the decks were continually machine gunned...During the attacks Australian nurses lay on top of children to protect them." In an attempt to boost morale and stem panic Prime Minister Curtin announced that "Darwin has been bombed but it has not been conquered." At this time the American flyers were yet to have any training on the combat aircraft they would be flying against the Japanese within the month!

On the 22nd of February George finally got his hands on a P-40. There were only two available and he spent his allocated two hours familiarising himself with the controls and instruments. Two days later he had an hour and a half of flying practice prior to heading off on the long trip to Brisbane to pick up a P-40 and bring it back to Sydney—where he put in as much time as he could. He was keen to pick up tips from the more experienced flyers. One of these was Captain Boyd D. "Buzz" Wagner, one of the first American aces of the war. George noted: "He knows airplanes and tactics inside out. He is certainly no lucky pilot—he's the real thing! I have a lot of respect for a guy like that."

By late February, thousands of refugees and evacuees from the southwest Pacific had passed through Broome south of Darwin before heading to the southern capitals. Early on March 3rd Zeroes of the 3rd Ku based at Koepang and Penfui airfields on Timor attacked Broome using intelligence from two predawn reconnaissance missions on the 2nd and 3rd. Knowing there would be no resistance, they destroyed every operational aircraft in Broome and killed at least 40 people.

Twenty three Kittyhawks and one B-17 set off towards Brisbane from Bankstown in Sydney on March 7th. Bad weather and low clouds forced the group to return after flying almost three quarters of the way. In early 1942 little more than 1% of Australia had been mapped; this combined with dodgy compasses in the P-40Es made a mother ship necessary for navigation. They finally made the trip on the 10th and spent some time in Brisbane for refitting before heading off across the country via the Brereton Route to Charleville on the 13th and on to Cloncurry on the 14th. After delays in Cloncurry they continued flying over land which George describes as being "so desolate it was scary" arriving in Daly Waters on the 16th. The inexperience of the pilots inevitably took its toll with only thirteen of the original twenty-five Kittyhawks arriving in Darwin on the 17th.

Following the initial raids, Darwin was a ghost town of empty streets, abandoned buildings and silence. When George Preddy and his companions of the 9th Pursuit Squadron arrived in Darwin, Marshall Law was in place and there was a sense of calm expectation. The remaining residents and armed forces had been resolutely awaiting the arrival of aerial defences, aware that the Japanese could land at any time. There was general excitement and a boost to morale when the American Kittyhawks began arriving. Aussie veteran Jack Mulholland said "They looked pretty damn good to us. We saw them as having come down under to save us from the advancing hordes and we were glad to have them with us!"

Ready or not, the 9th Squadron arrived in Darwin on 17th March. George wrote in his diary, "Finally got to Darwin with half the ships we started out with…this place really shows the affects of bombing, buildings and hangars are blasted all to pieces and big bomb craters are all over the ground." Some of the city was still intact as he also wrote: "All of the civilians have evacuated the town leaving swell hotels and apartment houses fully furnished. It is swell to get a place with running water, refrigerator and all the conveniences." They were soon in the thick of it, flying operations from Batchelor field about 50 miles south of Darwin. The first operational patrol flew out of Batchelor on 19th March 1942 and the young pilots quickly realised that there were fundamental flaws in the operation. There was no decent early warning system as George relates, "The alert around Darwin is lousy, we never get the call until bombing has begun, then it is too late." Flights often took to the sky just in time for an aerial view of the destruction. This, combined with the problems of taking a P-40 rapidly to altitude resulted in a great deal of frustration. The escorting Zeroes usually maintained an altitude advantage, knowing that pulling a P-40's nose up suddenly and firing the guns could result in a stalled aeroplane.

On the 22nd of March George recorded in his diary that "Harvey and Poleschuk shot down a Jap recon ship, the first for our squadron." 49th FG records show that a Type 97 "Babs" escorted by 3 fighters were tracked on the new radar that just come on line that morning. The Japanese had enjoyed a month of unopposed sorties over Australia but the newly arrived Americans had put an end to that. Although Japanese records indicate no loss for the 22nd March, they quickly realised that a change in tactics was required. Mission planners could no longer send bomber formations over the Darwin area without escorting fighters as they were now at the mercy of an enthusiastic if not particularly experienced group of young fighters. All the flyers were under 24 years of age, most, like Preddy had no combat experience at all. March 22nd also saw the arrival of "three old pilots who have seen action in the Philippines and Java, (Lt. William Irving, Captain Joseph J. Kruzel and Lt. Andrew Jackson Reynolds) it is nice to have someone around who knows...how the Japs operate. When our meeting was held tonight all the squadrons were changed around, forming eight flights." Along with Lt. Jack Donalson, the three new arrivals were the only seasoned pilots in the squadron.

Pair of P-40s from Capt. Joe Kruzel's flight sits in the open on a Northern Territory air strip. Kruzel scored six confirmed victories during WWII. After leaving the SWP, he became CO of the 361st FG and retired as a major general in 1970.

The Japanese launched strikes against the RAAF airfield in Darwin on 28th March with seven unescorted bombers arriving only to be

sent on their way with one shot down and one damaged. George wrote, "[T]he boys at Darwin intercepted (seven Jap Bombers) on the way back to Koepang and knocked down two for sure and possibly disabled two more." Throughout the history of aerial combat over claiming has been the rule rather than the exception and the skies over Darwin were no different. The Japanese changed tactics and on 30th and 31st March there were Zeroes in escort for the first time." The 9th PS was there to challenge them but the fledgling P-40 pilots were only able to claim one Zero in the subsequent engagements. On 30th March Lt. Preddy and his mates took off from Batchelor and patrolled for a while before returning to refuel only to be scrambled almost immediately to intercept an incoming attack. They arrived in time to intercept the Bettys but were bounced by Zeroes. George says ruefully of the engagement "I had a perfect shot at a Zero, but missed by not turning one gun switch on." On the 31st George saw "three bombers for the first time at night. Hollywood producer Lucien Hubbard visited Humpty Doo in 1942 and wrote "The Japs (came) again and again, in Zeroes and Bombers. The green kids went up to meet them."

On April 4th the Japanese returned with a force of six Bettys and six escorting Zeroes attacking the civil 'drome at Parap, only two and a half miles from central Darwin. Crews of the Australian 14th Heavy Anti-Aircraft Battery had their range and altitude dialled in and shot down two of the bombers before the 9th Pursuit Squadron transmitted a "Tally Ho" over the R/T, the pre-arranged signal for the ground gunners to hold their fire. Diving through the Japanese formation, the P-40 pilots claimed two Zeroes and six Bettys. George relates the day's events… "We gave the Japs the prettiest surprise today. When they came over with seven bombers and three Zeroes at 23,000 feet, we had a flight of seven P-40s at 26,000 waiting for them. When they left there were only two bombers and one Zero. Reynolds is becoming quite an ace, adding a bomber and a zero to his credit. Gardner was hit by our own ack-ack and bailed out. Livingstone overshot a landing on a forced landing after the combat and was killed when his engine cut out."

The airfield where Lt. Livingstone crashed was known simply as 34-mile airstrip, (some time later it was renamed Livingstone Field) it was one of four paved air strips in the immediate vicinity of Darwin built alongside the main north-south road between Darwin and Adelaide, over 3,000 kilometres to the south. George and the 9th PS set up digs there on April 14th "We have a camp built and are dispersed in the woods." By this time George had been assigned to Kruzel's Flight

Left to right: Preddy, Deams, Taylor, and Clay Tice pose at 34-Mile Strip in March 1942.

and said of the change "I believe I will like flying under his command."

Monsoonal Darwin is oppressively hot, from October until April with cloying humidity and temperatures hovering around 35+ Degrees Celsius in the shade, punctuated by dramatically violent electrical storms. The heat is relentless, famously rendering the inhabitants "troppo" during this time of the year. Prickly heat and fungal infections were rife. George said "The mosquitos are so bad you have to crawl in bed under a net after supper to stay in one piece." The mid year "Dry" replaces the mugginess with cooler, dry, dusty conditions. Despite the difficult conditions the pilots did their best to make themselves comfortable in Australia's northern combat zone. Their camps had volleyball nets, board games and furnishings requisitioned from Darwin buildings, including the Hotel Darwin, in a village of tents shaded by gum trees.

The Japanese attacked Darwin three more times in April, the biggest raid coming on ANZAC Day, the 25th with a force of 27 bombers 15 fighters and one reconnaissance plane despatched against the RAAF airfield on the outskirts of town. This was the first time that all three squadrons of the 49th Fighter Group operated as a unit. The 49th claimed a total of ten bombers and two fighters destroyed and one bomber damaged against no P-40 losses. The Japanese returned two days later and although at the time the 49th claimed victories over three bombers and four fighters, four P-40s were lost. George was thrilled to be getting some combat experience and was pleased to be able to make damaged claims for a Zero and a Type 97 Bomber, though still chafing from the lost opportunity to down his first plane. Three P-40s from the 8th Squadron were destroyed as was one from the 7th Squadron. There is some discrepancy as to actual numbers of enemy attackers during these battles as well as the number of kills and the above scores have been taken from George's diary entries for the days of the battles. George wrote of the engagement on the 28th..."Captain Strauss and a lieutenant are still missing. Andrew and Martin were both shot down and one jumped while the other made a forced landing in the water. Fish was shot through the head and was last seen going into the bay with his engine wide open and all guns firing."

April finished on a sad note with a runway collision which claimed the lives of Brigadier General George, war correspondent Melville Jacoby and Lt Jasper of the 9th. Their transport was hit by a P-40 which ran off the runway on takeoff on the 29th. General George survived the crash but died the following day in the 135th Med. Regt. Hospital. George Preddy had a great deal of respect for the General and describes him as "Probably one of the best pursuit pilots and authority in the world."

Lieutenant George Preddy stands beside his assigned P-40E after being assigned to Capt. Kruzel's "D" "Dragon Flight."

A rare in-flight photo of Lt. Preddy at the controls of his P-40E No. 85 prior to it having the distinctive dragon nose art. The name "Tarheel" appeared on both sides of the Warhawk at this time.

A seven week respite followed after the April raids. During this time George and his squadron mates settled into a routine as they continued to fly patrols out of Livingstone Field at Humpty Doo and perform training missions to hone their skills. By May the humidity was decreasing with the onset of "the dry" giving everyone a much needed break.

On 8th May the battle of the Coral Sea took place. Meanwhile, at Humpty Doo they were concentrating on necessities as George wrote, "Captain Kruzel and several other pilots went buffalo hunting and shot two large ones and a calf. We will probably be eating buffalo for the next month." There is no mention of the Coral Sea until the 12th when George says "There has been a lot of talk about the big naval battle off New Guinea where the U.S. Sank several Jap warships and troop ships. It is believed that this was directly responsible for stemming an invasion of Australia." There was also an official announcement by the War Department of Doolittle's raid.

The quiet time they were experiencing was the result of concentrated attacks on Port Moresby in New Guinea and also further south. On 31 May 1942 HMS Kuttabul lost 21 sailors during a raid by Japanese midget K-IX submarines in Sydney. Six Japanese submariners lost their lives. They were discovered to be Kamikaze and were buried at the time respectfully with military honours despite grumblings from the press who were busy with propaganda of their own. (Since the war the ashes have been returned to Japan and the men have posthumously become national heroes.) In June Midget submarines went on to attack inside Sydney Head whilst beachside suburbs were shelled by larger submarines.

The relative peace was shattered on June 13th with an attack on the RAAF airfield in Darwin. The following day 27 Zero fighters strafed the city. On June 15th the Larrakeyah Barracks and Stokes Hill were targeted, "Heavy bombers, about 27 of them, dropped their load on the town at noon. Our flight went into the attack just before they reached their target but we were attacked from above by several Zeroes. We were forced to dive out but Manning's flight caught them from above. After the enemy turned west, Manning and McComsey got one Zero each with Peterson and Fowler getting one between them. Taylor damaged one and the 7th got another. My ship was hit in three places. Otherwise, the squadron received no damage." As a parting gesture on the 16th 27 bombers escorted by 27 Zeroes returned to bomb what was left of Darwin town. There were no more raids that month. The box score for the four day killing spree was more or less even with 11 Japanese fighters claimed destroyed for the loss seven of P-40s with an additional three crash landings.

For the Americans it was a matter of urgency to log as many hours as possible. They were up against seasoned veterans some of whom had seen years of combat in China. In the absence of enemy activity the pilots were in the air as much as possible. Preddy and his mates were flying the Curtis P-40E Warhawk which was well armed with six 50 calibre Browning machine guns and provided the pilot with ample protective amour. Compared to the nimble A6M2 Zeroes that their opponents of the 3rd Kokutai were equipped with, the P-40 was more cumbersome. The Zero could easily out-turn a P-40 so the Americans avoided dog fighting when possible. They preferred to dive through the enemy fighter and bomber formations, lining up on a target, firing a burst or two and then continuing on below the enemy. Once below, the P-40s could use the speed gained in the dive to climb back above the Japanese formation for a second attack.

Flight of 5 P-40s from the 9th FS, #74 Lt S. Woods, #70 Lt E. Ball, #77 unidentified, #72 Lt J. Watkins and on the end , #85, Lt. George Preddy

George had settled into a routine of sorts, dogfighting, attending lectures, censoring mail, having "bull sessions" with the boys with the occasional bottle of whiskey and working with the others to set up camp and make it comfortable. He also did a lot of laundry, which he mentions in his diary with great regularity! Lucien Hubbard describes the "Dewy idealism…of our own 'few' who are doing so much for so many…", the Boy Scout camp where "monopoly was played by landless youths who were giving their best years to the acrid business of killing and being killed."

The last week of June, General Brett and three other Generals paid a visit. They admitted that decisions made in December of 1941 to send inexperienced pilots into combat had been a mistake. Relieved to have survived this period George felt by now he was "a little better prepared having learned from actual experience how to take care of myself up there, there is something to be learned on each mission and I am just a beginner." With the Japanese concentrating on territories around New Guinea, the pilots of the 49th continued training in relative peace, they busied themselves doing odd jobs around the camp such as making an oven out of an anthill, building a volleyball court and the general feeling was that owing to the change in direction of the Japanese onslaught away from Darwin their group would be moving on soon. George summed it up in his July 4th diary entry, "Tojo usually furnishes the target for our fireworks on holidays but he didn't comply this time." George was "still sweating out a ride south" hoping to get leave before they moved on, possibly to New Guinea.

Late in the afternoon on 12 July two flights of 9th squadron P-40s took off on a training flight over the Manton Dam south of Darwin to practice diving attacks on enemy formations. Preddy was flying on Lt. Richard Taylor's wing and the two were pretending to be the Japanese targets flying at 12,000 feet. Lt. John Sauber and Lt. I.B. Jack Donalson climbed to 20,000 feet to make mock diving attacks. Sauber flying in No. 87 went first and dove on Preddy's P-40. Jack Donalson recalled the moment "Sauber...looked at me and signalled he was starting his pass and I could see clearly that he was too close to have room to manoeuvre...before he could break off the pass he collided with Preddy." He severed the tail from Preddy's P-40, No. 85, "Tarheel" just behind the cockpit. Preddy was seen to egress his stricken craft but Sauber went down with his ship.

Back at Livingstone Field the crash crews raced against the rapidly setting sun to where Lt. Clay Tice was holding the position of the spot where he had seen Lt. Preddy's white parachute

Lt John Sauder was a popular member of the 9th FS before being killed in the mid-air collision with Lt. Preddy on July 12 1942

canopy float into the bush. Approaching the spot where Preddy landed, according to Lucien Hubbard's rather dramatic version of events he and Bill Irving ran through the bush for about a mile, Hubbard "caught a glimpse of white half way up the ridge, and made for it…it was Preddy. He was lying on his parachute. "Gee, fellow, I am glad to see you!" he called out…"Is my leg bleeding much? Is the artery ok?" He had bandaged it himself from his first aid kit. Falling through trees his body had broken through a big branch, gouging a deep hole in his hip and breaking his leg. He seemed more concerned about a thirty-foot termite mound nearby, not the best place to kip down for the night! When Irving arrived Preddy grinned. "You owe me dough boy," he said. "I bet I'd be the first to date one of those new nurses down at Adelaide River!"

With nothing to do but lie around in bed waiting to be shipped out to hospital, George updated his diary regularly, the day after the crash feeling rather sore and sorry for himself he recounts the fateful incident "…just lying in bed cut all up and stiff as a board. At least, I am thankful to be here at all. Sure sorry about Sauber. My right hip is cut about off and there is a big gouge in my left calf. Also my right shoulder is cut up. Sure lucky!"

Shipped off to Melbourne via Alice Springs to recover, George found himself in a bed near Major Robert Van Auken who was recovering from burns received when baling out over Melville Island during the raid on 13th June. He had been rescued by an Aboriginal known as "Old Johnny" and canoed back to Darwin for treatment. George and Van Auken had come down from Darwin in the same group. George recounts his first meeting with Joan Jackson, the girl he would later get engaged to. "Major Van Auken has been blessed with two beautiful female visitors today…" Thereafter the girls visited often, bringing reading matter and treats.

George kept abreast of the happenings of the war from his hospital bed. He was jubilant that a new early warning system had finally been created and the now seasoned combat pilots were setting upon the Japanese invaders with alacrity and sustaining fewer losses than ever. George's diary entry for August 25th says, "The Russians are sure catching hell around Stalingrad…News came today that Japs raided Darwin Sunday with 27 Bombers and 20 Zeroes. We got 9 Zeroes and 4 Bombers without a loss. They can come every day if our boys can knock them down like that!" George mentions in the same entry that his mate Jack Donalson had been awarded the DSC for service in Bataan. During his stay in hospital George was interviewed by ABC Radio, his interview was transmitted back to the States on August 30th.

He was falling in love with Joan and describes her as "…what I have been looking for, beautiful, smart, nice family and good fun to be with." Before long Joan was wearing George's wings. They spent plenty of time together and particularly enjoyed visiting Luna Park at St Kilda, going dancing and riding the rollercoaster. From their first meeting on the 23rd July until George departed three

Lt Clay Tice who held his position over the crash site enabling Preddy's buddies to successfully locate him

months later the two became increasingly close as his diary reflects. "Had dinner at Joan's house ...it was a very enjoyable evening. Her dad is a famous golfer (Gus Jackson) and a very nice fellow...I think I could love Joan." When Joan waved George off on the Brisbane train on the 20th October 1942 she couldn't know she would never see him again. George wrote, "I just realized how much Joan cares for me and means to me after seeing how hard she took my departure."

Arriving in Brisbane George expected to head on to New Guinea. Instead he was greeted with "two sets of orders within one hour, one set promoting me to first lieutenant and the other sending me home." He departed Amberly Field for home in a B-24 on October 23, 1942.

Leaving the 49th behind to continue the push in New Guinea without him, George headed back to the USA, where after some leave and time to see his family and have some southern fried chicken, he returned to service, continuing his training until he was called to the First Air Force on the 30th December 1942. On December 31st George reflected on the past year, "Well here it is, the last day of 1942, a year I will not soon forget. I am on a train setting out on my sixth crossing of the country since this time a year ago. Trips twice across the Pacific Ocean and up and down the continent of Australia have given me some great experiences. Hope by the end of next year I can have done more for my country than I have this year."

Although George did not score any personal victories over the Japanese while defending Australia he had learned many valuable lessons. He had looked in the face of death as well as love. George Preddy has been described by contemporaries as having "A core of steel in a largely sentimental soul." In a war that was all about the business of "killing and being killed," the experiences of the young George Preddy, his account of the situation in Australia and the daily details he shares reveal an intelligent and thoughtful man who would indeed take the experience of combat to a new level when he entered the European Theatre in 1943.

During WWII, Captain Eddie Rickenbacker's WWI score of 26 enemy aircraft destroyed became the benchmark of the exceptional American fighter pilot. History shows that although he did not survive the war, George went on to be one of the few legendary aces that equalled or surpassed that total and became the top scoring P-51 Mustang ace of all time. *The Greensboro Daily News* would say of George: "His aim was deadly, his flying skill remarkable..."

The complete article is available online at www.stardustudios.com. Photo credits 352nd FG Association, I B "Jack "Donalson, Robert Alford Collection.

Lieutenant Preddy's P-40 engine, propeller and spinner are visible in this photo taken at the Manton Reservoir crash site.

George Preddy's fiancée Joan Jackson of Melbourne, Australia.

"Open Mouth, Insert Foot"
by BIG Richard Arlen, USAF (Ret.)

On arrival in England, Flight Surgeon Richard Arlen was temporarily assigned to the 352nd FG at Bodney, England. In a book he wrote while on duty in the ETO about his wartime days, we found the following story.

I traveled to England in an old banana boat that would do about eight knots in a large convoy of ships. I was not assigned to any group or squadron at that time, but had the responsibility of watching over twenty-eight "Georgia Peaches" who had married some of the RAF pilots who were getting their flight training in south Georgia. These ladies wanted to be near their husbands in England and were en route to do so, all of them frightened to be on the Atlantic in a convoy which was subject to German sub attacks. We had our daily scare of U-boats and it was reported that our escort corvettes sighted one and sunk it. These women looked up to me like a big brother and later, I met a couple of them in London and learned that some of them were already widows.

I was soon assigned duty on an old RAF airbase at Bodney, in East Anglia, famous for its achievements during the Battle of Britain, England's darkest hour. Now, a new organized and trained American P-47 Thunderbolt fighter group of the 8th Air Force, fresh from the States, had taken over the field for operations.

I met the pilots assigned to me and they appeared to be somewhat quiet and dull. I had just come from a local club of which I was a member, and returned in somewhat good spirits. They were, at that time, billeted in Clermont Hall, a former British mansion secured under the Lend Lease Act, but later were moved onto their airfield to live in Nissen huts. Anyway, there they were in this large living room and one of tbe pilots was playing the organ and some were writing letters home and there were four senior officers in their pajamas, all very quiet and serious.

I remember speaking to one of them, a tall man with a red mustache and I started blowing off steam saying this place reminded me of a morgue. Then I said that "if I was the C.O. of this outfit I would throw out the snooker table and the organ and put in a bar, have a piano, radio, ice cream machine, popcorn and keep the kitchen open all night and throw a dance at least once a week." The red-headed officer looked at me and asked me my name and never said another word. I soon retired to my bunk and never felt better.

The next morning the Executive Officer, a Colonel Clark, asked me whether I had read the base orders of the day and I had to tell him I had not. I soon managed to read them and found my name, Capt. Richard Arlen, appointed president of the Officers' Club by order of Col. Joe L. Mason, C.O. That's when I realized the fellow I had spoken to the night before was the head knocker of the Group, Colonel Mason. He had me meet him in his office and assured me that he had enjoyed our previous night's discourse so well that he had decided the job was mine.

So, to liven the old place up I went to London with my men and went to the Carlos Club in Grosverner Square and looked up tbe owner, a Mr. Miller. And I asked him how I would go about buying a night club. He told me I should get in touch with the owners of the New York Club which had just been closed by the British government.

I did, and found it was nicely furnished with mirrors, bar, and everything we needed and we arranged to buy it. Calling back to the base at Bodney, we arranged for a five-ton truck and some GI's and in a few hours, using their crowbars and other tools, we moved these things to Bodney. We reached the base with our spoils and all the officers were amazed.

"What? Two thousand dollars for this? Are you in your right mind?" I told them lumber was expensive, that it was even necessary to get tbe King's permission to cut down a tree.

Anyway, I got the base carpenters to install the nightclub equipment and all the pilots got into the act. We now started to restore the ninth wonder of the world. In a few days the place started to take shape. The boys wanted a real drain in the bar so we hired a Cockney plumber from the village to do the job. However, to put in the drain, he had to pass it through tbe storeroom of the Lord who owned the mansion.

The caretaker of the estate told the real estate agent the Americans bad gone through tbe Lord's storeroom and the plumber was arrested and jailed and the agent was waiting for me when I returned. I decided to go on the offensive and told him that if he didn't get the plumber released I, as a Flight Surgeon, would condemn the estate on the grounds tbat the water was polluted by the cesspool and he would no longer enjoy his commission as the agent. I also told him we were holding a dance in a few days and the bar had to be ready for that. He quickly got the plumber out of jail and in a few days the bar was ready.

Officers' Club Interior View

On Saturday night we had about seventy-five English lassies and a number of American nurses there for our dance and we included some RAF pilots as guests, too. Since we had no money for booze I picked up some medicated alcohol from nearby bases and added fruit juices that made a palatable, but potent, cocktail. The pilots later dubbed it 'Thunderbolt Joy Juice.' Few if any ate anything, and everybody got happily and noisy drunk.

But after the dance, the battle royal began. The morning after I took inventory. The kitchen was cold, no breakfast, no coffee. We had to resort to K-rations and the Colonel was really pissed. None of the girls had gone home. We soon got fed up with the normal rations and since there were lots of deer on this large estate, we shot a few to enjoy a big dinner of fresh venison. Here we were in England living on this beautiful estate with all kinds of wild game including the King's deer and pheasant. We had guests at this dinner and one happened to be the Game Warden for the estate. He told us that his father and his grandfather were also wardens and they never hesitated to shoot anyone they caught hunting on the forbidden "King's Ground." He thanked us for the nice evening, saying he appreciated our not bothering the King's deer. It was a subtle warning and we realized that "a word forewarned is a tombstone unworn."

Three weeks later the 8th Air Force Inspector General sent two colonels to investigate our party. Colonel Mason sent for me and I explained to the Colonels that the party was good for morale, which it was. I told them that as their Flight Surgeon it was my duty to stimulate their morale and therefore improve their flying abilities.

It was with this fighter outfit that I really learned to understand the fighter pilot. They do not seem to age as fast whereas the rccon pilot is more of a lone ranger. The squadron leader of the fighter pilots is one of them. To see a squadron leave on a mission and then anxiously wait for them to return is a tedious and spectacular event. In the distance one will see what appears to be a flock of birds getting closer and larger, begin to hear the roar of the engines grow louder and more distinct. Then, as they fly over the field, each crew cbief is looking for his pilot. Yes, and some are missing. Some may do a victory roll and then we wonder. Who was left in the fields across the Channel?

They don't talk about that, but the empty seat in the mess hall is very conspicuous. Death or the loss of a comrade is not openly expressed. But the sentiment of their loss is sincere and genuine. There is that intrinsic sympathy and outward equanimity which is the antithesis in our normal social order, That day when a pilot gets a letter from a loved one, but he is not there to read it.

These men have learned to take adversities for they are continuously undergoing a bitter tempering as they grow to be silent and serious and more aware of life's values and realities. After dinner they silently sit down and read their mail, if any, and in comes the Flight Surgeon and starts a game of poker.

Being in the service during these difficult days the boys have learned the true meaning of friendship and give it a new interpretation, really learning who their friends are and the values of that relationship in their uncertain lives …

Editor's note: This story was extracted and rewritten from an unpublished manuscript of Dr. Richard Arlen, who served as a 352nd Group Flight Surgeon during WWII.

Wartime photo of the driveway approach side of Clermont Hall. The kitchen wing to the left was removed post-war. Note the English lorry and US Command Cars. (Jack Donalson)

Officers McCarthy, Lindsey, DeBruin, Zimms and two unidentified officers being served by Sergeant Feldman in the first Officer's Club — "The Thunderbolt Inn."

PFC Mitchell Simmons of the 486th operates the ice cream making machine at Bodney. (Sheldon Berlow)

Sergeant Nilan Jones of the Support Group 17 was an accomplished artist, as can be seen by the voluptuous Vargas nude on the ceiling of the Officer Club Nissen. Signature reads "Vargas by Jones." Nilan was responsible for many of the 352nd FG nose art, especially in the 486th Squadron, where Richmond's "Sweetie" was a standout. (Sheldon Berlow)

"Roll Me Over"
by Marc Hamel

Not long before his death in 1998, 17.25 victory ace John F. "Smiling Jack" Thornell, often referred to as "Direct-Line" Thornell because he always knew the latest rumors before anyone else in the squadron, shared this story with me. Jack always had that quick smile, a mischievous sparkle in his eye, and a ready spirit. Conversely, Group Commander Joe L. Mason was the typical rough and tough "regular Old Army" type of guy whose lighter side was seldom seen. Jack's story went like this:

"We had a lot of fun in the 328th and early on, before our Nissan huts were built and Colonel Joe made us move into them, we had some pretty wild parties. At that time we were still living in Clermont Hall, a big mansion our government had acquired under the Lend Lease Act. It was located east of our airbase.

"I came out of the Officer's Club one night after one of these parties and said to myself, "Man, you've had too much to drink. If you get on that bicycle you will probably spin-in and kill yourself. It's pitch-black dark and I can't walk back so you've got to find another way to get back to Clermont Hall.

"I looked out front and there was Colonel Joe's staff car and I thought, 'Hey, maybe I can borrow that for a little while and make it home!'

"Luckily, the keys were in it so I got in the car and started it up. As I went out the gate I returned the salute of the airmen guard who obviously recognized the Colonel's vehicle and passed me through.

"The road from the O Club to Clermont forks near the entrance, with one lane going around the back and the other around to the front. I took the wrong one and the next thing I knew I had rolled the car over into the ditch along that lane, making a complete rotation and coming up on its wheels. Believe me, that sobered me up in somewhat of a hurry and I opened my door to get out.

"About that time the back door of the car opened and out steps Colonel Mason. He had been 'sleeping one off' in the back seat and I hadn't noticed.

"'Oh, no, I thought, here it comes. What am I in for now?' The Colonel looked at me and said, 'Man, that was some rough sleep!'

"Many years later, I worked for Colonel Mason in England when he was Chief of Staff of the Third Air Force, and I was in Atomic Fighter Plans. We had a great many good laughs telling about the night we were 'rolling around England' in his staff car."

Submitted by Wayne Stock, Col. Mason's Crew Chief

"Wot A Way To Run A War"— Characters and Places
According to Ted Fahrenwald

Don McKibben of the 486th begins,

"Having flown most of our combat missions together, Ted Fahrenwald and I stayed in close touch after the war. After Ted's death in 2004, I received a note from his daughters that they had uncovered a collection of his wartime letters written to his mother, sister, and brother. He wrote frequently, starting in April 1943, and ending June 4, 1944, four days before his last mission. The typed transcripts fill 223 pages. Ted was not writing for the printed page. This was Ted with his colorful, and sometimes quirky, conversational style, speaking to the folks back home. Telling it like it was—at least most of the time. Sometimes, the chronology of events is treated loosely. "The other day" could have been yesterday, or last month. Who cares? And sometimes he stretched the facts for the sake of a good yarn. No harm done. As he warmed to his subject, some of us could smile, knowing that if Ted were telling the story in person, his eyes would be twinkling, because he would know we knew. Here are just a few selected excerpts."

Ted narrates,

"Our fighter Group, you know, is made up of three squadrons. Two squadrons' pilots (the 328th and 487th) and half of our own 486th live in a fine old chateau nearby which used to belong to some millionaire duke or something of the sort. Clermont Hall they call it, and quite a layout it is. A beautiful place with miles and miles of grounds around it, including forests of giant elms and large grain fields. It also has a shooting preserve—the whole works complete with a multitude of fat cock-pheasants…upon which I have my eye.

"Now what do you think of a place like this? Many rooms and baths with rose colored tubs. A master bedroom in which any number of cats might be swung. Servants quarters, gamekeeper lodge, greenhouse, thick forest well stocked with grouse and pheasants. Well, anyhow that's where this gang is now ensconced. We moved out of our temporary 'sewer pipe' *(Nissen-author)* several days ago

Great casual photo of certified comedian and pilot Ted Fahrenwald. His crew chief Robert German stands by. The 486th Operations Nissen is at back left. (Harold Stanfield)

Stephen "Andy" Andrew, A-Flight Leader. An unsure gunner, but one helluva pilot, diminutive Andy was known for pulling right on an enemy's tail before unleashing his .50's. He was very successful with this approach.

and into this castle where we intend to live for a long time....in the manner in which we ain't accustomed. From where I sit I can look out thru three bay windows and take in a very expensive sight indeed. And expansive. Restful to the eyeballs. Below the window are 500 feet or so of green lawns studded with giant elms and oaks, then a solid background of deep and dark forest. Mostly firs and evergreens of one kind of another passed by winding graveled roads. Hi ho, the gamekeeper hasn't been keeping the partridges out of the orchid garden. At this point I am settin' by an open window. Cool breezes float in. Little bunnies are hoppin' to and fro chasing each other in circles, with what in mind we all know. Gray and white clouds breaking up now and again to let the sunshine thru in nice slanting columns. Ruff way to fight a war, is it not?

"Do my flying from the field down the road, which is reached by Jeep in several minutes when there is flying to be done. Don't envy the boys doing the fighting from the dirty end of the war. I was lucky to be stationed in England, for here we live the happy life in our spare time. Just hop in the cockpit, careen around the uncivilized skies for a while, and then land back home again. Step out back into luxury, comfort, and within range of queens and good whiskey. I reiterate—Wot a way to run a war!"

A few months later the Officer's quarters, mainly metal Quonset-type huts called Nissens, were completed on the northeast corner of the field. The pilots relocated there, and Ted's letters continue from this new form of housing....

"Say, I neglected to say that we've moved out of our castle and into a quaint old corrugated tin Nissen Hut. Matter of fact I like it better. C-flight lives together again: Gerst, Rauk, Green, McKibben, Northrop, Heller, MacKean and me. Now these huts are nothing but an overgrown section of sewer pipe with the ends boarded up to keep folks from wandering through on their way to town. We looked forward to moving into them as we were getting fed up with leading the life of luxury. A little bit chilly sleeping these nights but I don't mind that. This morning at four a.m. I awoke to hear fleet after fleet of our bombers forming up overhead, which means we go out pretty soon.

Pipe smoking, booie-woogie piano playing pilot Al Marshall poses with his P-47 "Sweet Chariot." (Al Wallace)

"We just have a tiny stove in our hut, two feet high and eight inches in diameter. It burns coal, wood, shotgun shells, coke, high octane gas, etc. It wasn't throwing enuf heat so Rocky *(Rauk – ed)* and I removed the firebrick lining with an axe. The cockeyed thing got cherry red from the floor clear up the stovepipe. No one could get within six feet of it. That fixed 'em, alright. Now the stove, having been red-hot ever since, is melting down gradually. So we've been shopping around the field looking for a new one to swipe. Our hut is quite a homey little joint. Six feet away from the stove the temperature takes a slide down to around 35 or 40 degrees, so we wear flying jackets and fur boots when not in close vicinity of the heater. Sometimes we even get a thin, high layer of alto-stratus cloud hovering about the ceiling. But we have a peachy time.

"McKibben just came in from test hopping his ship after some repairs. Fifteen minutes ago a ship come over our Nissen hut, clearing the roof by about ten feet and the suction about lifting the bunks off the floor. I remarks to Greene, 'There goes Mac.' Naturally, it wuz him. He comes in grinning, 'Howdja like that 'un?'

"Had a few beers the other night at the Club and walked from there into A-Flight's hut. Had only been there a short time when I saw a duck waddle across the floor. From one bunk, across the floor, and under another bunk. I looked around, careful like, but nobody else seemed to notice this phenomenon. So I kept my mouth shut and shrugged it off as just another symptom of flak-happiness. Just before I left however, I poked my head back into the doorway and casually asked if they kept a duck in their hut. They said yes. Thank God!

Stan Swenszkowski, pilot Donald "Mac" McKibben, Dick Linn and Luman Morey pose with their oft-photographed P-51B "Miss Lace." Notice the grey painted 75 gallon tank. (Dick Linn)

Leo Northrup and unidentified crewman with Northrop's P-51D "Donna Dae", named for the wartime movie starlet who was Leo's childhood friend. (Earl Martin)

"Inasmuch as the poker game is stalled here for a while, I'd like to introduce you to some of my pals in this here outfit. Sure want to bring some of these boys out to the house some day. You'd like 'em all. Oh me, what a gatherin' we'll have one day around the ol' homestead.

"One of my good friends in the Squadron, Captain Eddie Gignac, is really a character. He had quite a writeup in last month's Esquire magazine as regards his sports and war record. He was a national collegiate ski jumping champion and is highly decorated for his feats in combat flying. Holds the Silver Star for gallantry. He's a little guy, built like a keg. Face all scarred up from a deal in which he was knocked down by Japs in New Guinea. Crashed in the jungle and fished out by natives after a week or so. Anyway, Gig is a natural born clown and is full of the wildest lying stories I've heard in a long time. He's the funniest guy I ever knew...a born comedian. A wild man in an airplane...unpredictable, violent maneuvers. I know, for I've flown alongside for a few 'rounds.' I think he's the best-liked feller in the squadron.

Captain Henry Miklajcyk displays a Polish flyer's insignia on his RAF style goggles. Miklajcyk was a real aircraft buff, and wanted to know everything possible about how his aircraft was constructed. He was KIA in November of '44 on his second tour. (Jim Leber)

"Well the other night he goes down to the kitchen here at Clermont *(English manor house, temporary quarters – editor)*. This in itself requires great skill, as Gig and me and one or two others are quite happy from having cooked up a batch of joy-juice. But Gig comes back with an armful of sugar, cocoa, butter and vanilla and states that he is going to make us some fudge. So he cooks up a batch on an electric plate and soon has all our mess-kits full to brimming with boiling, sticky, highly elastic fudge.....which we 'et' in about two minutes.

"The new cribbage board has been doing double duty ever since it hit the ETO. I got whupped today by Gignac by about 40 pegs, but have cleaned him at countless checker games.

"Gignac tells a droll story of fighting the war in New Guinea. When he wuz there they had no ice and so had to drink their beer warm. That is until he began taking a case along with him in the cockpit every time he went up to altitude to intercept the Jap bombers. The beer would be thoroughly chilled by the time he landed again. Novel, that.

"The other night about the time we were ready to hit the hay, Pappy beats on the door with a club and reels in, full of cheer. He keeps us laughing for a couple of hours with his amusing tales of skiing and flying. Gig's best story, tho, is the mournful telling of his bitter childhood days. He tells it with such a straight face, even in tears sometimes, that we almost believe him. Right now, he's growing a fierce mustache, that's sprouted red hairs in all directions, for his ground crew told him they were going to let their beards grow until he knocked down a Messerschmidt fighter, and they gave him 'orders' to follow suit.

"Lt. Colonel Luther Richmond flies at the front of our squadron and is responsible for a couple of million bucks worth of planes and pilots whenever we take off. He figures tactics and gives the orders over the radio transmitter. A calculating, conservative leader, he knows all the angles and is decisive. He's hard to know and stands off a little ways from the boys. He has much time in the air from running a training command outfit in Texas. He's absolutely fair and square and we never underestimate him, which would be easy to do. Still water running deep applies to Luther.

"We sure like to fly behind the Colonel... couldn't have a better C.O. He's a fine pilot and a cagey tactician. And as we say....he looks after his boys! I remember one day we were back over England after a mission and were plenty low on gas. So Gerst tells the Colonel that he has to land. Luther tells him 'Break away from the squadron, and let down through the overcast. You ought to find a good field right below us...and be careful now, Joe!' We got a kick out of that. Besides that, Luther leads us on all of the tougher missions.

"Of late we outlaws have gotten together in the evenings and held sessions of delightful music. My gittar...Northrop twanging away on a Jew's Harp. Now and then Luther drops by with his 'harmonicy' and adds to the general uproar. He knows fifty-three verses to 'Ol Chisolm Trail' of which fifty ain't fit to print. You know how that one goes...sing a verse alone and then everyone in earshot chips in on the 'Come a ty yi yippee yippee yow yippee yow, come a ty yi yippee yippee yay' part.

"Willie O. Jackson is scheduling flights and leading one himself. Willie's a sharp customer, a swell flier and will take a chance. He's aggressive and spoiling for a fight with any German, any time. Flew with him today, and he leads a smooth flight. Sure like this gent...a real friend. One of the bunch and has earned our respect and admiration both at the bar and in the air. Generally he's got a big twinkle in his eye. That's Willie O.

"Henry (Black Mike) Miklajcyk is a screwball professional avaitor. Our Engineering Officer as well as an excellent combat pilot. Lives, breathes, and talks nothing but flying, and is delighted to be over here chasing around the stratosphere in a hot ship.

"Al Marshall was, I think, a law student before the war. Fine conversationalist and loves to argue, which he does with high logic. Drives home his points with triumphant smiles. Smokes a foul pipe and plays a nice piano, as well as holding his own at the bar.

Martin "Corky" Corcoran of B-flight in a Stateside posed wartime portrait. That grommet would not have been in his officer's cap in England. (Larry Corcoran)

"Here's Ed Heller, and his name fits. Biggest gent in the Squadron, maybe 200 pounds and six feet tall. Ex-Pennsylvania State Trooper, and nicknamed nacherly 'Flatfoot.' A hot pilot and we argue now and again that his flying is no brains and all muscle, strictly, as compared with mine being a case of mind over matter. Powerful big guy and has the heartiest belly-laugh I ever heard. Good friend.

"Corky Corcoran is a peach of a guy. Great, tall, lanky bag of bones and another good pilot. He has a subtle humor and is the only one of the more reckless types who's married.

"Stephen Andrew....what a story! Captain Andrew really did some gory work, and the simple word 'Hamburger' will always refresh my memory, and I'll get to that. Andy is quite a gent. Shot down off New Guinea by Japs, swims five miles or so to shore, and hikes back to his field. The most precise and deliberate pilot in the squadron...cool like a cucumber and very analytical. He's the most curious guy I ever knew, and likes details. He inquires in his mild voice about every little encounter he hears about. For instance, should I tell Andy that I've just killed a rabbit with a rock, he might ask: '(1) Just where did you first see this rabbit, Ted?; (2) Was he hopping, running, walking, or sitting?; (3) How big a rock did you throw and what shape was it?; (4) Did the rabbit see you? (5) How fast was he going and what deflection did you allow and how much did you lead him? (6) Was the rabbit killed instantly or did you have to throw a second rock? (7) Did you examine the body to see why the rock killed him,' etc.

"You can figure out what happens to any enemy ship Andy meets up with. It's as if he gets one a day. He came back home mad the other day, saying he'd gotten too close to this '109 when he pulled the trigger. The enemy plane sprayed oil all over the captain's ship, thus blinding him. He was unable to shoot at the second Jerry ship, and that's why he was sore. He hates to let one get away. Next, he sez, he will shoot from a greater angle so as to let the oil spray miss his own ship. Andrew is a quiet mild-mannered guy on the ground, and a perfect officer and gent indeed. But in an airplane, the guy goes nuts. Completely.

So Much for Protocol

352nd C.O. Joe Mason and Executive Officer "Pop" Clark spent some time coaching their staff car driver, Corporal Ed Kerrigan, who hailed from Sunnyside, New York, the proper way to treat any VIPs who visited Bodney as they occasionally did, VIPs like Secretary of State Stimson and General Doolittle and others.

Kerrigan did real well with such niceties as opening the car doors For them, etc., but once he got under the wheel he would turn to his Passengers and say, "OK, where do youse guys want to go?"

Col Joe L. Mason's staff car.

"So this Andy character spots a Focke-Wulf 190, gets on his tail and starts shooting. Can't hit the damn thing, so he closes up fast. He overshoots and flies tight formation with this jerry fighter. Andy waves at the other pilot, drops back, and starts shooting again…this time with accuracy. The Jerry pilot bails out and is swept back squarely into Andy's propeller. Splat. Usually a blow of such intensity would knock down a fighter plane, but this Jerry hits squarely in the hub with an equal strain on all parts of the propeller. Well, pal Archy *(Lloyd 'Rocky' Rauk – ed)* is flying alongside, sightseeing, and inasmuch as this is a gory sight and Arch's stomach is full of party-type butterflies, he pulls off one fur flying boot and upchucks into it. He tosses the works overboard. On the way home, when over the Channel, Archy figures that one boot is no damn good so he tosses the second one out and into the briny deep. Andrew gets his ship home intact, lands, crawls out of the cockpit, walks to the propeller, picks a scrap of blue flying suit from spinner bolt and pockets it. 'Good souvernier,' he sez, and walks away.

"We shore had a time the other night. Went through the list of songs we used to know and love so well. Red River Valley, Walking Cane, and we are just in the middle of Colonel Richmond's bawdy version of some old Texas song when I get rambunctious and bust a couple of stings on my 'gittar.' A peachy evening. Things boomed most of the night. Some party in his hut, Captain Andrew to you, decides he wants the lights out so he can get some sleep. But his pals want to play cards. So Andy grabs his .45 and 'Boom, boom, boom!' the lights go out and three holes are punched in the tin roof. Nice quiet lot of fellows in this squadron alright."

Close Encounter
by William Neely

The mysterious thing that happened to Lieutenant Colonel Brown over Bremen in 1943 sent the pilot off on a quest that lasted his entire life. Finally he found the answer. It had been worth the wait.

In December 1943, Capt. Charles L. Brown flew his first mission over Germany as aircraft commander of a battle-weary B-17. What happened that day is an extraordinary untold story of World War II. Recently I sat with Lt. Col. Brown, USAF (Ret.) in the yard of his Florida home. His keen memory supported by a diary, he told me this tale.

The target was Bremen, Germany; the specific objective, a Focke-Wuif plant in one of the city's outlying districts. During the preflight briefing at his base in Kimbolton, England, the intelligence officer pointed out the flak areas to avoid. Bremen was protected by more than 250 guns manned by the best artillerist the Germans had. He told the pilots they'd be subject to attack by more than five hundred German fighters in the area. American and Royal Air Force fighters were scheduled to escort them to and from the target.

The group combat formation was to consist of the lead, high, and low squadrons, each made up of a three-ship element followed by a second four-ship or diamond element, for a total of twenty-one aircraft. Brown's ship was to fly on the far left of the second element, or low squadron, the slot known as "Purple Heart Corner."

Lt. George Arnold

After the briefing, Brown and his crew made the short, cold truck ride to "Ye Olde Pub" their B-17F. "As I stood there," said Brown, "I suddenly experienced a quiet, almost tranquil feeling. My thoughts wandered. I had just celebrated my twenty-fifth birthday two months earlier—actually my twenty-first—but to impress my crew and give them some confidence in my ability, I had told them I was twenty-five. The tranquility left as quickly as it had come."

The first of a series of signal flares arched through the ground haze, indicating that it was time to start engines. "Takeoff in any aircraft is an exciting moment for the pilot, but taking off for the first few times with a full bomb and fuel load on a combat mission is nothing short of awe-inspiring," Brown says. "This was no training mission; the guns, bullets, and bombs were real.

"Takeoff was at 0842 and by 0940 we had formed the group at 8,000 feet. Two other groups completed the wing formation of sixty-three aircraft. The cloud cover over the Continent was scattered to broken with most clouds topping below 10,000 feet.

"Our fighter escort, mostly P-47s, were doing their job perfectly until 'Ye Olde Pub' reached the initial point for the bomb run at 1132 when we were at 27,300 feet. During the ten-minute run we would cover more than 30 miles in a straight line. This gave the lead bombardiers time to setup their bomb sights and correct for wind drift and to check smoke and cloud obstructions. It also gave the German defense units time to determine their flight path and altitude.

All Brown could see in front of him was an oily black carpet of flak bursts. He remembered a veteran combat pilot telling him that you were in serious trouble when you were close enough to see the flames in the heart of the bursts. "About two minutes before 'bombs away' I was seeing, immediately in front of us, what appeared to be fantastically beautiful black orchids with vivid crimson centers."

"We're hit!" Two voices yelled over the intercom simultaneously. The nose section was partially destroyed and the oil pressure dropped on the number two engine. Shutdown procedure began immediately while Brown attempted to coax more power out of the remaining three engines. The B-17 lurched skyward as the bombardier called out the welcome "Bombs away!" With the sudden shedding of three tons, Brown thought perhaps he could remain in formation, but the number-four engine was also in trouble, making it impossible.

A sister ship on the left burst into flames and spun earthward, and suddenly 'Ye Olde Pub' was alone, a cripple with a feathered prop and another engine working only sporadically. Either condition normally attracted German fighters like blood draws sharks.

"Enemy fighters at six o'clock," yelled the tail gunner, and the intercom came alive with frantic calls: "Bandits at twelve"; "Six 190s at three o'clock high."

Every crew member could hear Frenchy Coulombe open fire with the twin .50-caliber guns in the top turret, and then Doc joined in with the nose gun. Both were trying to shoot down two 190s that were approaching in a coordinated attack from the ten and twelve o'clock positions.

"I saw the wings of the first fighter light up with machine-gun and cannon fire," said Brown, "and for a fraction of a second I was mesmerized. It looked like all the movies I had seen back home. Then it struck me: This is no movie; we're up to our ass in trouble.

"I pulled up and headed directly toward the attacking fighters. I thought it might scare at least one of them. It must have worked because both planes broke off by rolling over and diving. I heard Ecky screaming in the tail, 'Fighters attacking at six o'clock level.' And then he yelled, 'Get 'em somebody, my guns are jammed! Jesus Christ, they won't fire!'

"I heard the machine-gun fire from the fighters, and I felt the vibration as the bullets and then cannon fire struck us aft. I was scared and I don't give a damn who knows it. I got on the radio: 'Denver One! Denver One! Mayday! Mayday! This is Goldsmith under attack south of Wilhelmshaven. Need assistance!' It was all I had time to transmit on the fighter frequency.

"The next wave of German fighters hit our radio room. They also shot away the controls to the number-three engine. But even with the controls gone, the oil pressure and engine temperature remained stable and number three continued to put out a little more than 50 percent power. Had the engine failed totally, it probably would have been the end of the line."

The only guns still working were Frenchy's twin .50s and the single gun that was left of the nose. As new waves of fighters came at him, Brown turned his plane to meet them, using the battered B-17 as a two-and-one-half engine attacking fighter plane. This threw the German pilots off their routine causing them to close faster with shorter aiming and firing runs. Nevertheless, the American bomber was hit hundreds of times and sixty-degree below zero winds swept through the opening in the nose.

"While I was trying to determine the full extent of our damage," Brown said, "I glanced out the window. Just a few feet from our wingtip was an Me-109. For a moment I thought the heat of battle had been too much. I closed my eyes, figuring it would go away, but when I opened them again, he was still there. I nudged Pinky and pointed to the German plane. His mouth dropped open."

There was something different about this particular Messerschmidt. It was solid black; it was a night fighter.

"We assumed it was only a matter of time before the German pilot came in for the kill. He looked relaxed and confident, and with only one of our eleven guns operating, he had every reason to be. He nodded to us, but we didn't return the greeting."

Brown and his crew figured it was all over. Then the German saluted, rolled, and was gone, putting an abrupt and curious end to one of the oddest encounters they could imagine.

The ravaged bomber left the enemy coast and headed for home. The number-two engine was out, number three still operating at half-strength, and number four damaged by flak and trying to run away at every whipstitch while the copilot worked frantically to keep it functioning.

The rudder did not respond, and the elevators were very slow to do so. With the gaping hole in the nose the aircraft seemed to be swimming through heavy air, grossly overweight for the available power, barely answering its controls. Brown found that by dropping the left wing a few degrees and using trim tabs, he could maintain a relatively straight flight path. The question was, Could he remain in the air long enough to cross a couple hundred miles of North Sea?

"Much to the crew's relief, a pair of P-47s appeared and flew at each wing tip while keeping Air-Sea Rescue informed of our flight path and position. As the 'Pub' gradually lost altitude, the men began throwing out excess weight—guns, ammo, ammo cans, everything loose that wasn't essential." They dropped to 500 feet and still had no sight of land.

Finally, struggling just 250 feet above the water, Pinky called out, "There it is!" (the coast of England). And even more welcome, an air base. The crew lowered the landing gear manually as Brown took 'Ye Olde Pub' in with only one engine operating at full power.

"We had no effective brake or rudder control," he said, "but by a minor miracle, the aircraft remained relatively straight as we stopped, still on the runway."

Every part of the ship had been badly damaged. One wing showed a hole the size of a bushel

George Arnold's P-47 under wraps

basket where an 88-millimeter antiaircraft shell had passed through without hitting any thing strong enough to explode it; the vertical stabilizer was gone, the elevators were inoperative; the radio compartment was destroyed, whole sections of the skin had been torn away, and the hydraulic system was junk. One onlooker described it as a "flying wind tunnel that looked like a piece of Swiss cheese."

For years Brown tried to learn more about this unusual encounter but all his research went for naught. Then, in the July 1988 issue of the 8th Air Force News, he saw a letter in the "Mail Call" section which read:

BOMBER CREW SOUGHT – On either 13 or 20 December 1943, my flight of 12 P-47s came upon a badly crippled B-17 being shot up by five or more Me 109s. We had altitude on them, surprising them and destroying all five. Due to our fuel shortage, we had to leave them and return to Bodney when they were over the North Sea. Is there anyone who might have been on that plane? George A. Arnold 352nd FG, 10426 Brookside Dr., Sun City, AZ 85351.

An elated Brown immediately phoned Arnold, and the pair compared notes on the events of that day more than four decades earlier. Arnold identified the P-47s as part of the 487th Squadron of the 352nd Fighter Group. Official records show the destruction of the five German fighters.

Then, on January 18, 1990, Brown's hundreds of inquiries paid off. He received a letter from British Columbia from a man who had been told of this enigma. It read, in part:

Dear Charles,

All these years I wondered what happened to that B-17... did she make it or not? As I am a guest of the American Fighter Aces group, I inquired time and again, but without any results. I have been a guest at the 50th anniversary of the B-17, and I would still find any answers, whether it was worth a court marshal. I am happy now that you made it, and that it was worth it.

I will be in Florida sometimes in June as guest of the Am. Fighter Aces and it would be nice to meet and talk about our encounter. By the way, after I left you I landed at Bremen Airport with a bullet in my radiator...

Franz Stigler

Needless to say, Brown answered the letter from Stigler within a day of receiving it. In further correspondence he learned that Stigler was then 75, that he did fly on the bomber's wing on that date, and that he had been a highly successful German pilot with 28 Allied victories and had been shot down numerous times himself.

Finally, Brown and Stigler got together and Stigler told him the "B-17 was the most respected airplane he flew against." He said, "There was always a wall of bullets coming at me and I never came home without holes in my airplane. But," he continued, "I was on the ground when I saw this single bomber flying over at a low altitude. It was apparently badly damaged, so I rushed to my plane and went after it," the old ace told him.

"When I got near it, I could see that there was much damage to the nose and tail sections... I flew behind the plane, and I could see the gunner lying across his machine guns. There was a huge hole in the side of the fuselage, and the rudder was almost blown away. It was in very bad shape."

Speaking clearly as he remembered that day, he continued, "I could tell the pilot was in bad shape. I didn't have the heart to finish off this wonderful machine and its brave men. I flew beside them for a long time, trying in some way to help; they were trying desperately to get home, so I was going to let them do it.

"The short way to safety was to turn right and fly to Sweden, so when they banked left and headed for England, I thought, 'You crazy people. I hope you make it.'"

Stigler touches the arm of Charlie Brown, who sits next to him. Both men are silent for a moment. Then Stigler speaks, "I couldn't have shot at them," he says. "It would have been the same as shooting at a parachute. I shot down eleven B-17s, and I always waited to see how many chutes appeared. The more I saw, the happier I was."

Brown smiles slightly at the man he hasn't seen for forty-six years. "If you made a practice of this," he says to Stigler, "you would have been using a parachute yourself."

"I did," replies the German. "Six times."

William Neely is an amateur pilot and author of numerous books on auto racing. This article is adapted from his book, Pilots, published by Simon & Schuster.

Razorback Ace

by Tom Ivie

Moments after the Boeing B-17s unloaded their deadly cargo of high explosives on the airfield at Bramsche, Germany, sixteen pilots of the 487th Fighter Squadron, 352nd Fighter Group, headed for the deck to strafe the ravaged airfield. As the Mustangs streaked over the runways on their firing pass they received return fire from light flak and machine gun positions on the field. The official report described the ground fire as "light and inaccurate," but it was intense enough to deprive "Meyer's Maulers" of their most gifted pilot, Captain Virgil K. Meroney.

Thunderbolt Ace Virgil Meroney cheerfully poses for a publicity photo at Bodney in late Winter 1943-44. (Sheldon Berlow)

The 487th Fighter Squadron's most gifted pilot? Some readers might wonder about that statement and say, "What about John C. Meyer, George Preddy, or William Whisner?" On that, John C. Meyer himself had this to say. In his farewell speech to the enlisted men of the 487th, Meyer talked about the squadron's combat record and some of its notable pilots. In that talk he stated, "Captain Meroney was one of the most natural and complete fighter pilots I had ever seen or flown with. He could shoot well and fly with the best."

The records back up John Meyer's statement. When Virgil Meroney went down on 8 April 1944 he was officially credited with nine victories and he was pulling away from his nearest contenders within the 352nd FG. His closest rival in the Group at the time was Captain Donald S. Bryan of the 328th FS who had 3.0 while Meyer and Preddy were tied for fourth place with three kills each. Captain Meroney had personally accounted for nearly one-third (9 of 29) of the 487th Fighter Squadron's victories and had outscored the entire 486th FS. All of his victories were accounted for while the 487th FS was still flying P-47 Thunderbolts. He went down on his third Mustang mission.

It's difficult at best to talk about what might have been, but the odds are that if Virgil K. Meroney had been given the opportunity to fly the Mustang in combat as long as the highest scoring aces with the 352nd FG, he undoubtedly would have emerged as the Group's top ace. It's a known fact that none of the 352nd's scoring leaders really hit their stride until the Group converted to the Mustang. Meroney's Crew Chief and long time friend, Albert F. Giesting, put it this way: "If the flak had not shot him down so early in the war, he would have been the top scoring ace within the 487th. I have often thought about what fun it would have been to watch the scoring race between Meroney, Meyer, and Preddy. But Fate has its ways."

Virgil Meroney's military career was a long and successful one. It began in May 1937 when the 16-year-old native of Pine Bluff, Arkansas enlisted in the Arkansas National Guard. He was called to active duty as a Sergeant in 1940 and entered flight training in January 1942. After graduating in October 1942, Lt. Meroney became one of the original pilots assigned to the 487th FS of the 352nd FG. During the training days that followed he met and mastered that seven-ton monster from Republic Aircraft called the P-47 Thunderbolt. By the time the 352nd FG arrived in England and began combat operations in September 1943, Virgil Meroney felt quite at home in the P-47's cockpit. He had also honed his shooting skills razor sharp. Al Giesting recalled with vivid detail Meroney's marksmanship on the skeet range snd the high percentage of bullets he put through the aerial target sleeves while in training.

Close friends Al Giesting (Crew Chief), Virgil Meroney (pilot) and Jack Gillenwater (Assistant Crew Chief) were on a roll as a team in the 487th's Thunderbolt era. (Al Giesting)

The 352nd Fighter Group began combat operation on 9 September 1943, but nearly three months would elapse before Virgil Meroney would be able to test his skills against an enemy pilot. The date was 1 December 1943, and the Group's mission was to escort the bombers of the 1st Bomb Division back to England. Rendezvous was made over Holland in the vicinity of Rheydt and the 352nd FG's P-47s positioned themselves over the bombers for the flight home. The main force of bombers was in a good tight formation and was being left alone by enemy fighters, however a number of stragglers flying below and behind them were not so lucky. German fighters were bearing in on the crippled Forts and firing away. The 487th FS was the first to spot the German fighters and dove to the attack. Red Flight, led by George Preddy, got there first and Preddy blasted a 109 out of the air at 24,000 feat. Seconds later Lt. Meroney, leading Green Flight caught a pair of Me-109s at 20,000 feet and bounced them. He reported his first victory as follows, "...I passed over the rear box of bombers at 30,000 feet and saw two Me-109s attacking a straggler behind the bombers. We dove on the E/A and I fired a short burst out of range at 40 degrees deflection, observing no hits. The right E/A broke right. The other E/A rolled over and I fired a short burst at him aiming under him at about 20 degrees at a range of 300 yards. The E/A did not "split-S" but fell off to the left and the pilot bailed out. I then noticed the plane smoking a lot. Claim one Me-109 destroyed."

Lt, Meroney had no sooner pulled up from this encounter than he observed an Me-210 flying along on a straight course at about 10,000 feet. He quickly positioned himself up-sun from the unsuspecting German and then dove to the attack. After two passes Meroney had seriously damaged the plane's left engine and its cockpit, and then his wingman, Lt. Richard Grow, hit the 210 from the right side. Moments later the crew bailed out and the Messerschmidt headed straight down with both engines burning. With the destruction of this enemy plane and his own guns empty Lt. Meroney headed back to Bodney. His score now stood at 1 1/2 victories.

Three days later on 4 December 1943, the 352nd FG headed back to Holland. This time the Group was on a fighter sweep and was looking for enemy fighters to mix it up with. The Luftwaffe was quite accommodating and met the Group shortly after it made landfall over the Dutch coast. The 328th FS engaged the Germans first and the skirmish ended in a scoreless draw. Now it was the 487th Fighter Squadron's turn and White flight led by John Meyer bounced the 109s.

As Meyer disposed of one of the Messerschmidts, Yellow flight was in a dive to join in the action. Lt. Meroney, flying as Yellow 3, saw two Me-109s heading toward them and turned to counter the German attack. One of the Germans put his plane in a dive and Lt. Meroney followed. He closed to 250 yards and cut loose a couple of short bursts. His gunfire set the 109 ablaze and it dove into a farm house far below and exploded. As Meroney and his wingman, F/O J. L. Sweeney, broke off from this engagement they realized that they had become separated from their squadron so they joined up with Blue flight of the 328th FS. In an almost simultaneous movement, the flight was attacked by a lone Fw 190. As the German positioned himself on the tail of Blue 4 he failed to notice that Lt. Meroney and F/O Sweeney had pulled in behind him. Lt. Meroney delivered the first blow and mortally damaged the Focke Wulf before breaking off his attack. Then Sweeney swooped in and finished it off. The 190 was last seen headed straight down in an uncontrolled dive trailing flames and smoke. With these victories Virgil Meroney had increased his score to three and he was now the 487th FS's top gun.

With three victories in his last two missions, Virgil K. Meroney had shown that he was a fighter pilot headed toward acedom. He had proved all he needed was the opportunity to engage the enemy and, if he did, the chances were good that the Luftwaffe's inventory would be reduced.

However, the opportunity to engage the enemy disappeared after the mission of 4 December 1943 and would not reappear until 29 January 1944. In fact the period of 5 December 1943 through 28 January 1944 was an extremely slow period for the entire 487th FS. The squadron was only able to claim 8 ½ victories during that time.

The 352nd's most prolonged period of uneventful missions during this time frame was in January 1944. In the ten missions that the 352nd flew between 1 January and 28 January 1944, the Group encountered enemy planes on only two occasions and claimed one victory. This scoring drought seems odd in a period when 8th Air Force fighters were knocking down enemy aircraft in record numbers, but the reason was VIIIth Fighter Command's new escort tactics.

During 1943 a fighter group would rendezvous with the bombers at a certain point and fly with them until relieved by another unit. Now a Group would be assigned a particular area of the bomber route to patrol and would maintain the patrol until the bomber forces passed through. In

Meroney's P-47 "Sweet Louise" displays nine victory crosses and numerous "fighter sweep" brooms in this view. (Al Giesting)

other words it was just the 'luck of the draw.' If the Luftwaffe decided to make a stand in your sector there were encounters; if not you rarely saw an enemy plane.

The 352nd Fighter Group's luck finally changed on 29 January, and during the next three missions the Group claimed a total of sixteen confirmed victories, one probable, and five damaged. Of that total Lt. Meroney was credited with three victories and one damaged. The first of these victories took place on 29 January when he destroyed an Fw-190 without firing a shot. In that violent dogfight Meroney's tactics and maneuvers caused the German pilot to lose control of his aircraft and spin out. Seconds later the Focke-Wulf exploded into the Belgian countryside.

On 30 January, the 352nd was again providing withdrawal support for the bombers and their sector for this mission was between Ruhlertwisc and the Channel. As they made rendezvous with the bombers an air battle was already in progress. A mixed force of 20 single- and twin-engined fighters were nipping at the bombers so the 352nd FG rushed in to break up the attack. In the ensuing melee the Group claimed 9-1-4 and at no cost to themselves. One of those victories was claimed by Virgil Meroney and it was an historic victory. It was his fifth kill and with it he became the Groups' first ace. He described the day's activities as follows, "Leading Crown Prince Blue flight I turned into 12 plus Me-109s approaching from 7 o'clock at 30,000 feet and made a head on pass at the lowest one firing a short burst from 400 yards and 10-15 degrees deflection. I observed a few hits but did not see what became of him.

"I then made a fast 180-degree turn and chased a flight of four of the E/A. I closed on the nearest one firing short bursts as I closed from 400 yards to 150 yards. My last burst was as he was going straight down at 10,000 feet. The right wing came off and the E/A completely disintegrated. I had no trouble staying with the Me-109 and could over take 'em at will. I pulled up and joined some other P-47s as my flight had become separated."

Lt. Meroney's actions broke up the enemy formation and disrupted their attacks on the bombers. His wingman, Lt. Robert I. 'Iron Man' Ross also downed one of the 109s and raised the 487th's total for the day to 2-0-1.

After its very successful mission of 30 January, the 352nd FG was pulled off operations for the next three days. The break in the action gave the pilots a chance to relax and the ground crews had ample time to give the P-47s a thorough maintenance check. By the time Lt. Meroney and his P-47D 'Sweet Louise' (named for his wife) were ready to return to action, S/Sgt Al Giesting had also added the two new German crosses under the canopy and the plane now displayed five victory symbols.

The stand-down ended on 3 February and the 352nd FG took to the air on an escort mission to Germany. Rendezvous with the B-24s was made over Texel, but the bombing mission was aborted when they were confronted with a 10/10 overcast. The 352nd, however, continued on inland and finally encountered three Fw-190s near Aurich. The 190s were poised to attack the 487th FS when Meroney cut them off and fired at the two lead planes during a head-on pass. He then pulled up and did a tight wing-over which put him right on the tail of one of the Focke-Wulfs. The German turned and rolled desperately but the massive P-47 stayed right with him through every maneuver. Then the enemy pilot made a fatal error and tried to outdive the P-47. Meroney followed the Fw-190 and closed rapidly. At 200 yards he opened fire and then watched as the E/A burst into flames and plummeted to earth.

After his sixth victory Lt. Meroney was given a few days off and he did not return to combat operations until 8 February. The mission of 8 February was uneventful, but when the Group returned to Holland on 10 February, the Luftwaffe was up in strength. The 352nd FG engaged enemy fighters in a series of air battles, but their success was limited because of confusion in the battle area. The Group reported the day's activities as follows: "Group just off English coast when controller advised of change in mission. Bombers sighted mid-channel immediately thereafter. Landfall Zandvoort at 25,000 feet, 1028 hours. General area support furnished to vicinity Vechta. Close support difficult due to large number of fighters with bombers entire time. Nine Me-109s sighted Oldenzohl 30,000 feet at 1110 hours. Combat ensued until P-51s joined fight. During heat of battle R/T 11 identifying P-51s caused our Group to cease firing result-

Showing typical attention to detail, Al Giesting insures his ace pilot Virgil Meroney is set for another mission. The 487th's Blister hangar is in the background. (Al Giesting)

ing in Me-109s getting away with claims of only four damaged. Had P-51s remained away from combat, those engaged believe further victories would have resulted. White and Blue flights of 487th FS attacked three Me-109s, Deventer area at 1110 hours. Prior to this combat in same locality, one a/c seen to explode in field and two a/c seen spinning down, one exploding in mid-air. Twelve to fifteen Me-109s seen S/O Enschede, 30,000 feet. 1110 hours with British markings, repeat, British markings. R/T poor, intense jamming. Intercom satisfactory. Numerous P-51s seen flying singly, bounced by this Group as identification difficult until within gun range."

In spite of the confusion the 487th FS was able to claim two confirmed victories and the 486th FS claimed the four damaged Me-109s. One of the two Me-109s destroyed was Lt. Meroney's seventh victory.

In the sixty days that followed the mission of 10 February, the 352nd Fighter Group saw few encounters with enemy planes. During this period they destroyed 23 enemy planes, and the bulk of those victories (14) went to the 328th Fighter Squadron. The 487th Fighter Squadron was credited with three kills, the first of which was scored by Lt. Sanford Moats on 24 February. The remaining two came in March 1944 and both were scored by Captain Meroney.

The newly promoted Captain's lull in action ended on 8 March during a mission to Germany. The day began badly as six of the Group's planes were involved in mid-air collisions shortly after takeoff and Lt. Bond of the 486th FS was killed. The bad luck continued as the enemy was up in force, but the Group missed the bulk of the action. The fighter groups which had accompanied the bombers of the 1st Task Force through the penetration and over-target portions of the mission had encountered hundreds of E/A and claimed a total of 37-4-8 victories.

By the time the 352nd arrived on the scene to provide withdrawal support to the bombers the action had somewhat subsided. The Group did, however, engage the enemy fighters in two skirmishes. The first occurred near Dummer Lake and Captain Gignac of the 486th FS claimed an Me-109. The second occurred approximately 10 minutes later over Meppen. The 487th FS was being relieved of its escort duties by another unit just as the three enemy planes broke out of the sun and attacked the rearmost bombers. Captain Meroney called out a warning to the squadron and then led his Blue flight in pursuit of the enemy fighters. Meroney chose the leader of the German formation as his target and followed him down. The chase went down to tree top level and Meroney fired a series of short bursts and registered numerous hits all over the ETA. Moments later the plane officially identified as an Me-109 or 209 burst into flames and went down. The pilot tried to escape by parachute. Chances are victory number eight for Captain Meroney was an FW49OD (long-nose) rather than an Me-109. In his encounter report he described the plane as follows:

"I overtook him and as I pulled up on his right wing he jettisoned his canopy. I was sitting right on his wing and got a good look at him. The nose was extra long, and big, so it may have been an Me-209. It was painted in the usual colors with dark slate top and light undersides, with crosses on both wings and fuselage. In front of the cross on the fuselage was a dash and then some black chevrons pointing towards the nose."

His description of this plane, which was still relatively new at this stage of the war, outlined the silhouette of an Fw-190D "to a T." The markings carried on this Fw-190D indicate that its pilot was at least Geschwader Kommodore.

Three days later the 352nd FG embarked on a low-level mission which some former "Bluenosers" described as their most dangerous mission at that stage of the war. The mission was to sweep the Pas-de-Calais area and determine the strength of German flak batteries in that area. As the Group made landfall it split up into squadrons and each squadron headed toward its designated target area. The P-47 pilots found out immediately what a hornet's nest they had entered. They actually started receiving ground fire three miles off shore and by the time they were over land it was virtually impenetrable. None of the squadrons were able to strafe their assigned targets due to the ground-fire, and they suffered damage. The 487th was hardest hit and suffered one KIA, one WIA, and a number of its planes were damaged. The Group did manage to strafe a few flak and machine gun positions along the coast though and did inflict some damage and casualties on the enemy. Captain Meroney included two flak towers and some machine gun positions in his strafing runs. He damaged one gun position and set a second gun silo on fire. The total cost of the mission was two pilots dead, several wounded, and three planes were lost. Numerous other P-47s suffered various degrees of flak damage.

Having determined that the Pas de Calais area was alive with flak emplacements, Eighth FC returned the 352nd FG to its normal escort duties on 13 March. The mission of 13 March was totally uneventful, but two days later the Luftwaffe made its presence known and suffered by doing so. On the mission of 15 March the 352nd encountered numerous Me-109s and downed six of them. Five of the victories went to the 328th FS and the other was claimed by the 486th FS.

The enemy was again encountered on 16 March, although in much smaller numbers, and

"Miss Josephine" / "Hedy" artwork on the crew chief's side of Meroney's "Sweet Louise." (Al Giesting)

the 352nd FG claimed two more kills. The day's victories were scored by Lt. Wendall Parlee of the 328th FS, his first, and Captain Virgil Meroney.

Captain Meroney's ninth and last aerial victory took place over Bar le Due. France. He was flying as White 5 when an Me-109 made a pass at the flight. Meroney turned to meet the German's attack and the 109 pilot made the error of trying to outdive a P-47. Within seconds Captain Meroney had closed the gap and commenced firing at the E/A. One burst blew a number of large pieces from the 109 and smoke and fire began trailing out of its engine cowling. The enemy pilot then tried a loop through a cloud bank but never made it. The 109 stalled out and went straight down trailing flames and smoke until it crashed.

Just a little over three weeks after this action Captain Meroney climbed into his shiny new North American P-51B coded HO-V, and headed out on his last mission of World War II. Over Bramsche Airdrome a chance hit by flak accomplished what the Luftwaffe's finest pilots had been unable to do. Captain Meroney would spend the next year as a guest of the German government.

He was down, but in his own mind, certainly not out. It took him nearly a year, almost to the day, to make his escape. Freedom came on 7 April 1945, during a forced march from one Stalag to another. During the march Meroney and a friend bolted away from the column and took off. Twelve days later on the morning of 19 April they heard a column of armor moving along the road in front of their hiding place and peeked out to see what it was—an armored column of the United States Army's 12th Armored Division. The two tired former POWs rushed out to meet their liberators. The War was now over for one of the "Bluenosers" greatest pilots.

Just what personal qualities made Captain Virgil Meroney the outstanding airman he was? The remarks of Col. John C. Meyer were echoed and amplified by some of his former squadronmates. Fred Allison, who served in Meroney's flight and later replaced him as flight leader, had this to say, "I knew Virg well, and we flew together on many occasions. He was a natural pilot and extremely aggressive in combat. When the enemy was in sight Virg threw all caution to the wind and went after him."

Al Giesting added, "In addition to his skill as a pilot Captain Meroney was an outstanding leader and the other pilots followed his example." When asked about Captain Meroney's exceptional skill at aerial gunnery Al Giesting stated, "He had all the qualities of a born hunter. His keen eyesight, quick reflexes, and familiarity with his equipment and its capabilities were, in my opinion, the reasons for his success. His years of experience as a hunter and skeet shooter back in Arkansas also came into play. It was a real experience to watch Captain Meroney on the skeet range. He was almost casual in his firing stance and yet he knocked one clay pigeon after another out of the sky and rarely missed. Captain Meroney also knew the value of taking care of his equipment. When he was assigned P-47 #42-8473 he told me, 'Sergeant, when you say it's ready to fly, I'll fly it and take good care of it.'

"He had total faith in his ground crew and made it a team effort even down to the naming of the plane. He told us the left side of the plane was his and the right side was for us to name. He was very particular about his 'Sweet Louise' and wanted to fly it on all of his missions. All of his victories were scored in it." Al then added. "It was all of these traits combined with his pilot skill that made him so successful in his air battles."

Virgil Meroney remained in the USAF after the end of World War II and served in a number of command and staff positions. He remained an active fighter pilot throughout the rest of his long career. During the years since WWII Colonel Meroney participated in a number of historic events. He was a member of a flight of four F-84Gs that established a world record flight from Japan to Australia in 12 hours and 15 minutes. The trip was a 5000-mile hop and required three in-flight refuelings.

He also was involved in the first flight made by humans through an H-bomb test conducted at Kawajalein Island. In addition to these accomplishments, Colonel Meroney was also named as commanding officer of the first F-100 Super Sabre squadron based in Japan.

He saw combat action in both the Korean and Vietnam wars—a three-war veteran. In both conflicts he flew fighter-bomber missions. Even though he was assigned to SAC during the Korean War he managed two TDY tours in Japan, where he flew missions over Korea in an F-84 Thunderjet. These sorties were flown with the 27th FBW.

In 1968, Colonel Meroney went to Thailand as Deputy Commander for Operations of the 8th Tactical Fighter Wing, and within six months he was named Vice Commander of the Wing. While stationed at Ubon, he flew 81 missions in F-4 Phantoms over North Vietnam. During that time he took his son, Captain Virgil K. Meroney III, on his first combat mission, and some weeks later Captain Meroney took his father on his last combat mission. Tragically, the younger Meroney was later lost during a mission over North Vietnam and is listed among America's 2500 MIAs.

Colonel Meroney retired on 1 January 1971 after 30 years of outstanding service to his country. For his wartime services the Colonel was decorated many times. His medals include the Silver Star, the Distinguished Flying Cross with four Oak Leaf Clusters, the Air Medal with 14 OLCs and the Purple Heart.

After his retirement Virgil Meroney and his family returned to their native state of Arkansas where he passed away during the summer of 1980.

The author would like to thank the following individuals for their generous assistance in making this article possible: Mrs. Louise Meroney. Mr. Albert F. Giesting. Mr. William H. Kohihas, Mr. William N. Hess. Mr. Dwayne Tabatt, Mr. J. Griffin Murphey III, Mr. Fred Allison, Mr. Sheldon Berlow.

Pre-Briefing Sessions by Ray Mitchell

I volunteered to represent the 328th Squadron, along with Dick Gates, our Operations Officer, at the early morning pre-briefing sessions with Colonel Meyer and "Stormy" Gysbers, our weatherman, and representatives of the 487th and 486th Squadrons, and I will never forget those briefings.

Colonel Meyer would huddle over the table before us and with his pronounced squinting, study the map very carefully. On an occasion when "Stormy" would advise him that the weather in the target area would not permit us to escort bombers that day, I knew what that would mean.

Colonel Meyer would go to the phone and contact 8th Fighter Command Headquarters and ask permission to do an alternate "strafing mission" on one of the many German airfields he had previously spotted on escort missions when we could not leave the bombers to make attacks on these fields.

Strafing German airfields was particularly dangerous and we lost a number of pilots making these attacks during the war. The leading flights crossing the fields had the element of surprise and usually made it with less enemy resistance. However, by the time the following flights came across the enemy ground fire was ready and usually put up intense and accurate ground fire that took its toll. Nevertheless, these were important missions that contributed greatly to the destruction of the Luftwaffe and our gaining air superiority over the Luftwaffe.

Ray Mitchell taking off in his "Carol" at A-84, Chievres, Belgium.

Gig's-Up!

Edward Gignac: The 352nd's All American Flyer

"Leading Purple Flight shortly after rendezvous with bombers, I saw a straggling B-17 at about 5,000 or 6,000 feet. At approximately 1530 I saw an E/A approaching the B-17 from 6 o'clock. I immediately called in the bounce and started down. The E/A scored several hits on the B-17 before I could get to it. As I closed in on the E/A he broke off his attack on the bomber with what appeared to be a violent aileron roll. The E/A then pulled up in a vertical climb. I reefed back violently and took a short burst. My closing speed was very high so I could not follow him up. Even though my burst was very short and at a great deflection, I claim destruction of this Me 109 as Lt. Heller, flying Green 2 saw the pilot bail out just after I fired at him, and he also saw the E/A explode as it hit the ground."

These are the words of Captain Edward Joseph Gignac, USAAF describing one of the high points of his outstanding flying career on March 8, 1944. An excellent Flight Leader and pilot, Gignac never attained the exalted title of "ace" with the Army Air Forces. However, his awards, accomplishments, and the respect garnered from his peers clearly marked him as an outstanding pilot.

This All-American flyer's love of the air, competitiveness, and daring manifested itself in another way early in life—on skis.

"Eddie" Gignac, was born on West Street in Lebanon, New Hampshire on September 7th, 1918. A quiet boy from prime skiing country, Eddie was on skis by the age of three wearing boots tailored to fit his diminutive feet. "Rather on the short side," as a friend later described, Gignac's stature grew slowly, as his skiing prowess advanced rapidly.

Gignac in top ski jumping form, Berlin, New Hampshire, 1940 (Marilyn Gignac)

Around age nine "Eddie" tried leaping from the ski jump in his hometown, and found it to his liking. With encouragement from the local ski coach, jumping became his serious avocation by 7th grade. Soon, a high-school-aged Gignac was making his presence felt in Northeastern competitions, winning several prominent winter meets and perfecting a somersault leap off of ski jumps.

Gignac entered Kimball Union Academy on scholarship in 1936, and gave notice of arrival to the ski world. After major Class B skiing victories in 1936, "Ed" qualified for Class A (top level) ski jumping competitions with a second place at Brattleboro VT during the '37 season (in February of 1938). This set up one of the biggest upsets in ski jumping to that date.

One week later, on February 27, 1938, Gignac ran away with all of the marbles at the U.S. Eastern Amateur Ski Jumping Championships at Gilford, New Hampshire. His unprecedented debut in Class A was marked by outjumping the top Olympic, National and Eastern title holders of the day, giving him the Eastern U.S. jump title.

In the fall of '38, he entered Middlebury College in Vermont and continued his winning ways despite a college football-related knee injury. During the '39 ski season, he joined a national invitational exhibition tour with other top U.S. ski jumping stars, and claimed Lake Placid's Ski Meister Trophy for all-around skiing prowess. Though his nagging knee injury cost Gignac a place on the '40 Winter Olympic team, he bounced back to take the College Class jump title at the Nationals held in Berlin N.H. An ace jumper and competitor in anyone's book, Ed began looking for another challenge for his considerable talents.

With enough college credits under his belt and rumors of war about, Edward Gignac enlisted in the US Army Air Corps in the spring of 1941 after his mother's death from cancer. Technically too short to qualify, 5'3" Gignac wrangled himself into flight training, starting at Darr Aerotech in Albany, Georgia. Surviving upperclassmen, washout checks, poor food, and scrubbing the wind tee for an improper takeoff at Augusta, Georgia, he was soon off to Pursuit School at

The 40th FS's Operations Hut in New Guinea. Gig"is seated fifth from right, facing camera, in a ball cap. America's top early war ace "Buzz" Wagner is standing fifth from left. (Carlos Dannacher)

Gignac's P-400 "BW-117" after after an engine failure and crash landing at 30 mile strip in New Guinea. This crash resulted in the prominent scar seen in later photos across Gig's eye and cheek. (Phil Shriver)

Craig Field. "Learning how to effectively take lives and destroy property" was his joking description of this last stop before graduating Class 41-I as 2nd Lieutenant Edward J. Gignac. This was December 12, 1941, and America had been at war for five days.

With our country in dire straights in the Pacific Theater, Gignac arrived in Australia with the 7th Pursuit Squadron to defend freedom in early 1942. The month of May saw "Gig" transferred to the "Red Devils" of the 40th Pursuit Squadron who were then off to defend Port Moresby, the last Allied holdout in New Guinea. With little experience and inferior P-39 Airacobra aircraft (and rejected lend-lease export versions designated P-400), the "Red Devil" squadron pilots went to battle against overwhelming odds daily.

The following Silver Star Medal writeup describes what June 18th, 1942 held in store for him:

"For gallantry in action over New Guinea, on June 18, 1942. This officer was flying a P-39 type aircraft as part of a flight of three, which intercepted nine enemy bombers and eight enemy fighter planes. The enemy fighters were at a considerable height above the bombers, and when our planes attacked the bombers, they were met by a diving head-on attack by the Zeroes. Lt. Gignac selected one of the bombers and continued to press the attack in spite of the fact that a number of Zeroes were firing bursts into his plane. After the first pass, he chandelled in front of the bombers and, though slightly wounded, he managed to damage another bomber. After the encounter, Lt. Gignac succeeded in flying his crippled plane back to the home base and landing it. His persistence and fearlessness are highly commendable and are in keeping with the best fighting traditions of the U.S. Army Air Corps."

July 11, 1942 brought a winning day for the embattled 40th Pursuit Squadron home team, with five hard fought victories over the Japanese. Gig was not so fortunate however, as he experienced an engine failure just prior to the attack on the enemy formation. Fellow Airacobra handler Philip K. Shriver remembers,

"From 15,000 feet with a glide angle of a flat rock, he managed to cover a considerable distance and crash land alive on a makeshift strip. From that day forward, he carried the mark of every pilot that crash landed a P-39...an imprint of the gunsight on his forehead."

Though carried on the 40th Squadron's roster, extended recuperation from his injuries prevented further combat flying before rotation home in the late Fall of '42.

Gignac, promoted to 1st Lieutenant, arrived home for a well-deserved leave in November. Posted stateside to the 320th Squadron of the 326th FG (an OTU or Operation Training Unit for P-47 Thunderbolts), Edward noted that, he "wanted a crack at the Germans next." This OTU Provided the original cadre of pilots that formed the 352nd.

Gignac poses for publicity photo in Joe Mason's first P-51B. Mason's aircraft were often used as photo props due to their proximity to 486th Operations. (Sheldon Berlow)

Nice profile of Pappy's "Gig's-Up II" P-51B ready for the June '44 Russia Shuttle Mission two weeks after Gignac's death. Note the 108 gallon paper tanks seldom used on 352nd P-51B/C's. (Cy Hall)

Fidgeting at the lack of combat activity, Gignac took a Thunderbolt home for a memorable "buzz" of Lebanon and Kimball Union Academy before being assigned to the 21st Squadron of the 352nd Fighter Group.

Attaining Captain rank in March of 1943, Gignac's experience found him well suited to Flight Leader status in the 486th (as the 21st was renamed in May). The 352nd shipped for England on the Queen Elizabeth, and arrived at their Bodney airbase home in July. Operating from this base, the 352nd produced some of the top scoring ETO aces, and later boasted the nickname "Bluenosed Bastards of Bodney for their blue painted cowlings.

Adapting to ETO missions pounding the Luftwaffe's fighters, now "Pappy" Gignac put his first mark in the victory column with a shared kill over an Me 110 January 30th, 1944 while flying his P-47D-2 Thunderbolt (aptly named "GIG'S-UP"). While the nickname "Pappy" was commonly applied to older experienced pilots, Assistant Crew Chief Art Nellen remembers that Gignac's sobriquet was from his chin's resemblance to the "Lil' Abner" character "Pappy Yokum's" chin. Only war could advance "Little Eddie" to "Pappy" in a few short years.

Undoubtedly this chin was carried high as Gignac was awarded the DFC in February, less than a month before the March 8th Me-109 victory described at the start of this article. This victory was the first using a P-51 Mustang for the Group, which was transitioning from the shorter-legged P-47s in early spring (see the chapter entitled "Raining Thunderbolts" for more on this eventful mission). Though people close to Pappy noted that combat was taking its toll on him, he was advanced to Assistant Group Operations Officer in April. Promotion to the rank of Major in May prompted another new job, acting Group Operations Officer. Gignac proudly wrote in a letter home, "I'm no longer a mere cog, I'm a wheel."

June 6, 1944 is a date that Americans remember with pride. D-Day! The 352nd FG was busy shaking the earth bombing and strafing in France to prevent movement behind the beaches. The following day brought more of the same, as the Luftwaffe was not able to contest the skies.

The second mission of June 7th found Gignac leading a tactical assault flight of Mustangs over France. Pappy's Mustang, "GIG'S-UP II," wasn't available for this mission, so he was flying his CO Willie O. Jackson's P-51B "Hot Stuff." After bombing a marshalling yard at Trappe, a group of prime movers, fueling trucks, and personnel carriers, were spotted near Voisin-le-Bretonneux at 1400 hours. Quickly setting up a "race-track" strafing pattern to prevent mid-air collisions, gunnery passes began. After the second set of passes, a large amount of light flak began reaching for the 486th pilots. Asked if he was going to make another pass, Gignac radioed, "I can't—I'm on fire." The Mustang, hit in the right wing ammo bay and streaming smoke, pulled up to 2000 feet as Pappy tried to gain altitude for bailing out. The plane was then seen to nose down sharply, and explode in mid-air as the right wing fuel tank exploded. Pappy was killed instantly.

Assistant Crew Chief Art Nellen at left checks the rigging of a 108 gallon paper tank under Pappy's P-47D "Gig's-Up." Their revement was immediately east of 486th Operations. (Cy Hall)

Typical wartime publicity cartoon covering Gignac's exploits as pilot and pre-war skier. (Marilyn Gignac)

The Germans in the area removed his dog tags and ordered the French Mayor of Voisin-le-Bretonneux to bury his body, and assist in salvaging pieces of the aircraft. While at first blush the removal of the tags sounds barbaric, in fact this was common practice as the Germans forwarded dog tags to the Red Cross for notification. Sadly, Pappy's tags never made it to the Red Cross, causing him to be listed as Missing In Action (MIA) until much later. It is presumed the Germans with the tags were later killed in the hedgerow fighting in France.

At great risk to himself, the mayor secretly recorded the remnants of the aircraft's serial number on the fin as well as the "Hot Stuff" nose art during the salvage operation. As the graves registration personnel moved into the Voisin area later in the war, the Mayor came forth with this crucial information he had recorded, allowing Pappy's identification to be established later when lost-aircraft records were researched in 1945.

An interesting fact to note is that the French citizens of Voisin-le-Bretonneux never knew the identity of the heroic American aviator that was briefly interred in their cemetery in 1944. In 1999 Gignac's sister Marilyn, using the author's research records, walked into the Mayor's office in Voisin-le-Bretonneux to inquire about her brother. Now a French-speaking Catholic nun, she was greeted with welcoming arms once the official realized that Pappy was their "unknown hero." On June 7th, 2000, the Town of Voisin-le-Bretonneux dedicated a permanent memorial to Gignac, recognizing the sacrifice he made to the liberation of their country.

The author Brewster, writing for the Kimball Union Academy Alumni Bulletin in 1947 eloquently remembered Gignac:

"He was a great little guy; his heart as large as his body was small, and his physical courage was boundless. He flew, wearing the colors of Uncle Sam in his last great event with the same fullness of endeavor that so marked his life. After a desperate losing battle in which his body was riddled and scarred by the bullets of the Japanese, he returned as indomitable as ever to go forth to the European front and there in his last great contest, ride his last plane with its symbolic name, 'GIG'S-UP.'"

Fellow flyers and friends contributing to this article remember Edward as "rather quiet, unassuming, a rock of stability, well-liked, respected, a fine man, excellent pilot, peerless leader, fair, trustworthy, decent, a great gentleman, and one Hell of a man." While a man could have much worse words said of him, they also miss the slightly wild edge that led this All-American to fly from ski jumps and airfields around the world. A great remembrance from his hometown sums this up well: "He was a hellion, in the nicest possible way."

Edward Joseph Gignac is buried in the American Military Cemetery located in Epinal, France overlooking the Moselle River. His awards include the Silver Star, the Distinguished Flying Cross with one Oak Leaf Cluster, the Air Medal with three Oak Leaf Clusters, and the Purple Heart.

Acknowledgments: The author would like to thank Helen Gignac McCaffrey (sister), Maria Downey (cousin), Al Heigh, Joan Bishop (KUA), Kim Ehritt Smith (Middlebury), Carlos "Dan" Dannacher and the members of the 40th Pursuit/Fighter Squadron, Bob Powell and the members of the 352nd Fighter Group, Sam Sox Jr., Tom Ivie, the U.S. Air Force Museum, the Town of Voisin-le-Bretonneux, and especially Robert L. Hamel, for keeping "Eddie's" memory alive for the last 50 years.

Flightschool mates George Preddy, Howard McClatchey (96th BG pilot), and Eddie Gignac show the effects of Snetterton Heath's huge party thrown in November of 1943. Emblazoned on the deuce-and-half trucks sent to pick up English girls, this was billed as the "Red Dress Party," and the 96th BG was reportedly stood down for several days... and the *last* girl left the base three weeks later. The 8th AAF boys certainly needed something to help unwind and maintain morale in late '43. (Joe Noah)

Breaking In the Mustang

By Marc Hamel

It seems impossible to scan any aviation magazine today without seeing the familiar image of a P-51 Mustang against a blue sky. If a survey were done, it is not hard to imagine the top answer to the question, "If you could only have one airplane, what would it be?" A Mustang. With such fame it is easy to forget that this distinctly successful bird was not always a known quantity. It may also startle the novice to learn that these Mustangs did not come out of the Inglewood and Dallas plants wearing blue, red, green, yellow, or checkerboard noses.

The 17th Service Squadron repairs two early P-51B's in the 328th area in April of '44.

This writing will share a little of what the transition was like in early '44 at a small grassed Eighth Army Air Force base in East Anglia. The 9th AAF's Pioneer Mustang Group, the 354th FG, got the first of the Merlin P-51B/C's into the air around December 1st, 1943. By late February of '44 the 357th, 363rd, and the 4th Groups had received and transitioned to the Mustang. The 352nd FG was notified that their first P-51's would be waiting at a depot for them to pick up on March 1st. The first transitional Mustangs were from the 362nd FS of the 357th FG who had been flying them since mid-February.

In the 352nd, it was decided the first Mustangs would go to the 486th Squadron. A few years ago the author asked the 486th's C.O. Luther H. Richmond why his squadron received the first Mustangs versus the 328th or 487th? He replied, "Because Colonel Mason *(the 352nd's C.O. – author)* kept his plane with our squadron, of course." Wayne Stock, crew chief for Joe Mason recalls, "I got the first Mustang, and it was a green one with the throw-over canopy."

Veteran pilot Chester V. "Chet" Harker was notified that he would be leading the small group of pilots to pick up the aircraft. Chet shares with us, "When we heard that we were transitioning from P-47's to P-51's, I managed to scrounge a Tech Order about the plane for our squadron. Like all T.O.'s, it spent most of the print with descriptions of the parts and particles, and very little about flight performance and in-flight characteristics. About the only thing we managed to get out of the manual (between preparing for, recuperating from, or flying the combat missions) was how to get the

Mid April of '44 photo with Rocky Rauk's PZ-R and Ed Heller's soon to be famous "HELL-ER-BUST" being serviced in the 486th flightline. (Cy Hall)

thing started plus the location of such things as landing gear and flap handles, trim tabs and fuel selection levers. We didn't have a loaner P-51 to practice on before our transition, and I don't believe the other squadrons had one until after we had picked up the first batch.

"We flew down to the depot in our UC-64 Noorduyn Norseman (I think they were Canadian built). I don't recall the name of the base—we just thought of it as the 'depot.' I do remember that it was located south of Bodney and N.E. of London. This base was pushing them out as fast as they could unload, tack on wings and give a cursory check. We checked in to Ops at the depot, signed for some planes by numbers, and were told, 'They're all yours now. Saddle up and clear the field.' When we got to the planes and first looked into the cockpit, the only instrument gauges that stuck out were the kerosene float compass and the needle and ball instrument. Other than that there were a helluva lot of black holes where instruments should have been. We were missing, among other little niceties, an altimeter and airspeed indicator. With English weather, it was nice to know how close you were to the ground. Luckily weather was excellent (for England). As I recall we had at least a 1500' ceiling and 2 or 3 mile visibility. This was fortunate because our communication gear consisted of a high frequency 'coffee grinder' radio—no VHF and no IFF on board.

"Since we had no guns, no external tanks and no communications and navigation aids to speak of, the planes were light and able to take off on a minimum amount of runway. If my memory is still with me, we came back in pairs (at least I remember I had a wingman en-route that I could hand signal to). I had managed to borrow a roadmap from one of the lads at the depot so we used it and the seat of our pants for navigation. Fortunately the railroad tracks we located eventually took us to Norwich where we picked up a compass heading for Bodney. My first impression, after flying 40 some combat missions behind the Thunderbolt's R-2800 'armor plate' was, 'Will this sewing machine fly?' It was a dream after I got used to the change in the sound of the engines. Landings were quite a bit different from what we were used to. The plane was light, to start with, and much looser on the controls than the 'Jug' (P-47D). In addition, it would ground-loop if you didn't stay right with it until you were down to taxi speed.

"After a plane was flown to Bodney, the 17th Service Squadron stripped and checked the P-51 before returning the plane to the squadron. The squadron level crews and armorers then installed the communications equipment, fitted the guns, loaded ammo in the wing bays, etc. When everything was ship-shape the planes were returned to us for test hops and combat-ready acceptance checks. I didn't have any trouble with guns jamming except when, on two occasions, enemy fire or flak exploded in the ammo trays in the right wing. The guns didn't jam—the ammo just wouldn't feed."

Groundcrewmen admiring Woody Anderson's new artwork on "Texas Bluebonnet," PZ-Abar. April 1944. (Earl Martin)

Due to the canted mounting of the four .50's in the wings of the P-51B/C's, some units in the winter '43-'44 had problems with guns jamming under high "G" turns in combat. This was due to the sharp break-over of the ammo belts entering the gun breaches, and was solved in some units by adding small electric booster motors to pull the rounds from the trays and into the breaches. Armorer "Stan" Stanfield related that the 352nd cured the problem by simply being meticulous in hand-loading these belts to insure a stray out-of-alignment round didn't cause a jam.

Early shot of Henry Miklajcyk's P-51B "The Syracusan" having its guns harmonized. (Earl Martin)

As we shall see, Chet Harker wasn't the only one to find that landing the new planes was "different." Martin "Corky" Corcoran was an original member of the 486th Fighter Squadron like Chet. They both joined the unit while it was still designated the 21st FS back during Operational Training in the northeast United States, and even lived in adjacent houses as they were both married. Corky was in a flight of pilots ferrying Mustangs on March 4th, and drew P-51B number 43-6569 (an O.D. aircraft with a birdcage canopy).

Corcoran wrote this delightful letter home to his wife a few days later. Colorfully reflecting the flavor of the times, it reads, "March 6th—Well Honey, I'm out of the hospital again. I don't know how long my poor old head will hold out, but there are probably several good bumps left in it yet. This stay in bed was a very short one. The other day I was checking out a new plane. I was sent to another base to ferry up an airplane I had never flown before. I was very much sold on the plane all during the flight. Then I came in to land and I was just a little long. I probably could have gotten it in but I was playing smart—I was going around to make a good approach.

"Well, I got part-way around in nice shape, but then the engine and I began to have differences. It decided that was a heck of a good time to take a rest, and I (thinking entirely differently) started to pull and push everything in that cockpit that would push or pull. At the same time I tried my darndest to reef it in back into our field and land it on the wheels, but it took only a moment to see that idea was nothing but a wild dream of mine. I had neither the speed nor altitude to make it.

"That left me with not a heck of a lot to do but pull up my wheels and look for a decent field to belly land. And there just ahead to my right was a beautiful plowed field very well suited for my purposes. I headed for it and just as I was about to settle into it the engine started to turn over in great style—I thought I had won my argument. Within three or four seconds the engine began to cough and spit. It had caught only just long enough to drag me out of a nice field and into a no good rutted holey hilly junk patch. Of course, I had to be so bright I plowed into the ground downwind with no flaps. All these brilliant little combinations brought me into the ground at approximately a hundred and twenty.

"Honey, if you ever meet anyone in doubt about it, you tell them you have some first-hand information that coming in contact with old mother earth at that speed provides quite a jolt. On first impact I started to have more troubles. First the tail figures it's grown up and decides to start out on its own (see photos). Next, just as I decided very suddenly to lean forward, the gunsight decides to be where head is heading. My head proved just as hard as the steel on the sight, but my skin gave a little.

The P-51B that attempted to kill Corky Corcoran in March 1944 as described in this chapter. The cleanup crew from Bodney looks on from their truck. (Larry Corcoran)

"When the cockpit (about the only part left with me) came to a stop, I crawled out took a look at the multitude of pieces of airplane lying about. At the moment I wasn't a bit worried about this horrible representation of the taxpayers' hard earned dollars because I was sure this cut on my head was costing me no less than a pint of blood per minute.

"The old meat wagon was very much on the ball that day because I had walked hardly fifty yards when I saw it coming. I walked towards it and climbed aboard, whereupon they wheeled me very rapidly to the dispensary on the field.

"In the dispensary the doctors appeared to look upon my wound much more lightly than I. I was still afraid to move my head for fear it would fall off. I swore that much blood couldn't come out unless my throat was slit clear back to the vertebrae. The docs very calmly proceeded to clean out and sew up what I soon learned was nothing more than a two inch gash over my right eye. Then I, who had been treated with such loving care upon arrival was now looked upon with disgust because I could complain of no other aches or pains in other parts of the body.

"I had figured they would turn me loose so long as my injury was so slight. Ah! How foolish I am. I should have known army medics better. Here, in me, they had a wonderful opportunity to slap a guy in bed and keep him there. After transporting me over to a hospital to have my noggin X-rayed, they brought me back and tied me in a bed. As you would expect, the X-rays of my head came back blank.

"As a result of my confinement I missed our squadron party, which I had planned on so much. And all for a little cut on the head. Yesterday evening they turned me loose. I would have written last night, but I was so busy answering a million questions about the accident that the boys were asking that I didn't have the time.

"Today I went in to see Richmond *(Squadron CO Luther Richmond – author)* to talk to him about the wreck. I figured I'd have to meet the board today or tomorrow. The board is made up of all the wheels and meets to place blame for accidents. I was just happy as hell that the board had met on the deal and came out with a verdict I never heard of them giving before. The pilot was considered blameless, and the cause was placed as 100 percent material failure. Nice, huh? I must have told them some fancy story while the docs were patching me up because that's the only pilot's statement they had. That's the way to do it I guess; just demolish the thing so they have nothing to investigate and tell them a good story. (I did tell the truth though, honest I did.)

With his head bandage visible under his officer's cap, Corky Corcoran chops wood to feed the small stove in his Nissen in March 1944. (Larry Corcoran)

One of the Norseman hack planes that served Bodney.

"Today were received cigars from Sontag's *(486th Intelligence Officer)* new assistant. He heard today that he became a daddy last Thursday.

"I'm sorry I have scribbled so long on one subject. It is too late to write more, but maybe tomorrow I can think of something else to write about. Good night, darling. I love you ten thousand times more each day. Mart." We are fortunate that Larry Corcoran, Corky's son, thought to keep the letter and the photo of the wreck he has shared with us here (opposite page).

Chet adds about the party Corky mentioned, "I recall the party. Our liquid refreshment consisted of port wine (N.A.F.I. ration). Unfortunately, one of the pilots in the squadron (we won't name him as he might not like this passed around) had just received a 'Dear John' letter so several of us joined in by tossing off quadruple port wines. We found this pilot a bit later in the middle of the airfield in a somewhat dazed state. Another of my flight was found wandering around down in the swamp near the officers club. I think it was Don Higgins who helped me round them up. I sure had one heck of a headache the next day and haven't touched a glass of port wine since."

Fellow founding 21st/486th pilot Donald "Mac" McKibben adds, "No, I did not pick up any aircraft from the depot, and nothing stands out in my memory about transitioning from the P-47, except that I felt right away that the P-51 was built more to my scale. I liked the closer fit of the cockpit. The other thing that impressed me was that the slender Mustang was 'slippery' compared with the round-nosed Jug, and it took a little time to adjust to this in formation flying." Mac flew on March 8th, 1944 in the first "mixed" P-47 and P-51 mission the Group flew, with the first seven checked-out Mustangs participating. It was flown in intensely bad English weather, and a collision in the overcast resulted in a reduction of the 486th's Thunderbolt strength by four aircraft.

This mission was detailed in Air Classics issues Volume 36, Number 1, January 2000 and Volume 36, Number 2, February 2000.

The total transition took a month and a half in the 352nd, with the 487th and 328th squadrons receiving their planes (in that order) after the 486th, with the 328th flying their first all-Mustang missions in mid-April of '44. Fortunately, the 328th had the benefit of an early production loaner P-51B for the transition (coded G4-D, serial number 43-12456, from the 362nd Fighter Squadron of the 357th Group). Historian and Editor for the 352nd FG Association, Robert "Punchy" Powell allows, "As I recall, my first real knowledge of the '51's was that I heard some of the 486th pilots talking about them. Then, one day when I came in from a P-47 mission I was told (by Bob Lyons, I think) that we 328th pilots were to get 30 minutes in a P-51 parked on our side of the field. I'm not sure if it was a loaner from the 486th or where it came from *(see above – author)*. I do not recall seeing any Tech Order over in our squadron on the plane until much later. I do recall asking our Engineering

Loaner Mustang G4-D from the 357th FG, 362FS, used at Bodney to transition pilots in the 328th from the Bolt to the Mustang.

Officer (Gustavson) what he could tell me about the aircraft settings, etc. but he said he had no T.O. and knew little or nothing about it.

"I do know that when my turn came to get my 30 minutes I took it up to 5,000 feet plus and did a couple of chandelles, then some good hard vertical turns and slow rolled it to get the feel of it. Then, knowing I had to land that critter, I put the wheels down to see how it reacted to a stall and then brought it back in for a landing. It was either the next day or a few days later that I flew my first mission in one.

"I remember that, since we were not familiar with the aircraft, some of the 328th guys aborted their first mission in one because they said they had a 'rough engine.' Then we were told to change the settings slightly and it would smooth up. Never heard this complaint again.

"I don't recall anyone giving us any instructions on flying the aircraft, however. I don't remember how soon after we got the Operating Instructions for the aircraft, but we did and I read the parts which seemed most important to me at that time." *For more on Punchy Powell, see Air Classics issues Volume 29, Number 6, June 1993 and Volume 41, Number 11, November 2005.*

Other problems with the new Merlin Mustangs were encountered and solved by ingenious engineers and crews. North American solved a radiator clogging problem apparently due to dissimilar-metal corrosion, cooling leaks were discovered and overcome, and windscreen defrosting was improved. Fouled sparkplugs were to remain a constant maintenance headache for the tireless ground crews, but refined ground handling by pilots helped this somewhat. However none of these vexing teething problems did anything to diminish the sense of security the P-51B/C provided to bomber crews as they were escorted deep over Germany. Of course, the Luftwaffe knew nothing of these issues either as they discovered with dismay the sleek single-engined fighters over terrain no escort had ventured before.

This article is dedicated to my son and "Junior Bluenoser" Andrew Carleton Hamel. My sincere thanks go out to Larry Corcoran, Chet Harker, Tom Ivie, Mac McKibben, Punchy Powell, Sam Sox, Wayne Stock and our late friend (and 8th Air Force Historian) Roger Freeman.

A nice photo from the time of this story. Al Wallace's P-47D and P-51B in his revetment at the same time. Both were named "Little Rebel." (Al Wallace)

Bryan's Anson Story
by Don Bryan

Aviation author Marc Hamel wrote to 328th Ace Don Bryan asking him to tell him about a crash experience he had in April '44 and this is the reply he received in Don's own words:

"I'm not sure just when this happened but I do know that it was when the 328th Squadron was about to get P-51 Mustangs. Not sure either who all the others were who were involved but I was the one who was ordered to pick some pilots to go wherever it was to get those airplanes. The three pilots in my flight that I picked because I thought they were the best were Alm, Nussman and Quinn. I'm not positive, but I think there were some other 328th pilots involved, too.

"Anyway, a pilot from 8th Air Force Headquarters flew into Bodney in a junkie British twin-engine airplane known as an Anson.

So, we lugged our parachutes out, climbed in and started putting them on. The Anson pilot informed us that we didn't need to wear them in that aircraft so we took them off.

"Since I was the senior officer in the crowd, I got to sit in the right seat (the co-pilot's seat). The others all sat in the back of the plane. As I recall, there weren't any real seats or seat belts in the rear. If there were, nobody was wearing them on takeoff.

"I got into the right seat and started strapping in.

"The pilot said, 'You don't need to fasten the seat belts in this plane. We never do.'

"Where the hell he got that from I don't know. After this incident I found out that his flight to Bodney was his first flight in the Anson.

"We lined up for takeoff, to the NW, I think. At least it was toward the 487th Squadron area. When we got clearance for takeoff, he started to roll—without checking the mags."

"'Aren't you going to check mags?' I asked.

"His answer? 'No. We never do.'

"Well things were going well — until we broke ground. That's when the left engine snorted and the prop started windmilling. There wasn't a thing I knew how to do. I sat there and watched the left wing slowly (it seemed like an age) lower down and start to drag on the ground. It was kind of fascinating the way the wingtip started curling up. After what seemed like a week or so, the nose started down and the tail came up. About that time I said 'to hell with it' and grabbed hold of my seat. Those were pretty good seats. You could reach down and grab the seat bottom and get a real good hold.

"As the nose started to drag, the end tore off and you could see the ground. The nose kept getting shorter and I was afraid that my feet would get caught so I pulled them back under it to keep them away from the ground. At about that time the pilot flew out of his seat and struck me on the left side. It's amazing how strong you can be in times like this. I didn't have my seatbelt fastened and he flopped into me and I never moved from my seat — but the seat moved. When he hit me it tore the seat loose from the cockpit.

"Suddenly things stopped moving and it got real quiet. My first thoughts were 'What the hell, I'm still alive,' and I started trying to get the pilot off of me. When I did, we took off through a large opening on the right side.

"Today, wearing seat belts in cars and aircraft saves lives, but in this case, *not* wearing one evidently saved the pilot's life. One of the props was embedded in his seat and he wasn't in it!

"You can imagine the condition of that plane. A partial list follows.

The left wingtip was curled up.

The left engine had come partially loose and the prop had cut through the cockpit and imbedded in the pilot's seat.

The nose was gone.

The right wheel was missing.

The right engine was completely gone and had rolled about 75 yards down the field.

The entire right side of the plane from the cockpit to just in front of the tail was gone (the three pilots in the back of the plane had exited the A/C through the side onto the wing. The fuselage had broken at about three feet in front of the tail and the tail was twisted 90 degrees to the right.

"Years later when I was in the Training Command I was responsible for reviewing all aircraft accidents in the Command. I reviewed more than 200 major accidents and never once saw a picture of an accident in which the plane was so torn up as the Anson without any serious injuries.

"I think the only one of our guys that was hurt was Lt. Nussman. He got bopped on the head, but by the time anyone got to us he was OK. All of us 328th pilots gathered around the front of the plane. Everyone was OK and when we got in front of that plane and saw what was left of it, we all started laughing and hooping and hollering like we had just survived the sinking of the Titanic.

"Anyway, we were all out in front of it when our Group C.O., Col. Joe L. Mason, came screeching out in his staff car. He stomped up to me and said, 'Bryan, what the hell have you done to MY plane?'

"We didn't stop laughing and I told him, 'It's not your plane, it's an 8th AF Headquarters plane.'

"Then he said, 'Oh, is anyone hurt?'

"When we told him no, he kind of stepped back, looked at the wreck and said, 'By golly it *does* look kind of funny.'

"I never heard another thing about it and I did not get to fly out to pick up the P-51s for the 328th. If they had sent another HQ pilot to ferry me I'd have told them to stuff it."

Don Bryan's youthful look seems at odds with his rapidly climbing victory tally. (Robert Powell)

William H. Alm of the 328th.

Harold Nussman relieves himself before a long flight (known as the "last minute pee") while listening to flight instructions.

The Day It Rained Thunderbolts

by Marc Hamel and Simon Dunham

The 352nd Fighter Group flew an unusual mission out of Bodney on 8 March, 1944, not unusual in its intended goal, but rather in the composition of the Group on the mission. Using a mixture of seven P-51B/Cs and 50 P-47D's, Mission 69 was a success in one respect for the 352nd, and an unexpected, hair-raising tragedy in another. Victories over the Luftwaffe were chalked in the win column, though these were tempered by foul weather and falling fighters.

Donald "Mac" McKibben poses with dwarf "Sneezy" nose art with crew chief Luman Morey below. (Earl Martin)

The Eighth Fighter Command began taking delivery of the Merlin-powered P-51B and C models in early 1944. Older groups such as the 4th FG commanded by Col. Don Blakeslee got the first Mustangs and the 352nd had to wait a few months before their first Mustangs nosed their way onto their field. March 1st saw the arrival of seven of these new fighters, two of which were assigned to the Group C.O., Col. Joe L. Mason, and Lt. Col. Luther Richmond, C.O. of the 486th Squadron. Colonel Mason stabled his mount, "This Is It," with the 486th.

Using a loaner P-51B for training in February, Colonel Richmond had his pilots ready for action in early March. The 486th would fly the March 8th mission with seven Mustangs and 16 Thunderbolts—an unusual and unorthodox procedure due to the different performance characteristics and fuel consumption variance between the two fighters.

John "Curly" Edwards' P-47D HO-E being harmonized. This plane was involved in the mid-air with Flaps Fowler.

The 8th of March dawned ominously with typical English winter weather—a full overcast from 700 feet to 3,500 feet. The pilot briefing revealed the 352nd was to provide withdrawal support for a mixed force of Second Division B-24s and First and Third Division B-17s comprising an aerial armada of some 600-plus bombers returning from striking the VKF ball-bearing factory near the Spree River in the heart of Germany, in Berlin—a "maximum effort" mission. The 328th

Pappy Gignac's soon-to-be "Gig's-Up II" P-51B gets initial maintenance from Cy Hall, crew chief in March 1944. The plane is in its revetment, photo looking north towards 486th Blister Hangar. Note loaner P-47 in background as was used by Henry Miklajcyk in this chapter. (Cy Hall)

and 487th Squadrons each launched 17 P-47s into the gloom. In view of their longer range capability, the Mustangs lifted off first followed by the P-47s, forming up in their usual four plane flights and heading into the dense cloud overcast.

As it was common practice, only the flight leader would fly instruments through the overcast. He would maintain the flight path by careful attention to his instruments as he had no visual clue of attitude in the dense clouds. The other members of the flight had to maintain visual contact with the leader, flying formation on him spaced as close as possible to keep visual contact, usually only a few feet apart. This could become quite hairy if vertigo reared its ugly head.

Vertigo is a sense of disorientation and dizziness brought on by the inner ear sensing

Ted Fahrenwald's Caterpillar Club membership card with original pin. Mac McKibben and Henry Miklajcyk certainly earned theirs. (Ted Fahrenwald)

Caterpillar Club
Certificate of Membership
LT. T. P. FAHRENWALD
is a member of the CATERPILLAR CLUB having saved his life by parachute.
SIGNED
HON. SEC. EUROPEAN DIVISION.
IRVIN
IRVIN

Ted Fahrenwald gets his parachute harness adjusted by beloved 486th parachute rigger Mike Sandorse in front of 486th Operations. (Tom Ivie)

false movement when the eyes have no visual reference. It can easily happen when flying in a dense overcast or dark conditions. The pilot can be in a deadly spiral but his sense of balance in his inner ear doesn't recognize this without some visual reference, in this case the flight leader had to rely on his flight instruments, monitoring them with great concentration and faith, completely disregarding the sensations that he was turning.

Taking off in the first flight of P-47s, Lt. Stan Miles of the 486th, flying his "Bundle of Joy" P-47D, found himself flying in extremely poor visibility. At times his leader and those in tight formation with him disappeared momentarily from view even though they were just off his wingtip. The flight led by Captain Henry Miklajcyk was following close behind in near zero visibility. Captain Don McKibben, flying his P-47 "Sneezy" in that flight, relates this experience.

"Our squadron, comprised of four flights of four aircraft each, was ascending through the thick overcast following the usual procedure entering the overcast one flight after the other. Miklajcyk was flying on instruments and Lt. Bond and I were flying visual contact on him and the fourth pilot in the flight was flying visual contact on me. Suddenly suffering 'vertigo,' he had to pull away and go on his own instruments. Therefore, he was not involved in what happened.

"Apparently, Lt. Miles inadvertently became separated from his flight while trying to fly visual on his leader. Suddenly, I became aware of the form of the shadowy form of an airplane where it shouldn't be, above and to my left, and numerous pieces of aircraft tumbling about.

"I reacted with a violent move up and to the right. Normally I would have recovered and continued on instruments. However, my gyro horizon was tumbling uselessly due to my sharp maneuver and I knew there wasn't enough room under the overcast in which to recover visually. I opened the canopy and started out on the wing. Halfway out I discovered that I was still attached to the plane by my oxygen hose. I jerked off my mask, got out on the wing and jumped.

"I tumbled through the air for a second and pulled the ripcord. It turns out that I was upside down when the chute opened and the opening shock dislodged my escape pack from the pocket inside my jacket. It came up and hit me in the face, giving me a 'shiner' later. I didn't have much time to look around as I floated down since the ceiling

Promotional photo of Lt. Colonel John C. "Curly" Edwards. (Sheldon Berlow)

Ace Henry Miklajcyk on the wing of his P-47D "The Syracusan" after a mission.

was so low. I tugged the risers a little to maneuver into a plowed field instead of a road and saw the flaming wreckage of my plane nearby. On the same side of the road as my crashed plane was a house and a nice lady came out and offered me some tea. On the other side of the road was a thatched roof house that had been set alight by the burning gasoline thrown from the wreckage and some smoldering trees in a churchyard. That bailout earned me membership in the famous "Caterpillar Club" and I still have my caterpillar pin somewhere."

To observers on the ground it seemed to be raining Thunderbolts. Mr. and Mrs. Lemon lived near the site where "Sneezy" plummeted from the sky and exploded. Mrs. Lemon recalled, "We heard a loud clattering noise and then there was silence." Mr. Lemon added, "I was working at Hampton Hall, heard the noise and knew something was wrong. Seconds later a plane came hurtling earthward so I grabbed my bicycle and hurried to the scene of the crash. A few moments later a pilot parachuted down and landed about 100 yards from the wreck. Debris and burning fuel was flung toward a thatched cottage which eventually burned and the elderly lady owner, Daisy Moss, was trapped in her outside toilet by the propeller. My wife managed to remove it so she could escape. Ten or so fire engines from Hethel and Hardwick came to fight the fire but a nearby straw stack also caught fire and burned as well. A sycamore tree in the churchyard 250 yards away was also badly scorched." The gas-filled Sneezy certainly caused a stir when it returned to earth.

Back at Bodney that evening, Mac stopped in at the Officers' Club. His best friend, Ted Fahrenwald, was there and Ted recalled, "I was there when Mac came in for a few drinks. After he relaxed, the delayed shock hit him and he turned white as a sheet. I ended up escorting him back to his bunk."

NO BUNDLE OF JOY

Stan Miles confirms this traumatic incident with these words, "The day was solid overcast. I'm not sure if it was raining but I estimate the ceiling of about 700 feet. I was not flying on instruments climbing into the overcast but flying visual on my flight leader. I lost sight of him and was soon hit from underneath and behind by Henry Miklajcyk, flight leader of the flight behind ours, pushing me over into a dive as a result of the collision. My engine was knocked out and hydraulic fluid and oil came back all over the windscreen so I couldn't see in front of me. I pulled out of the dive right at the treetops and out the side window I spotted a large base with a beautiful runway below. I dropped the gear as it did not depend on the hydraulic system to be lowered and landed on that base. I rolled to a stop and it turned out to be the bomber base at Hethel which had an exceptionally long and wide runway unlike our field at Bodney. Fortunately, I did not try to bail out since, when I got on the ground, the canopy was stuck shut from the impact of the collision.

"I recall seeing a couple of columns of smoke at the time, presumably Bond, Miklajcyk and McKibben's crash sites. A lot of Hethel's personnel met my plane as I stopped. They got me out and took me to base operations where I waited until someone from Bodney came to give me a ride back. I never went to the hospital and I certainly didn't visit the Officers' Club as I was greatly saddened and depressed by the events

That pilots' pilot, 486th CO Luther Richmond relaxes in his Jeep in March 1944. (Rad Choate)

of the day. I was actually in a state of shock for days."

The USAAF "Report of Aircraft Accident" details the damage as "Cowling torn up; engine oil cooler line broken and further damage; and prop tips bent. The plane was left at the 389th BG's base and later sent off for scrap."

THE SYRACUSAN JUMPS TOO

Miklajcyk was intently flying his instruments when the collision occurred, undoubtedly preventing him from seeing Miles' aircraft. With a now crippled aircraft at low altitude, he quickly decided that discretion was the better part of valor as well, and took to his chute. The "Report of Aircraft Accident" flatly describes the P-47 as: Completely Demolished," a simple way of saying

"Sneezy" PZ-Y and its handlers – Stan Swenszkowski (Assistant CC), Dick Linn (Armorer), Luman Morey (CC) and Joe Miraglia (Radio). (Dick Linn)

the big Thunderbolt augured into a hedgerow between Wreninghham and Flordon, almost burying the aircraft.

A piece of Mike's plane landed just a few feet from farm worker Phillip Taylor. "I was busy muck-spreading some 500 yards from the crash site and the exploding shell noise startled my horse. My brother Leslie was working in the field nearest the crash and rushed to the scene. He was also scared by the exploding shells and beat a hasty retreat. He then watched the fire burn up the plane from a safe distance but did not know what happened to the pilot." Miklajcyk soon landed safely in his chute without injuries, also qualifying for Caterpillar Club membership. On hitting the ground and gathering up his parachute he was greeted by a forward young lad from Flordan who asked, "Can I have your parachute, Mister?"

LT. EARL BOND

The report for Lt. Earl H. Bond recounts, "Lt. Bond was flying Red Two position when his flight leader collided with another ship in the overcast. It is believed that in pulling away from the debris of the collision, Lt. Bond lost his orientation and spun in." It also stated that his aircraft was "Completely Demolished." At 1425

A mix of transitioning P-47s and P-51s in the 486th area in March 1944. Andy Andrew's PZ-A at left, with Ted Fahrenwald's Bolt PZ-F in center. (Al Wallace)

hours the aircraft impacted at the edge of a wood one half mile from Margate Hall, Bracan Ash. Earl Bond was killed instantly and his aircraft totally demolished.

After the recovery of Bond's body, the remains of his plane, "The Bid," remained undisturbed for 46 years, partially hidden by a bean crop, dense woodland and ground ivy. The boundary hedge still bears the scars of the crash and the large oak tree remains blackened and malformed in its upper branches. Local English farmer, Rex Webster, recalls the collision. "I remember it being around lunch time and hearing what sounded like aircraft in the clouds fooling around and then some noise followed by three aircraft diving out from the cloud in violent spins. Seconds later there were three separate explosions as they hit the ground, one on the boundary of my father's land in Braconash." Another similar account from Phillip Taylor adds, "I had just returned from dinner about 1330 and was muck-spreading that afternoon on a horse-drawn wagon when the incident occurred. I could hear the noise of aircraft above but could not see them as they were flying in the clouds. Then all of a sudden there was a sound like smashing crockery and the planes came diving out of the clouds. One crashed at Hapton by the houses, another came down near the sheds at the back of John Bett's land in Braconash, and one landed just across the field from me on Webster's meadow. I think another came down on the airfield at Hethel. I remember seeing the pilot floating down up there at Hapton. He looked a bit like one of those thistledown seeds. Now the pilot of the one on the low meadow…I don't know where he came down as the wind must have carried him away from the crash. I never saw him."

A mix of 486th Bolts and Mustangs in front of the Firing Butts mound in March 1944. Stan Miles' P-51 PZ-S in foreground with Earl Meyer's PZ-E "Bonnie Lee" P-47 behind. Photo looking north from field. (Ted Fahrenwald)

Meanwhile, the remaining 352nd FG fighters assembled on top of the overcast and proceeded on their mission to escort the withdrawing bombers after their Berlin attack.

And, of course, the surviving pilots gave great credit to parachute-rigger Michael Sandorse of the 486th, known as "Sandy." He was lauded for his record of 25 out of 25 chutes performing when they were needed during the war, giving the pilots great confidence

McKibben's "Sneezy" crashed threw burning debris onto this home in Flordon, England, which set the thatched roof alight. The locals formed a chain and removed the furnishings before the house burned down. (Sheldon Berlow)

A fantastic find – a photo of Miles' P-47 on the ground at Hethel after being sliced under the cowling by Miklajcyk's Bolt. The shine is due to oil. (Sheldon Berlow)

William "Flaps" Fowler, who tangled with Curly Edwards in the fog. (William Fowler)

that their chutes would save them if need be. Two chutes were issued to each pilot and these were unpacked, dried, inspected and repacked every two weeks or even more often if the weather was particularly humid. Pilots like Luther Richmond, Henry Miklajcyk, Stephen Andrew, Gus Lundquist, Mac McKibben, Ted Fahrenwald, Jim Gremaux and many others owed their lives to Sandy's capable hands and dedication.

486th Thunderbolt pilots at Goxhill gunnery school, Winter '43-'44. Back Row, l to r: Tom Colby, Stan Miles, "Andy" Andrew, Henry Miklajcyk. Front row, l to r: Earl Bond, Earl Meyers, Chet Harker, Bob Babbitt (Chet Harker)

Any Old Port in a Storm
by Charles Rogers

On one mission I will always remember, we were escorting a group of bombers to their target when we got a call from the lead bomber stating their primary target was canceled and asked if we would escort them to their secondary target, which was agreed to do.

I don't remember if we stayed with them until they reached their secondary target or not, but at any rate, when we left the bombers we took up the heading we were given at the prefight briefing and started a let down for our return to Bodney. When we broke out of the overcast there was nothing but water around us and we were unable to contact the English controller for a corrected heading.

Someone volunteered to climb to an altitude where he could make contact with the controller to get and radio the correct heading home to the rest of us. At that time I called those in my flight to take up the new heading we were given and not to accelerate so as to stay in formation.

My altitude was probably about 900 feet or so and I flew for some time with my fuel gauge moving toward empty with some rapidity. Just about the time I was thinking about having to make a water landing (and P-51s didn't ditch very well), I saw a shore line. As I got closer there appeared to be nothing but forest straight ahead so I decided to follow the shore line and just as I made a right turn I was without fuel and the engine stopped. I made a U-turn and sat the plane down on the sand, wheels up.

We had been flying more than six hours, much of this time on the verge of panic, and I just got out of the cockpit and sat my butt down on the wing. While sitting there contemplating my situation I saw an Englishman approaching. He asked if I was OK and I assured him that I was. He commented that I was a very lucky pilot, and then told me that had my plane landed 25 or 30 feet closer to the tree line I would have been in a mine field.

He took me to a nearby RAF base and after a short debriefing there, they asked me for the Bodney telephone number, which I didn't remember. So, they pulled some strings and got the number and called our base. Sometime later, I was picked up by someone from the 352nd and returned to Bodney.

Lt. Charles Rogers

Pas-de-Calais Malaise
by Marc Hamel

The date of March 11th, 1944 was remembered by justly-famed General James "Jimmy" Doolittle for many years after the war. Why? Because it was the day he ordered 36 planes of the 352nd Fighter Group to perform a strafing mission on the Pas-de-Calais area, with the real intent being a "reconnaissance by drawing fire" to assess the German pre-invasion strength along the French coastline. It was one of the few mistakes the General made in his career, and it was an ugly trial-by-fire for the Group. Many years after the war at an 8th AAF reunion in Phoenix, Arizona, pilot Robert "Punchy" Powell had a conversation about this mission with Doolittle, one of Powell's true heroes, while waiting for Doolittle and Senator Barry Goldwater to place a plaque at the monument of World War I ace Frank Luke. The conversation went something like this:

"General Doolittle looked at my name badge and said, "Powell, what fighter group were you with in the 8th?"

"The 352nd Fighter Group, General," I replied.

"Well," he said, "I owe you and your fellow pilots a big apology."

"An apology for what, Sir?"

"Do you remember a low-level mission into the Pas de Calais area of France a short time before D-Day?" he asked.

"I sure do, Sir. We got the Hell shot out of us on that one and we lost some pilots and planes."

"Well," he continued, "That's why I owe you an apology. That was my stupid idea in the first place."

"Sir," I said, "May I ask what the purpose of that mission was?"

"Actually," he replied, "there were two reasons I ordered that mission. First, we wanted to continue our efforts to make the Jerries think we were planning the invasion for the Pas de Calais area, and secondly, we wanted to find out how heavily that area was being defended—where their guns were."

"Well, Sir," I said, "we certainly found out where their guns were. Almost every one of us took some hits from their groundfire."

Three 328th Bolts take off on the "short" south-north runway at Bodney. They are just clearing the treeline, heading north, at the 486th Operations Nissen. (Ted Fahrenwald)

Mission FO 265 was officially intended as a low-level "Rhubarb" mission to attack ground installations, taking off at 0808 hours and heading for West Detreport, Montdidier, Nieuport, and then home at approximately 1100 hours. We have many wartime Encounter Reports that give us a feel for this harrowing Channel-side mission. In addition we have a March 11th letter home from 486th FS pilot Ted Fahrenwald that reveals much more about how it felt to fly the mission than any typically restrained Encounter Report could relate. Ted was well known for his humorous speech and antics, and it is reflected in his writing. Heir to an alloy-manufacturing family in Chicago, Fahrenwald was quite eloquent, though he often spiced his writing "hayseed style" for effect. His plane wasn't named "The Joker" for nothing…

Perhaps the finest flyer to come out of America, Jimmy Doolittle commanded the 8th AAF at the time of this mission. (Sgt. Peterson, 4th Gp)

Ted begins, "Howdy! What a dull week. That's a lie—it's been wild and wooly from start to finish. Recently, in a half hour or so, I did more flying than I have done in the past six months. Here's how it came about, and to hell with the censors.

"About 6:30 am today there was a Group Pilot's Briefing, which was surprising since the field was buried in fog and the clouds were on the ground. So Colonel Joe L. Mason asks for 36 volunteers for a risky trip. We all wanted to go along, so the squadron commanders picked out twelve pilots apiece and the others are asked to leave. So we are briefed for a low-level attack on three airdromes in Pas-de-Calais area, Abbeville being the 486th Squadron's target.

"Pas-de-Calais is notoriously about the most heavily defended of all German-held territory. It's the stretch of French coast across the Straits Dover from Dover, England. It's the coastline where the Jerries sit with their fingers on the ol' trigger, sweating out the invasion. And it is fairly obvious that our High Command wants a bit of first-hand poop on the area. So our briefing consists of planning a surprise attack.

"We are to take off and hug the treetops across England, flying as low as possible so Jerry radar can't pick us up. Then we plan to buzz down the Straits of Dover, parallel to the French coast and half-dozen miles offshore. At a signal, the three squadrons will fan out abreast, make a 90 degree turn, and barrel in up and over the enemy cliffs…whereupon we hedgehop inland to our targets. And we're all flying Thunderbolts this trip, although the 486th had already partially converted to Mustangs beginning on March 1st.

"So I damn soon find myself skimming the choppy waters of the Channel, getting salt spray on my windshield, for the waves are just a few feet beneath us and the ragged clouds are just a few feet above my canopy. I'm at one end of a long line-abreast string of 36 ships, and they look mighty mean. All bobbing and jerking up and down slightly, for it's difficult flying in that little slot between waves and clouds. We dump our belly tanks at a signal and I see them bounce and splash behind the other ships. So we make our complicated swing towards the coast, en masse, and we're enroute to hostile country. It's all a big secret we think, but then in front of us we see many white plumes of water spout up, which is shellfire from cliff gun batteries.

"So I pull up over a cliff and head into the thick of it. I soon found myself clocking plenty of MPH in a crazy, twisting and turning buzz job. As you no doubt know, a buzz job is a zero altitude situation, wherein, if you find yourself looking down at the tops of trees, you get back down to where you'll stay healthy. And that level is about 3-1/2 feet above the local terrain. Some call it "contour-flying" and the British call it "rhubarb-raiding," but I call it one helluva fancy buzzing.

"So the instant we barrel up over the cliff, things go all to hell and it's every man for himself. The country is hilly and the tops of the hills are in the low-hanging clouds. It's a case of jerking the ship up a hillside into the soup, counting 'one, two, three' and pushing the stick forward and hoping you're over the hill. Off to my right I see a Thunderbolt explode and hit the ground. But we try to bore inland, individually. The Jerries cut loose at me with rifles, bricks, shotguns, water-pistols, bofor-guns, 88mm guns, rocks, light and heavy machine guns, old shoes, etc. I never knew before that I could fly with such skill and precision. Had everything up to the firewall (wide-open throttle) and never got above ten or fifteen feet...under high tension lines and telephone wires, etc. About this point the squadron leader hollers, 'Everybody out!' and I suppose each pilot thereupon tried as hard as I did to retreat back to the Channel. I couldn't turn around, for the bastards with their flak kept herding me inland. Saw many amusing things, not counting the tracers weaving their shiny trails around my cockpit.

Ted Fahrenwald with his 486th C-flight P-47 that did not carry a dwarf artwork from "Snow White," just PZ-F. (Ted Fahrenwald)

A typical, massive concrete German anti-aircraft flak tower on the Continent. (Earl Martin)

"Flew down a narrow gulch and I spotted the 20mm shells exploding on the bank beside me. So, I cruised thru a city to get away from the fancy shooting. I think it was Amiens or Abbeville. Flew right down the main street at the first-story level. Shops and houses and stuff going by lickey-split, to say the least. People dashing madly about.... wonder I didn't lop off a few heads with my prop. Then out the other end of town with tracers chasing me, and a skidding turn around a church steeple and they clobber me when I go past. I consider coming back and blasting same, but don't dare pull up for the turn.

"So, up another valley. Came up over a timbered hill and out across a plowed field.... couple of peasants cultivating with a two horse team, which immediately went hog wild when I went by. Last I saw of it the team reared up and the two farmers were hanging onto their cultivator for dear life. Over the river and thru the woods....came upon a farmer perched atop a haystack. When I emerged from the forest heading straight at him he takes a graceful nose-dive and huddles at the bottom of the stack. Went by him about ten feet away and he sure was terrified...I could count the hairs of his whiskers. Felt sorry for the old boy. Many farmers would wave back at me when I went by 'em, but a lot would just crouch by a road and stare. I would wave at one and all...friend of the people, that's me. Headed for a gap in a wooded hill and when I saw it was too narrow for my wings to fit thru, I went on thru up on edge, sidewise. And so forth, to add up to one of the damndest flying jobs I've ever pulled off.

"Did a little shooting too, and seemed able to hit things I aimed at. Much haze and low clouds too. So pretty soon I approach the coast and barrel down thru a kind of gully which led to the beach and then I was over the water again, happy to be there. I figured myself to be the sole survivor, thinking that nobody could fly as crazy and crooked as I'd done and get away with it. But I see a few tattered Thunderbolts emerge from the cliffs and then another, about mid-Channel, stagger around and then the pilot bails out into the water.

"So I fly back home alone and land, and I get out and sit on the grass and have a smoke while I ponder the devious ways of life. My propeller is stained green on all blades from chopping through the weeds and grain fields of la belle France. And I discover that my kidneys have functioned somewhere along the line, without orders from me. So 'hi ho,' I'm on the ground and we wait awhile and count noses. Of our 36, three are down in Pas de Calais, one is in the Channel, one has crash landed at Manston, and eight more are riddled with flak—three to the extent of being scrapped. My ship has only a couple of little holes in it, plus a branch of a tree being stuffed into the airscoop. Whew!"

While the terse Group's encounter reports are not as colorful as Ted's above narrative, they still give us valuable information about the mission from each pilot's viewpoint. Group Commander Col. Joe L. Mason was piloting borrowed 328th Bolt PE-Z, and reported, "I was leading the Group, flying with the 328th Fighter Squadron when we sighted land vicinity Le Touquet. Flying south, we entered the enemy coast at Le Treport on the deck. Taking a heading of 77 degrees, we flew into enemy territory, encountering ground fire from three miles off the coast until we crossed out. Necessary evasive action made it impossible to fly compass course and realizing we would not be able to attack our briefed A/D target, I directed attacks against all ground targets of opportunity. In the vicinity of Frevent, I shot up a combination observation and flak tower on a hill top, and on coming out the enemy coast, vicinity Cayeaux, I opened up on the gun installations along the coast."

486th FS C.O., Lt. Col. Luther H. Richmond, adds from his report, "I was flying White Leader, returning to the coast from inside France, when I had an opportunity to shoot up a High Tension Tower. I noticed a good concentration of hits. The exact place is doubtful, but could have been 10 miles east of Dieppe." Luther was flying his Thunderbolt "Sweetie" with the Nilan Jones eye-opening artwork in Vargas-girl style.

Bob Sharp in the 328th's P-47D PE-V allowed, "I was flying White 3. We made landfall at Le Treport at approximately 0858 hours and went inland on about a 70 degree course on the deck. I saw an army truck but was unable to fire on him. Later I saw a Flak Tower and fired on it with my wing man Lt. H. L. Miller, for about a four-second burst, seeing strikes all over it for the whole burst. Being unable to reach our intended target, we started out and on the way I fired a long burst on another Flak Tower, seeing strikes on it. At the coast I saw a machine-gunner standing by

Ray Phillips shows off the flamboyant artwork for his PE-N "Hildegarbe."

his gun, ready to fire I fired a burst at him and he was obscured in the dust from my bullets. I claim two flak towers and a machine gun emplacement damaged."

Wingman Harry Miller, following in PE-D, confirmed Sharp's report with, "I was flying White 4. As we were on a low-level attack mission we all were on the deck and were carrying a good bit of mercury. Shortly after crossing in, my element leader and I spotted a flak tower. We opened fire on it at about 1000 yards and continued to fire short bursts at it until approximately 100 yards distance. All this time I observed strikes. Shortly after this we spotted another flak tower. I opened fire again at approximately 1000 yards and continued to fire until I had to lift my wing up to miss running into it. Again I observed many strikes. On crossing out French Coast I saw many of the enemy who seemed more interested in ducking in a ditch than shooting at me." A close call with that second tower.

328th FS pilot Ray Phillips adds, "I was flying Spare and filled in as White Two. We made the enemy coast and turned in at the beach. We then flew up a valley where I saw many gun emplacements. The visibility was about one mile and the clouds were at about 300 feet. I fired at a number of what I thought were emplacements. When the Group Leader called 'everybody out,' we headed for the coast and close to the beach I fired again, but make no claims. Just after we crossed out I saw someone bail out and his A/C went into the water. His flight stayed with him. When we got back to England someone on the ground opened fire at us with a good many tracers, which came very close to us. Group Leader had Wing call them off and we came home without further incident." Ray was piloting PE-H.

"Slender, Tender & Tall" pilot Bill Halton details, "I was leading Red Flight on a low-level mission. We crossed in the French coast at zero altitude over Le Treport at 0580 hours, after paralleling the coast from Cape Gris Nex down. We went in over some cliffs about 200 feet high. I gave the rim of the cliff a burst just before I passed over, thinking there might be some small guns there. We passed over several small towns and some woods and after flying in a NE direction for about 10-15 minutes Group Leader called a left turn and in the turn I lost White 4 under my right wing and had to turn sharply left. When I did this I lost the squadron and flew SW for 10-15 minutes and broke out south of Le Treport. I observed ground fire from the woods, from the open fields and the towns. It was mostly light, intense, and accurate with occasional puffs of 20mm. The guns seemed to be dispersed all over the countryside. I attacked a tower which might have been a flak tower or water tank and I observed many strikes. The rest of my firing was purely at random. After crossing out I heard Red 2 calling in a Mayday for Red 3. I flew north along the coast until I found Red 2 and Yellow Leader and Yellow 3 circling the spot where he had gone in. We four stayed and circled until two more P-47's arrived to take over and direct the Air-Sea Rescue Service. I did not see any evidence of Red 3 except an oil slick where his ship had gone in. My return to base was uneventful."

Bill Halton with his P-47D
"Slender, Tender, and Tall" code PE-T

Ace Jack Thornell reported on his experience with, "I was flying Red 2. We continued on course to target at zero altitude. While on course I fired at targets of opportunity which included a gun position, a radio station, a B-17 Fortress, and enemy troops. The B-17 was lying in a plowed field with its wheels up from a crashed landing. It looked like it was in fair condition. Approaching it I fired a long burst at it. It had 1st Division triangle on the tail and wings. After passing this wreck we approached gun positions dug into a hill. It was at these positions that Lt. Schwenke, Red 3. was hit. I saw strikes on his engine but we crossed out together with his ship smoking. He called me and said, "I am bailing out; stay and get a fix for me." We climbed to 900 or 1000 feet and he went out. His chute opened immediately and he hit the water okay. I made a 360-degree turn and came back slow just as his ship sunk and I observed his chute go down but did not see any dinghy. We stayed out over the oil slick for 45 minutes, then came back home. A Walrus *(British amphibious air-sea rescue plane)* was over the spot when we left." Jack was piloting Bolt PE-I.

John Coleman in PE-L adds a little about Schwenke's disappearance with, "I was flying Yellow 3 when we entered the coast of France at Le Treport for a low level attack mission. Finding visibility to be low, about ¾ mile, our Group Leader ordered us to abort, strafing targets of opportunity on the way out. My targets were suspicious looking buildings inland and on the beach coming out at (approximately) Le Tourquet. One of our pilots, Red 3, bailed out nearly 8 miles off the coast at that area (Le Tourquet). So we spent the next 45 minutes circling the oil slick. No evidence of the survival of the pilot was noticed. Obvious to all was the preparation of the German Army for invasion. Flak towers, gun emplacements, gun batteries and gun trucks were ready-manned and fired at our approach."

Former 487th FS Operations Officer and Pacific Theater ace I.B. Jack "Medals" Donalson had just taken command of the 328th when this mission was laid on. Jack's attention to fine detail and operation shows in his report. "I was leading Yellow Flight. We sighted land about Le Tourquet. Flying south we entered the enemy coast at Le Treport on the deck. Taking a heading of about 76 degrees

Captain Donalson

we flew into enemy territory shooting at targets of opportunity. The visibility was about 3/4ths of a mile, ceiling 1500 feet. We encountered ground fire from the time we entered enemy coast until crossing out. I first shot up a heavy gun emplacement of about six or eight inch caliber. I then shot at a rifle range with soldiers practicing, and went on to fire at a camouflaged small caliber machine gun emplacement. On crossing out the coast I shot the area I was crossing out of, observing a heavy caliber machine gun mounted on the cliffs of the coast but no one manning the gun. On leaving the enemy coast I could see the water splashing around my plane from machine gun fire from the enemy coast." Interestingly Jack was still piloting his 487th P-47D named "Mauree" and coded HO-D. The information in Jack and John Coleman's reports must have been exactly the sort that Intelligence was looking for.

Lothar Fieg allows, "I was flying Yellow 2 on the low level attack mission. We made landfall, flew for a few minutes over enemy territory and as we were leaving the enemy coast I fired at a flak tower." He was jockeying his "Denali" coded PE-F.

Today, Punchy Powell recalls this mission like it was yesterday: "This mission was a nail-biter early on. Flying tight formation stacked up just above the Channel waves to stay under the German radar took a lot of concentration. Then, as soon as we made landfall we came under intense and accurate ground fire, sort of like sticking your head into a hornet's nest. Somebody was going to get stung, and we were it. "The visibility wasn't the greatest either, and at our low altitude it was difficult picking up targets but easy to be one. I know that the 'pucker factor' was very much in evidence throughout that whole sweep."

The day's official loss tally was four, not including the one scrapped upon arrival at Bodney. These included:

1st Lt. Harold S. Riley Jr. of the 487th FS, killed by flak near Calais (20 miles east of Pont Joie) flying HO-R, "The Kid."

1st Lt. William O. Schwenke of the 328th FS, bailed out into the Channel 8 miles west of LeTouquet after flak hit his PE-S "Kathie."

2nd Lt. Lawrence F. Allard of the 328th became a POW after being hit by flak near Essen.

Robert H. Berkshire of the 487th crash landed HO-U "Brutal Lulu" at RAF Manston.

And so ended a mission that the pilots of the 352nd, and especially those of the 328th Squadron, would never forget. The "Brass" did indeed learn of the massing of troops and guns at Pas-de-Calais instead of at Normandy. This was no doubt valuable "Gen" or intelligence, but the 328th boys have long questioned the cost of gaining that information.

Three Short Tales from Russ Liebfarth

Russ Liebfarth

"Up Through the Overcast"

It was SOP ***(standard operating procedure)*** in climbing through overcasts in formation for the leader to fly the instruments and the other planes in formation to fly tight formation, very tight formation. On one such flight I was leading the second element of a four-ship flight.

A minute or so after entering the overcast I felt uncomfortable and made a quick check of my flight instruments and discovered we were in a steep banking turn to the left. I quickly went on instruments myself and leveled off. My wingman and I topped out into the clear straight and level. Shortly afterwards we saw our flight leader and his wingman break through some distance to our left. Later, when we returned to Bodney, I asked him what happened.

He said he had been flying instruments using his artificial horizon and it indicated his wings were level and that he was climbing slightly but when he topped out he found himself in a steep left turn and realized he had never uncaged his instruments.

Flying does not allow many "free" mistakes, but that time, he got one.

Russ Leibfarth with his PZ-Lbar "Old Reliable," formerly "Cile II" of Chet Harker. (Leibfarth)

"The Tower Shoots Back"
Russ Liebfarth

This incident occurred on a dive-bombing/strafing mission in the Rheims area of France. We found an abundance of freight cars and locomotives in a marshaling yard and setup a traffic pattern attack to work them over. There was a water tower adjacent to the tracks and one of our strafing aircraft flew close by it and lower than the tower as he was shooting up some of the boxcars. Suddenly he pulled up and called out that he had been hit.

Actually, one of our planes ahead of him in the pattern had hit the water tower and as he flew by it the water was spraying out of the holes with such considerable force that he ***thought*** he had been hit, and he had...but ***not*** by enemy ground fire.

"Chandelles and Champagne"
Russ Liebfarth

We had been on an escort mission during which I had logged six hours and fifty minutes and I was forced to land due to a fuel shortage. The air controller advised me to stay south of the Albert Canal as that area was now friendly territory. I landed between some trenches in a grass field which had been used by the Luftwaffe just the day before.

After landing I turned around to taxi back to the point of touchdown where I found a group of people gathered and my engine died from fuel starvation at that point. There was a Polish unit encamped nearby doing maintenance and repair on their tanks who gathered around my plane.

The mayor of the nearby town invited me to spend the night in his home but one of the Polish NCO's advised me against doing so because there were German SS snipers in the area. So I stayed with the tank unit for three days, sharing a tarpaulin leanto attached to a tank with a Polish officer who was later shot and killed by a sniper about 100 yards from where I was.

On the second day they gave me a squad car, a driver and a submachine gun and away we went toward France to try to find some aviation gas. We hadn't driven far before we spotted some RAF types and asked them if they knew where we could find some av/gas and were told there was none in that area of Belgium, that we would have to go into France to find any. They were in the process of preparing the first advanced airfield for use by the Allies. So on to France we went and eventually found some av/gas. I was able to get 60 gallons in five-gallon Jerry cans and we started back.

The trip into France was quite interesting—there were burned out tanks and trucks and abandoned personnel carriers all along the way.

The day after I returned to the Polish camp we went out to my P-51 and strained the gas into the tanks—30 gallons on each side. We drove our vehicles to the takeoff area to determine how much room I would have for takeoff. I was concerned that it was marginal, but I was amazed at how quickly the P-51 jumped into the air with 10% flaps down, holding the brakes and powering up until the tail came up, and then releasing them and going to full power.

Before takeoff the local citizens presented me with two dozen roses, a magnum of champagne and about 100 personal letters to their relatives back in the States. They all seemed to have relatives in the Detroit area where they worked in the automotive plants since the Polish were well known for their tool and die skills.

All the troops and local citizens signed their names or wrote something on the fuselage of my Mustang and the Polish troops asked me to give them a show when I got airborne. I did what I could—a few low-level passes, a couple of chandelles and a loop. (You should try that sometime while holding champagne in the cockpit...)

"Big Miller's" Big Swim

by Marc Hamel and Fremont Miller

The "Bluenosed Bastards of Bodney" were fortunate to have the services of Fremont "Big Miller" during the war. A contradiction of sorts, this tall, easy-going beekeeper from Wyoming turned into a fierce fighter pilot when the war called. This is a look at an amazing story of survival.

Fremont Miller begins, "I joined the 352nd Fighter Group at Westover Field, Mass. After intensive training in P-47's there and at Mitchell Field on Long Island, New York, we shipped to England on the Queen Elizabeth, going directly to our base at Bodney, a grass airfield with no paved runways, but very satisfactory for a fighter base." The RAF had used it during the Battle of Britain.

Fremont flew with the 328th Squadron of the Group and did all of his missions flying the P-47's. "I named my Thunderbolt 'Red Raider' and, on the nose I had painted a large Wyoming insignia in red—the bucking horse and rider. This brought a letter of commendation from the Wyoming Governor, Lester C. Hunt."

Fremont Miller poses on his P-47 PE-W with the bold Wyoming artwork "Red Raider." (Fremont Miller)

First Victory

On February 4th, 1944, Fremont put his first mark on the scoreboard with a victory over a German Fw-190 fighter. A gaggle of 190s had initiated an attack on the bombers the 328th was escorting and Miller cut one off as it made a climbing right turn. His wartime encounter report reads:

"Here I closed in enough to take a few short bursts at the enemy aircraft. He half-rolled and started spiraling down. I gave him several short bursts and observed strikes all along his wing and fuselage. He quit spiraling and I got several good hits around the cockpit. The canopy flew off and the pilot looked as though he was going to jump out. I kept firing and saw my strikes going into the cockpit and off the wings. The e/a began smoking, and small pieces blew off the wings." Fremont then successfully evaded the fire of two other 190's that had come to the aid of their comrade.

Fremont saved ace Jack Thornell from a German's fire on March 15th, then proceeded to score an Me-109 fighter victory. "We each started after different bandits, but I saw one pulling up on Lt. Thornell's tail so I turned slightly and gave him a short burst from about 400 yards. I did not observe strikes but he rolled over and headed for the deck. I gave him another short burst just before he went into the overcast. I pulled up from this and was gaining altitude when I saw another Me-109 about 2000 feet below me flying at about 4000 feet. I went down on him and gave him a good long burst with deflection of about 45 degrees. I got very good strikes on him from the engine back through the cockpit. I am sure the pilot was dead for he took no evasive action. I last saw him spinning into the overcast."

April 8th saw Fremont gain a ground victory at Quackenbruck airfield in Germany. While ground credits are not currently recognized in a pilot's victory tally, they certainly were during the war. The ground strafing of airfields was one of the most hazardous undertakings of Fighter Command due to extremely fierce and accurate German anti-aircraft fire. Fremont continues:

"....I came in abreast of Captain Coleman and we each took an Me-210 and gave them a good long burst. Immediately they both burst into flames, burning with a dark red flame and black smoke which enveloped the whole front half of the plane."

On April 10th Miller again scored over another Fw-190, this time over Dormans, Germany. Fremont dove down to cover a P-47 that was attacking an Fw-190, and fortunately noticed two more 190s diving in for an attack. "They broke away without firing when they saw me coming. I followed them down and started firing on one at about 600 yards and saw a few strikes. When I was at about 400 yards I got a few more strikes and his right landing gear came down. The pilot then bailed out and his plane crashed into a little patch of woods."

Eighth Fighter Command selected April 15th for the first "Jackpot" strafing mission, planned exclusively to destroy German airfields. The mission put over 600 U.S. fighters into the air simultaneously to try to cripple the Luftwaffe's fighter force. While a rousing success, it was not without cost, and losses for the Bluenosers, including Robert "Iron Man" Ross of the 487th Squadron, the 486th Squadron's popular C.O., Luther Richmond, and very nearly Fremont as well. Fremont recalls that mission more than any other:

Captain John Colemen. (Sheldon Berlow)

"On April 15th, 1944, we were called out for an early morning briefing. The only difference with this mission was the fact that we would only be strafing. In fact, the entire 8th Fighter Command was attacking airfields in Germany. Our target was Diepholz.

"The takeoff was without incident. I was the flight leader of the fourth flight. The 'spare' plane was with me, making five planes in my flight. Normally, the 'spare' was sent home before we crossed the coast of the Continent if no one had mechanical problems and had to abort the mission. Since this seemed like a really fun mission, and he didn't want to miss anything, I let him tag along.

"The four flights separated several miles from Diepholz. My flight was to come across the field from South to North. I lined them up abreast and we were set to go across the field at a very low altitude. As there were no planes at my end of the field, I concentrated my firing on several buildings there. Just as I pulled up to go over the buildings, one of them exploded and I was caught in the blast. Some of the debris struck my plane.

"My left wing had a large hole in it. The metal skin was flapping in the breeze. Something had broken the oil line between the engine and the supply tank. The hot oil was coming out of the engine and covering my canopy instead of returning to the supply tank. Knowing that I would be in trouble after the 30 gallons of oil available had run through the engine, I took a heading for home. I looked back, before the back of the canopy was covered with oil, and I could see a huge black cloud of smoke. I don't know what was in those buildings, but it must have been bombs, ammunition, or fuel.

"I never was too great at flying on instruments but I had to rely on them now. I tried to open the canopy, but hot oil came in, so I had to close it. The Squadron Commander called me and told me to bail out, as he thought my plane would catch fire. I didn't answer him because I didn't want him to know I heard him. Anyway, he didn't know how hot that oil was. I would have been covered with it on the way out. The other members of my flight couldn't stay with me since I was leading them right through clouds. Also, it wasn't wise to be too close to my smoking plane! I finally got up to about 10,000 feet, a fairly safe altitude to cross the coast so that the flak gunners wouldn't be so apt to shoot me down.

"My plane wasn't flying too great. The motor was running rough and the left wing kept pulling down. By applying a little cross-controls (right stick and left rudder), I kept it heading in the right direction. I thought everything was fine until all at once my instruments started to go crazy. I was losing altitude very fast. When I pulled the stick back to raise the nose of the plane, the engine picked up speed. When I pushed it forward, I would be lifted off of my seat.

"Although by this time I knew what was happening, I couldn't believe it. If I had been doing aerobatics, this would have been called a 'Split S.' This maneuver is performed by pulling up the nose slightly, rolling the plane on its back, and pulling the stick back, adjusting the throttle as you fly through this power dive. We were advised never to try this under 10,000 feet in a P-47 that was in good condition and with good visibility. I had none of these conditions at the time.

"My altimeter was unwinding so fast I had little time to do anything. I kept trying to pull it through. After I passed my last 1,000 feet, the odds against me were pretty great. With the lag in the altimeter and the condition of my plane, I didn't think I could possibly do it. Yet I put all the pressure I could on that stick. About that time I blacked out. I thought I had died and gone to heaven. The feeling was great. I had the most beautiful visions, almost impossible to describe. When I came to, the motor had almost stalled out. I was sitting at about 5,000 feet, going straight up. I pushed the stick forward, trying to get leveled off before the motor stalled completely and made a 180 degree turn to get back on course and then climbed back to 10,000 feet.

"About that time I tuned my radio to the emergency channel and called the British Air Sea Rescue (ASR). Contact was almost immediate and very clear. I gave them a very brief description of my problems. They instructed me to keep my heading and altitude and stay in contact with them. They must have called me every two minutes to get a new fix.

"It wasn't long after that when the oil quit running over the canopy. I opened the canopy and looked out. All I could see was water. The engine didn't run five minutes more. It quit completely. I almost had to go into a dive to keep the plane from stalling.

"A little voice came over the emergency channel. 'Are you going to ditch or bail out?' I said I would bail out at 1,500 feet. The voice said, 'Keep talking so we will have a fix at the last minute.' I started reading off the altimeter. At the same time I loosened my safety belt and shoulder straps.

"As this was my first experience bailing out, I thought of all of the advice we had been given: 'Don't jump out or the tail of the plane will catch you. Crawl out of the cockpit onto the wing, sit down, and slide off the trailing edge of the wing.'

"When I reached 1,500 feet, I pulled the mike and earphone connections and crawled out on the wing. I sat down and slid off. Then I counted to ten. During the time I looked down to make sure I had the ripcord. I gave it a good strong tug. When the parachute opened, the buckle on the chest strap hit me in the mouth. I thought I knocked out a tooth. While I was checking to see which one was gone, I remembered to unsnap the leg strap before I hit the water. Before I could do that, I was in the water. Then I remembered to inflate one half of the Mae West (inflatable life vest), then get out of the harness and inflate the other half. By this time, the parachute had settled over me. I had to work to get it off of me.

"My one-man dinghy had been a pad on my seat-type parachute for over a year. The CO_2 bottle in the dinghy always seemed to be right on the point of my butt. There were many times I would have liked to trade it for a sponge rubber pad. At that time I was thankful I had not.

"The instructions were: 'Spread dinghy on top of water and inflate slowly.' This I did. Then, 'Take hold of handle on each side of dinghy and pull yourself over the small end.' This I started to do, but water ran into the dinghy. I thought I would jump over the big end. That didn't work. The dinghy flipped over and landed upside down on top of me. So I pulled myself in over the small end.

"I knew ASR would pick me up right away so I didn't even bail out the water. I recalled being told that if we could not get in the dinghy right away we probably wouldn't make it. Now the clouds were rolling in and the weather looked more threatening all the time."

Discussing this event after the war, fellow 328th pilot 'Punchy' Powell explained, "I remember being told that if we did not get into the dinghy within 30 to 60 seconds in the North Sea, which never gets much above 32 degrees, our muscles would be too cold to allow us to do so."

Don Bryan adds, "I was coming back later from this mission on April 15th, and I spotted someone in the drink. For some reason or other, I knew it was Fremont."

Fremont continues his story: "Within an hour or so, I heard planes circling overhead," Miller recounts. "Once in a while I could see one of them flash through an opening in the clouds. My boys had gone home and refueled and the ASR people had guided them back to the exact spot where I was floating in the North Sea. They couldn't see me through the fog and clouds, but they knew where I was. Before long, I heard a high-speed launch coming. It went by me, within a couple hundred feet. All I could see of it was just above the waterline, as the rest was obscured by fog. After it passed, it turned left and worked a square. When they got down to the middle of it, I wasn't there. By this time, it was getting dark. The planes went home and so did the launch. I wasn't discouraged because I knew they would be back tomorrow. So, I bailed the water out of the dinghy and settled down for the night.

"After dark, the wind came up and the waves were really high. I found that while riding in a dinghy you cannot keep your back to the wind. Every time I got to the top of a wave, the rolling top always hit me in the face and the water would run down my neck. I stayed soaking wet all night.

"The next morning, the 16th of April, the wind let up, but it stayed very dark and cloudy. Planes could not fly. Some kind of big black bird was sitting on a floating object, perhaps driftwood, about 100 feet from me. It looked like a big raven, only larger, maybe a turkey buzzard. I don't know.

"Water, water, everywhere, but I had no water to drink. I did have three small squares of butterscotch candy—about one inch square and a half inch thick—plus five milk tablets. Because of uncertain weather, I wanted to ration this meager supply of food. So I ate one-half square of butterscotch on that day."

On the 16th Don Bryan was itching to get out to find his friend, but the weather was awful over Bodney. "The day after Fremont went down I almost got four of us killed pulling a stupid stunt. I tried to get a flight of us out to go find him, even though the weather was lousy and we were stood down. I was trying to fly low under the weather and get out to the water. We took off in P-47's with the men echeloned off of me, and almost immediately ran into a snow storm. I happened to look down as I entered the snow storm, and saw a patch of ground below. Rather than go on instruments, I kept my eye on that spot, turned the flight 180 degrees out of the mess, and landed back at base. It was a foolish thing for me to try and fly a mission in that mess."

Fremont continues, "The second night was almost identical to the first. The next day, the 17th, was like the day before. Even the 'buzzard' was there. I don't know if he stayed all night or came back in the morning to see if I was still alive. I was determined to outlast him, so I ate another half square of butterscotch. My dinghy was getting quite soft, so I tried to put some air in it with a small bellows-type hand pump. I don't know if it did any good. The third night was very bad. The wind was terrible. The waves were higher than before, or so they seemed.

"When morning came on the 18th of April, the sky was clear and the bird was gone. During the day, I could see planes flying over. Bombers and fighters were going on another mission. I ate my lunch, another half square of butterscotch. I had a mirror in my survival kit, and used it to try to attract attention. I realized it was a meager effort, but I had nothing else to do.

"Late in the afternoon, a P-47 came over, very low. I waved my arms as hard as I could. The plane circled and climbed up to where he could contact ASR. I knew he was reporting my position.

"In the 1970's I learned that a 352nd FG flight leader, Don Bryan, coming back from a mission, called his flight and said, 'Let's go down and see if we can spot Big Miller.'"

Bryan was recently asked if he thought they would actually find Miller that day, and he replied, "Oh heck yes! As ornery as he was, I knew he was alive! Remember, I was pretty sure he was in his dinghy that first day."

"Miller continues, "The 16 aircraft of the 328th FS spread out at about 500 feet and searched the area they had searched three days before. Captain Bryan spotted me and called ASR."

Bryan adds, "We searched in a pattern that expanded in 30 second blocks. You flew 30 seconds one direction, turned 90 degrees and flew another 30 seconds, then turned again, and so on until you made a square. The next time around the square you expanded each square side by 30 seconds until you found the guy. The British ASR was awesome. They gave us the location to search, we started the '30 second' search pattern, I looked down, and there he was! Damn they were good. They had calculated the wind and water currents, and come up with the spot to search. Remember they didn't have computers. They did this in their head, and they put us right on the money."

Fremont continues: "In a very short time, the ASR launch appeared. It pulled up beside me. The planes then went home. The men on deck threw me a rope, and I climbed it hand-over-hand. I was never able to do this before.

"That was the end of 76 hours of sitting in my dinghy in the North Sea. The crewmen carried me down into a warm room and gave me a dry suit, similar to coveralls, made out of heavy woolen blanket material (or so it felt). After I got my wet clothes off and this woolen suit next to my skin, it felt just great. They gave me the choice of a shot of rum or a cup of broth. I chose the broth. It was so good they let me have a second cup.

"Then I lay down on a bed and was covered with blankets. They asked the identity of my group and squadron. This information was radioed to shore. My squadron was notified. The answer came back, 'We haven't lost a pilot for three days. It couldn't be one of ours.' They asked me again and I told them I had been out there for three days. This was the first time I realized that some of my outfit had given me up as lost. Never-the-less, Major Earl Abbott got into a little Tiger Moth and flew to the coast and was there to meet the boat when it docked. When he came through the door and saw me, he turned white as a sheet —just as if he had seen a ghost. He just couldn't believe it was me.

"They took me to the Royal Navy Hospital, somewhere in the Great Yarmouth area. I stayed there from April 18th until the second of May. I can't remember the doctor's name, but he was well up in years. He had a red beard, trimmed up in a Van Dyke style. He was a wonderful doctor, and the nurses were the most dedicated I have ever seen. They didn't have much to work with, but none of them ever complained. I do not know if there was more to the hospital than the ward I was in. It had about 30 beds, and they were filled most of the time.

"The doctor had my feet elevated much higher than my head, and there was a metal cage over my feet to keep the sheets and blankets from touching them. They had a fan blowing cool air over my feet to keep them from spoiling. They told me that the blood circulation had been cut off from my legs and feet while I was sitting in the dinghy and cold water for so long. My feet were swollen so much it looked like the skin would break, and they were almost black in color.

"A lot of doctors and other people came to see me while I was there. Most of them asked me the same question, 'How did you survive for so long?' Most of them concluded that it was my having lived my life in a climate where winters were extremely cold. They surmised that it made my body able to cope with the extreme temperature. Also, I was in very good physical condition. I was of large stature, weighing over 200 pounds.

"My idea of how I survived was somewhat different. I believe it was due to my frame of mind. I knew I wouldn't die after having gotten so far. If I was going to die, it would have been back in Germany. I couldn't help but wonder why the Lord had spared my life, and what he expected of me. I had been raised in a Christian home. This experience made a better Christian out of me.

"Some doctors suggested that my English doctor should take my feet off. He was a very stubborn man. He told me that he would take care of me as he saw fit. Thank Goodness! One evening, after he made a very careful inspection of my feet and legs, he told the supervisor that if the circulation didn't improve by morning, he would have to take my feet off. After she told me what the doctor said, my reaction was, 'Okay.' By this time I was thinking the same thing.

"The doctor came in the following morning to check again. While he was checking for a pulse just below my ankle bone, I saw a faint smile come over his face. He looked at me and said, 'Let's wait one more day.' We did, and the pulse got stronger and the crisis was over. Although it was very painful, I was very, very grateful.

"The food I had in the hospital was the best they had. It was well prepared—mostly fish with a vegetable. Every morning I had a fresh egg. I didn't see anyone else with an egg for breakfast. They were so good to me. The only nurse I can remember was Peggy Nugent. She was just like an angel. Seems as though she was working most of the time. She would stop by my bed at night to visit or bring water. I wasn't able to sleep very much because of the pain in my feet and legs.

"About the second of May, they transferred me to our base hospital at Bodney. I wasn't able to walk or even put weight on my feet. One of the enlisted men on our base, Sgt. George Wilcox, of Winstead, Conn., made me a special wheelchair out of bicycle wheels and tubing. It was on the order of a tricycle. I peddled it with my hands and steered the front wheel. I hand-peddled this contraption all over the base. Several times, going around the perimeter track, I turned it over. Someone in our control tower would usually spot me, call the 328th to tell them I had 'spun-in,' and they would come by and put me back on it.

"The boys in the 328th wanted to do something for the English nurses and Royal navy Hospital, so they had a party. Lots to eat and drink. Our music was on records, but they danced and had a lot of fun.

"I was transferred to the 65th General Hospital in England about the second of June, and back to the States on June 5th of 1944. They placed me in the base hospital at Mitchell Field, Long Island, New York for a few days, then sent me on to Fitzsimmons General Hospital east of Denver. I stayed there until the first of July. After a 30-day leave, I went back on active duty until the war was over."

Don Bryan recalls Fremont with affection today, "That guy was the biggest, strongest person I have ever seen. At the time I was one of the smallest pilots in the unit at 120 pounds. As a joke, Fremont would come up to me, grab me by my belt, flip me over, and hang me upside-down by my ankles. He would hold me out so I couldn't even grab his legs or anything. Then he would laugh and shake me until everything fell out of my pockets saying, 'Hey, Bryan, got any change?' He was that big and strong! Even though he would pick on me like that, it was all in fun. He is a fine guy."

352nd Historian "Punchy" Powell remembers Fremont for his western drawl, big smile and hearty laugh.

"One time we were weathered in and sitting around the 328th Pilots' Room and Fremont was reading the Stars and Stripes, the military newspaper. Suddenly he looked up and said, 'It says here that some young feller back in the States has set a new record for the high jump—something over six feet it says. That feller would be plumb hard to keep corralled, wouldn't he'?"

Powell also recalls a visit he and Betty had from the Millers some years later. Answering a telephone call, Punchy said this voice said, "I'll bet you don't know who this is." After Punchy said he didn't, the caller said, "Well, this is Fremont Miller from Crowheart, Wyoming."

"Where *are* you, Fremont?" Punchy asked.

"We're up here at a Shell station trying to find out where you live," Miller replied, so Punchy directed them to his home, only a mile away. It was the whole Miller family—his wife, Bernice, two sons and a daughter. The Millers were originally from North Carolina and had spent the holidays with some relatives there and Fremont had decided he wanted to visit his old wartime friend, Punchy.

Fremont Miller was able to hand-pedal himself around base on this trike after his North Sea adventure due to the efforts of S/Sgt's Wilcox and Hynds. (Fremont Miller)

Since this was just a few days after Christmas—turkey hash time—Betty pulled some steaks out of the freezer and fixed them a good dinner. Afterwards, Punchy said, "Fremont, you *are* going to spend the night with us, aren't you?"

"Wal," the big guy said, "I was kinda hopin' you were going to ask us." And they did.

Punchy recalls, "Although we had five bedrooms, Betty's mother was with us and we had to farm our kids out to the neighbors to make room for them. Fremont and I stayed up talking half the night when everyone else had gone to bed. I asked him if they could stay a few days so we could show them Atlanta, but he told me he was a member of the Wyoming Legislature and they had to get home before it reconvened and they would have to get an early start the next morning for home. Looking at the clock, I told him it might be a good idea to get some sleep if that was the case."

"He asked me to wake him up about 6 am. So, when that time came, I went to his bedroom and knocked on his door and announced it was 6 am.

"He opened the door looking very sleepy-eyed and said in that western drawl, 'It shore doesn't take long to spend a night at your house, now, does it?'

"I laughed all the way back to my bedroom," said Punchy.

Fremont Miller flew 58 combat missions for a total of 158 combat hours and received the Air Medal with three Oak Leaf Clusters, the Distinguished Flying Cross with two Oak Leaf Clusters, the American Service Medal, the European Service Medal, the World War II Victory Medal, and the Purple Heart. He was also made a member of the Caterpillar and Goldfish Clubs for using his parachute and being rescued from the North Sea. Fremont left the service as a Captain and remained in the Reserves.

After the war, he returned to beekeeping in Wyoming, and started raising a family with his wife Bernice. At the prodding of many that knew him, Fremont ran and was elected to public office in 1957. He went on to serve with Wyoming's House of Representatives for eight years, and the Senate for six years, during which time his level-headed work helped mold the future of the state, a state his parents had helped pioneer. Sadly, he died of pulmonary fibrosis on May 9, 2002 after Bernice preceded him in death.

There is little doubt, if any, that he was the most popular pilot among the airmen, ground officers and his fellow pilots in the 328th Squadron. Few men had his ready laugh, his happy enthusiasm for life, his quiet courage, his calm manner, or his quick acceptance of people for what they were. To meet him was to be his friend.

"Curly Who?"

"Bill" Wilson Edwards was a member of the 133rd American 'Eagle Squadron' consisting of American pilots who joined the RAF before the U.S. entered World War II. After Pearl Harbor, these pilots were transferred into the U.S. Army Air Corps and Bill Edwards found himself flying P-51s with the 336th Squadron of the 4th Fighter Group in the 8th Air Force out of England.

On July 13, 1944, then Major Edwards, piloting his P-51 Mustang on a mission to Munich, was shot down by flak and taken prisoner and taken to Oberurset, a Luftwaffe interrogation center near Frankfurt. There he met the infamous Hans Scharff, the top German interrogator.

Scharff, who was quite fluent in English, kept questioning him on one point, saying: "Why do you insist that you were a Major with the 4th Fighter Group when we know that you are Colonel "Curly" Edwards of the 352nd Fighter Group? We even have pictures to prove it."

For one month straight they tried to get Bill Edwards to admit that he was "Curly" Edwards. Finally, after pounding him over and over with this question, they decided that he really didn't look much like the pictures they had of Colonel John C. Edwards of the 352nd FG and sent him off to Stalag Luft 1 where he spent the remainder of the war.

(Submitted by the nephew of Major Wilson Edwards.)

"Jackpot" Over Vechta

by Marc Hamel in collaboration with
Maj. Gen. Luther H. Richmond (USAF Ret.)

"His tracers looked like red golf balls coming up, and normally I would have zigged and zagged a bit so that he would miss. On that day however, the temptation was too great and I held my aim steady as I could see my tracers hitting the flak site. I felt the ship get hit, and almost immediately a tongue of flame licked back from around my feet and burned my hands quite badly. 'You're on fire, Colonel!' my element leader radioed."

These words punctuate the abrupt transition of Lt. Col. Luther H. Richmond from aggressive leader of the 486th Fighter Squadron to a troublesome "guest" of Germany. Lt. Col. Lynn Farnol succinctly described Richmond in his wartime VIIIth Fighter Command tome *To The Limit Of Their Endurance* as, "That pilot's pilot." A better summation cannot be found, for Richmond's surviving squadron members praise his aerial leadership, foul-weather flying, and navigating skills to this day. These talents were tested over Vechta Airfield, Germany on April 15th, 1944 when it became the stage for a violent melee between the 486th Mustangs and the Luftwaffe's fighters and flak.

This photo looking north towards the firing butts shows Martin "Corky" Cocoran being readied for a mission by his crew. (Jim Leber)

Murdoch "Scotty" Cunningham with his P-51B "The Flying Scot II," which later went to Bud Fuhrman. Its code was, oddly, PZ-Ddot. (Chet Harker)

Early on the 15th, Lt. Col. Richmond pondered a peculiar order: "All of our Fighter Groups were ordered to strafe German fighter airfields! The 352nd FG was to hit three fields in northern Germany. My target (leading the 486th) was an Fw-190 field at Vechta, about 50 miles from Bremen." Noted aviation historian Tom Ivie adds that VIIIth Fighter Command selected April 15th for this first "Jackpot" strafing mission, involving over 600 USAAF fighters.

"We proceeded in group formation at about 25,000 feet, then split into squadrons at a predetermined point (near Dummer Lake). My plan was to pass from about 25 miles south of Vechta, to a point 20 miles southeast of it, then head for the deck toward the field with the sun behind us. This we did. I had a section of eight aircraft with me on the deck (White and Red Flights), and eight above providing top cover." This top cover consisted of Yellow and Blue Flights, tasked with carrying out diversionary simulated dive bombing runs and post attack photography.

The 486th FS pilots were flying their P-51B models (and one C) Mustangs on this mission, having started the transition from Thunderbolts beginning in the first week of March. The ships had yet to receive their famed blue spinners and sweeping cowling paint, which later led to the group's sobriquet "The Bluenosed Bastards of Bodney." Lt. Col. Richmond was piloting his assigned P-51B-10 Mustang coded "PZ-R" in natural aluminum finish with olive drab spinner and anti-glare panel. Richmond appreciated its bright aluminum sheen, and allows, "I wanted the Germans to see me, so maybe we could get together and sort things out. I figured that they would see the silver color from a long way off."

Richmond continues, "As we approached the airfield right on the treetops at about 1355 hours, I was preparing to pull up to 300 to 400 feet for the strafing run. I glanced up and saw about twelve Fw-190's circling the field at about 500 feet. I called my men and told them to forget the strafing run and engage the airborne aircraft.

Chet Harker listens attentively to updates from his crew chief and friend Rebel Harris, kneeling on the wing of "Cile II, Luck of the Irish" code PZ-H. (Chet Harker)

The Jerries apparently never saw us, as they were at 500 feet and we were on the treetops. They may have been watching our top cover at 10,000 feet... probably were!

"We soon had a number of Fw-190's burning on the ground. I remember they were in a rectangle as though the traffic pattern had been transposed to the surface. I had a beam shot at my Fw-190, and he half-rolled into the ground. I've never been sure whether I shot him down, or whether he was taking evasive action without realizing how low he was. Things happen pretty fast in a dogfight. Anyway, I received credit for the kill as my wingman reported it."

The Encounter Report of Lt. Chester V. "Chet" Harker, flying as White 3, confirms Richmond's victory: "Colonel Richmond opened up with a deflection shot with the result that strikes were soon all over the cockpit. The E/A then flipped over, spun, crashed into the ground and exploded." Strike one Fw-190.

Lt. Alton J. Wallace, flying in Red Flight, recalls: "Seeing the 190's, Richmond sounded the alert and immediately turned after them. Within seconds he had a 190 in his sights and sent it crashing to earth. As Lt. Col. Richmond pulled up, Lt. Henry Miklajcyk, flying his wing as White 2 in 'The Syracusan,' picked out another and sent it down with a few quick bursts of fire." Wallace followed this action from the cockpit of his 'Little Rebel,' one of the first seven Mustangs assigned to the 352nd.

Mike Miklajcyk described his Fw-190 victory in his Encounter Report stating: "I made about a 100 degree turn and went after the Fw-190 when he turned to the left and headed southeast. I used full throttle and closed in. I gave him a burst at about 450 yards so he would turn. I saw some strikes, and when he turned he almost lost control due to high-speed wing stall. He straightened out and I gave him another short burst. The canopy fell off and a few parts were flying around. He then turned again to the right and I gave him a two-second burst from 70 to 90 degrees, and saw hits in the cockpit. The ship blew up and went out of control, making a barrel roll just before it hit the deck and exploded. I crossed over the Vechta airdrome and strafed about four barracks northeast of the field and all of the barracks on the field. The flak was intense and accurate, and I used violent evasive action."

Hearing the Colonel's call of planes in the air, Red Flight joined the proceedings. Lt. Martin Corcoran, leading Red Flight in his "Button Nose" PZ-L, chopped two low-flying Fw-190's from the air in separate encounters. He reported: "Just as we came upon the field, White Leader *(Richmond)* called in 'planes in the air' and at the same time I observed Fw-190s going under me at almost 180 degrees to my course. I pulled up to turn around and see where my flight was. I couldn't see them, but saw a burning plane crash. I then started to turn and saw an Fw-190 in a right turn start across the field. I went down to get on him but he made an even steeper turn, so I took a two-second burst at nearly 90 degree deflection. Since he was hidden under my nose, I couldn't observe the results of this burst. I then picked him up again, and as he began to straighten out, I gave him another burst. He burst into flames, straightened out, and crashed into the ground sliding some distance before exploding. I then pulled up and observed another Fw-190 going away from the field. I don't think he saw me as I pulled up and gave him a short burst. He caught fire and bailed out. This action was from the deck all the way up to 800 feet."

Al Wallace went after four 109's that passed below at 180 degrees to his path. "I made a tight 180 degree turn to come down on their tails, but they turned sharply left, giving me about a 90 degree deflection shot," Wallace's report details. "I fired a one-second burst at the first one with no results *(more on this later)*. I then laid off a little more deflection on the second and observed many hits around the canopy. I did not see this plane crash, but Red Three (Cunningham) says it caught fire and crashed. The number three aircraft was not making as tight a turn as the rest and broke off when I fired at the others, giving me a chance to swing around on his tail. I fired a two-second burst at about ten degrees deflection with no result, probably because he was skidding and slipping. So I stopped firing and carefully slid over exactly dead-astern at about 200 yards. When I opened fire this time there were very many hits all over the plane and he burst into flames, crashing almost immediately."

After a close shave evading a ramming by one of Wallace's two crashing Fw-190's, Lt. Murdoch "Scotty" Cunningham (Red 3) pursued and flamed another E/A which tried to evade by hugging the tree tops. His Encounter Report details, "I continued after my 190. By this time he had some distance between us, but I soon closed, fired and observed strikes despite his violent weaving and turning. He stayed within inches of the ground, and I fired many short bursts from various angles of deflection, from zero to fifteen. At no time did he make a tight turn. He still had his belly tank on until he crashed. He burst into violent flames and spread himself all over the countryside. About halfway through my attack, I observed Red Two (Wallace) coming in and covering my tail." Fifty two years later Wallace added, "Scotty's 190 left this blazing trail across the field, whacked into the row of trees at the end, and went up in a ball of flames. Out of the whole mess came the guy's

Luther Richmond's PZ-R, the second Mustang delivered to the 352nd, poses with William Chamberlain's 388th BG B-17 at Bodney in early April, '44. Chamberlain flew over to visit with his friend Luther and try out the new Mustang. (Earl Martin)

belly tank. End over end over end, rolling across the next field. Of course the trees had stopped the rest of the plane, but 50 some years later I can still see that belly tank... end over end."

Richmond continues, "After initial contact, we were badly split up and had ballooned up to about six or eight thousand feet. I noticed a German survivor of the dogfight attempting to flee the scene at low level. I was diving to intercept this Fw-190 when a flak gun opened up on me. I could see his tracers coming up as I passed through about 1000 feet. The flak position was right in my sights, so instead of taking evasive action, I steadied on my target. Temptation overcame my normally good judgment and I got off a long burst."

As recounted at the beginning of this narrative, the flak site found its mark and flames leapt into the Mustang's cockpit and seared Richmond's hands. The Missing Aircrew Report for this mission carries Chet Harker's statement, "Anti-aircraft fire opened up in the southwest corner of the airdrome, probably 20 mm, and during the initial firing the Colonel's right wing tank was hit. Gas sprayed all over the plane, which continued about a fifth of a mile after it was hit." Richmond continues, "My hands were immediately burned quite badly, and swelled up like hams with huge blisters." To Chet Harker's warning of fire, he radioed, "I know it!"

"I was going pretty fast at about 350 knots, and took a few seconds to pull out right on the treetops deliberately to avoid the continuing flak." Lt. Thomas Colby (Yellow 2), flying "Lil' Evey," detailed that this flak was the most accurate he'd encountered outside of the Ruhr Valley.

Richmond continues: "Concerned about whether my 'hams' would be of use in bailing out, I struggled with my safety belt and harness and got it unbuckled. All that remained was to pull the yellow canopy release lever. I pulled up quickly intending to zoom up to four to five thousand feet, roll inverted, and fall free. I started up, pulling the canopy release lever with some difficulty.

"The canopy came off and there was an immediate concussion that sucked me half out of the cockpit due to my speed. My helmet came off as soon as I hit the slipstream, along with my goggles and oxygen mask. I wound up bent back over the turtle deck with my left foot caught around the throttle quadrant. The parachute straps across my rear were likewise caught on something. The airplane eased up to about 400 feet and then began a gradual descent with the engine running fine but the cockpit mighty hot.

"I managed to squirm down until I was hanging from the left side of the cockpit, and struggled to free my G.I. shoe. I thought of my wife and kids, and said a little prayer, all in a split second. Then I quit struggling, felt my foot slip loose and the heavy web straps rip from around my rear (this turned out to be the dingy, though I thought for a moment that the chute had gone).

"Free of the plane at last and floating feet-first on my back, I missed the tail I'd been staring at for a second. All I could hear were the flak guns going off all around me. The trees were going by fast. With little time left, I managed with my right 'ham' to get the ripcord out. The chute opened with a terrific jerk and the harness straps pressed my escape pack (in my inside jacket pocket) hard enough to crack 3 or 4 ribs. It was only a five or six second ride to the ground.

"The ground was hard and I hit like a ton of bricks, badly spraining both ankles. There was a burning Fw-190 about 150 feet from me, and a flak position a couple of hundred feet in the other direction. Soldiers in the flak position were glancing over at me, and with difficulty, I rolled up my chute attempting to be inconspicuous, but to no avail."

Harker's report back at base detailed, "The right wingtip of the Colonel's plane hit a tree, it cartwheeled and burst into flames immediately. It is my contention that the Colonel was hit." As can be gathered from these words, great confusion surrounded whether Richmond was able to get clear of the plane.

Richmond continues, "Soon the 'All Clear' siren sounded, and I was immediately surrounded by a dozen or more German soldiers with rifles pointed at me. My hands were very painful when hanging, so I tried to keep them elevated to reduce the throbbing. I then found that my left arm was practically useless, as apparently I had hit something during bailout. A German soldier returned my gabardine flight cap to me, as it had apparently blown out of storage in the cockpit when the canopy was released. As the soldiers prepared to load me into a truck, a German 2nd Lieutenant came up and started screaming in my face and patting his pistol. I thought he was going to shoot me, and so did the surrounding soldiers as they started backing off. I didn't understand what he was saying, so I could do nothing but stand up straight and look him in the eye. He finally wound down, and the soldiers moved back in and took me to the Vechta airfield doctor's office. Here

Respected 486th commander Luther Richmond presides over the payday in the Fall of '43. Before Nissen accommodations were built at Bodney, the men lived in tents as seen here. (Cy Hall)

the doctor examined my hands, punctured the huge blisters with a pair of scissors, and applied ointment and paper bandages."

Amazingly Werner Oeltjebruns (an avid researcher in nearby Wardenburg, Germany) located an eye-witness to this event during early 2000. Gottfried Pagenstert, now a Senior Farmer, was 13 years old in 1944 and was one of the first to Luther's crash site. He relates, "On one day in April 1944 I was standing in front of our farmhouse, and several German fighter planes were circling around the airfield at Vechta. Their altitude was not very high. Suddenly some U.S. fighters came down from above (he must have seen the top cover) and shot down seven or eight German fighters in a surprise attack. At the same time the flak also began to fire. During the second pass of the U.S. attack, I saw a single-engine airplane directly in front of me coming over from the northwest. The fighter was burning and leaving a trail of smoke. Just a few seconds before the fighter hit the ground the pilot bailed out. Just when the parachute opened the pilot was at the ground. A few hundred yards from the pilot the aircraft smashed onto a field and burned out."At the same time, a few people from the airfield were standing with a kind of little lorry not far from our farmhouse and went to go get the pilot for prison. Soon after, a fanatical German Flak Lieutenant came up riding a bicycle. The chain had been coming loose many times. When he arrived he shouted at the pilot. Another man, a courageous Sergeant of the Luftwaffe, asked the Lieutenant if he knew about the Geneva Convention, and think about what he would feel if he were the pilot. The Lieutenant backed off, and the injured pilot was advised to sit in the lorry and was moved away from the place." Today Luther adds, "I would very much like to meet that German Sergeant."

Tom Colby adds, "As I recall, there was much confusion back at base over whether Luther survived or not. It was weeks or months later that we had definite news that he was alive."

"Next stop was a small cell in the base jail with a plank cot along one wall and a couple of blankets," Richmond continues. "I hurt all over by this time, and lay down to try to forget what happened. About dark, someone woke me by entering the cell and I struggled to my feet. The jailer entered with a young soldier and said, 'Herr Oberst Lieutenant, this is the Soldat who you Abgeschossen Hat.' I figured this was the soldier who shot me down. Groping for appropriate words, I replied, 'Tell him congratulations; I hope he gets the Iron Cross.' The kid smiled and said what translated to, 'You were very brave and killed some of my comrades, but I was brave too. I stuck to my guns and got you.' I replied 'C'est la guerre,' and they both understood that. So much for sportsmanship. That night my nice warm leather jacket was stolen from under my pillow and I was without it for the duration. Northern Germany is cold in mid-April!"

Al Wallace readies for a mission. Note the Malcolm Hood canopy on Joe Mason's second P-51B, PZ-M. (Al Wallace)

Modestly, several 486th pilots have stated that perhaps it was a fighter training unit that was attacked that day, as the aerial action was very one-sided in favor of the 486th. The German records indicate otherwise. According to noted researcher Doctor Jochen Prien, the unit attacked was III/Group of Jagdgeschwader 11 who at the time was based near the City of Oldenburg. The Group was on a routine flight after earlier downing seven U.S. P-38s that day near Hamburg (most likely these were from the 364th FG, though the 55th FG also lost P-38's that day). Obviously these Luftwaffe pilots were no "beginners." In fact, one of their pilots downed that day was Major Anton "Toni" Hackl, who by April 15th had amassed a victory record of 141 planes. Badly injured that day, he recovered to raise his tally to 192 confirmed (including 34 heavy bombers) and 24 unconfirmed before the end of the war. German pilots with P-38 victories that day were Ofw. Schulze, Ofw. Zick (two), Maj. Hackl, Ofw. Laskowski (two), and Fw. Keil.

Other interesting material from Germany regards the Luftwaffe loss record for that day. It appears that eight German fighters (seven Fw-190s and one Me-109) were downed on the 15th, versus the 486th's claims of six. This is unusual, as in the wild melees over the Continent, it was easy to over–claim due to confusion. At present it is assumed that either Al Wallace actually connected with the first '190 he fired at, or two pilots bailed out deciding that "discretion was the better part of valor" (at low speed and low altitude versus the attacking Mustangs).

Four German pilots bailed out and survived, and four were killed in action. Those KIA were Uffz. Karl Blaha (the Me-109 pilot), Obfhr. Horst Binder, Uffz. Herbert Regel, and Lt. Willibald Kilian. Kilian crashed away from the field, and would undoubtedly be "Scotty" Cunningham's highly evasive victim.

Though precipitously plucked from his position as highly respected C.O. of the 486th Fighter Squadron, "Kreigie" Richmond was to carry on the war by plaguing the Germans manning his POW camp. Ultimately imprisoned in Stalag Luft I at Barth on the Baltic Sea coast, Richmond participated in an innovative escape plan and tunneling project, as well as passive resistance. Two hundred tunnels were initiated from this Stalag alone during the war, and although his escape attempt was unsuccessful, actions such as these tied up German manpower that could have been operational on front lines. Other noted pilot inmates at Stalag Luft I during Richmond's stay were Duane Beeson, Gerald Johnson, Gabby Gabreski, Loren McCollom, Mark Hubbard, Cy Wilson, Henry Russ Spicer, and Hub Zemke.

After release in early May 1945 *(see the story "Stalag Luft I" later in this volume)*, Lt. Col. Luther Richmond returned to duty in the USAAF, ultimately attaining and retiring with the rank of Major General. He served in commands in the U.S., Japan, France and Germany, as well as two tours of duty at the USAF HQ in the Pentagon. One of his more interesting assignments was with the American Embassy in Bonn, Germany. Here, from 1955 to 1958, he was responsible for rebuilding the German Luftwaffe. Richmond recalled recently, "A few years ago I made a list of aircraft I have qualified in, and as I remember it totals about 78 types from single engine to four-engines." This impressive total ran from the P-12 biplane all the way up through contemporary jets. He retired in 1970 after a long and successful career.

A fitting tribute to Colonel Richmond is the memory of the 486th FS's popular parachute rigger Michael "Mike" Sandorse. "Everyone knew that the Colonel was an 'ace' both in and out of his airplane. I do not specifically recall my thoughts at the time we learned Luther was missing, but I do remember telling the pilots that all of us would miss him very much. I kept asking the pilots after the incident if they believed Luther was killed, and they all said there was no way he could have survived the crash.

"When I attended my first 352nd FG Association reunion, I was told Luther was looking for me. I asked, 'Luther who?'

"'You know, the commanding officer.' I replied that I thought he was killed—then I got the rest of the story. Luther found me and hugged me and said, 'If it wasn't for you I would not be here today.' What a wonderful moment for me that I will never forget."

Luther Richmond smiling with P-47 PZ-M collar up

"Miss Lace's First Encounter"

by Marc Hamel and Donald McKibben

One of aviation artist Troy White's "Bluenoser" series of paintings is entitled "Miss Lace's First Encounter," depicting Donald "Mac" McKibben's April 19th, 1944 victory over an Me-109. This typically fine work by White was commissioned as a surprise by Frank McKibben for his father. Mac shares his thoughts with us.

"She was a beautiful bird, and I'm tickled pink to have Troy's faithful rendering. I never thought I would own such an impressive piece of art. Thank you Troy and Frank. What a thoughtful and generous surprise! I'd like to think that someday one of my grandsons will pass it along to his grandchildren.

"I am asked if I can tell all I recall about that mission. I remind you that the Encounter Reports, prepared by the pilots right after they returned from a mission, are likely to be more accurate than those retrieved from memory some 57 years later. Here is what the report said:

"I was flying Blue Three. As we approached the bombers' initial point, I noticed 15 to 20 Me-109's diving towards the box of bombers that had just turned to make its target run. I called these E/A's in over the R/T, but our squadron was in maximum climb to reach several contrails that were hovering in the vicinity. At that point I noticed a Fortress going straight down, and at the same time one of the contrails crossed directly over me and I identified them as P-51's.

"Then a lone Me-109 passed by me on the left, going the other way. I called and said I was following him and told the flight to follow me because it looked like all the E/A in the vicinity were on, or were headed for, the deck. As I wheeled onto the '109, he split-essed and I was able to keep a lead on him during this maneuver. I fired once in an inverted position, but saw I was under-leading him. He took no evasive action after the split-ess and I was able to close to about 100 yards at 2 to 3,000 feet. My last burst caused numerous pieces to fly off the airplane. I flew through the pieces and pulled up to avoid a collision. I looked back and saw the pilot bail out. The E/A glided, inverted, into the ground.

"I climbed back up to about 15,000 feet and joined four red-nosed P-51's (4th FG) and one

Donald "Mac" McKibben poses beside the original painting showcased in the paintings pp. 275 and following. (Mac McKibben)

red and yellow checker nosed P-51 (357th FG). The rest of my group was nowhere to be seen. I then noticed that I had accidentally hit the 'Off' button on my radio set, and that accounted for the fact that nobody followed me on my pass. I continued home uneventfully.

I claim One (1) Me-109 Destroyed.
Ammunition Expended - 680 rounds."

Commenting on this story, Don McKibben had this to say: "As war stories go, this one is not remarkable and certainly not enough to justify having a top-notch aviation artist capture it on canvas. But there was one element, not mentioned in the official report, that gave this incident special significance as far as I was concerned."

Here, Mac's modesty shows through. In the Winter of 1943-44, the 8th AAF was experiencing great frustration in dealing with German defense aircraft. The German fighters wanted to get to the American bombers that were doing damage to their homeland. They wanted no part of the roaming wolf-packs of 8th Fighter Command. These Allied fighters were causing no real damage to Germany at that time, but could easily deplete the German fighter stocks. So on-the-whole, the Germans simply stayed outside of the combat radius of the Thunderbolt, and laid-in-wait for the bombers. Therefore, many, many missions were flown by the "Mighty Eighth's" fighter groups with no combat to show for it. However, these efforts did limit the time the enemy fighters could attack the USAAF heavy bombers. This stand-off situation changed to Allied-dominance once the P-51B/C made the scene in the early Spring of '44, when the P-47 was plumbed for three external drop tanks giving much greater range, and when the bugs were worked out of the long-legged P-38 in the ETO.

The 352nd Fighter Group went operational on September 9th, 1943. In the seven months that followed, I put in about 100 operational hours, so-called 'combat time.' In all of that time, I never got close enough to an enemy aircraft for either of us to threaten the other. If my flight was protecting the rear end of the bomber stream, I could be pretty sure that any German fighters in the area would be concentrating on the lead bomber formations way up ahead. I spent a lot of time bird-dogging around in the wrong part of the sky. Of course, I am aware that sometimes our mere presence near the bombers helped keep the German fighters away from the B-24's and B-17's. After all, they did refer to it as 'escort duty.' Still, I felt I should be doing more, if only to balance the books with our taxpayers on two P-47's demolished while in my custody."

A late summer '44 view of Stan's P-51B under repair in the 486th Blister Hangar at Bodney. (Archer, UK)

The ever unconventional Cy Hall, crew chief of Gignac's "Gig's-Up II," gives pilot Frank Cutler a few pointers. Note Stars and Bars insignia on Hall's overseas cap.
(Cy Hall)

McKibben continues, "I'm glad the painting shows the beginning of the episode instead of the end. It doesn't show 'Miss Lace's' guns firing, or my adversary bailing out, although that might be exciting to the casual viewer. For me, the most dramatic moment was when I first spotted the enemy aircraft diving towards me, slightly to my left. This was it. Seven months of frustration had come to an end. After all of my griping about not getting the breaks, it was time to put up or shut up. It was time to see if 'Miss Lace' and I could do what we were sent there to do. That tension filled moment was what Troy White captured in the painting I call 'Miss Lace's First Encounter.'"

Diligent Luftwaffe research work by Troy White, Lynn Ritger, and Charles Bavaroise produced a listing of five German fighter crashes near Spangenberg for April 19th, 1944. Of these, the events surrounding the actions indicate that Unteroffizier Rolf Lohmer of JG2 was Mac's opponent. Lohmer was killed-in-action later in the war in a dogfight over Ruhlerstwist.

You may wonder how "Miss Lace" got her name? After his P-47's demise in a mid-air collision on March 8th, 1944 (see "The Day It Rained Thunderbolts"), Mac pondered a name for his new mount. The siren's call was Milton Caniff's daily cartoon strip, "Male Call" featuring the fantastic blonde, Miss Lace, who reminded the men just what they were fighting for. Caniff's other strip, "Terry and the Pirates," provided alternative artwork for the fuselage side...the equally curvy and luscious Dragon Lady. Why were they on the right side of the aircraft, traditionally the crew chief's domain? The P-51B was parked along the west edge of the perimeter track bordering the field's west side at Bodney. The aircraft usually faced north, so the crew chief's side was exposed the majority of the time. No need to waste seductive Nilan-Jones-painted artwork on the trees of the "Great Wood" at Bodney, and that's why.

"Major" Trouble

This story comes to us from 486th Armorer Harold "Stan" Stanfield (right) of North Carolina.

"One night at Bodney, two other armorers and I were loading and inspecting .50 caliber belts getting ready for a mission the next day. We were in the Nissen near the Pilots' Hut, and we were pretty tired, but were helping each other out to share the load. All of a sudden this major in sharp Class A's came in.

At Bodney we didn't salute much, unless it was our CO or Joe Mason. Most of the Stateside procedures were overlooked while we were in England, and we were all left to get our job done. A few officers were demanding about saluting and all, but us enlisted men called that "chickens#%t" at Bodney.

Well this major starts screaming at us because we didn't salute, stand at attention, and pay him proper respect. He yelled and yelled at us, and I thought he was going to back us up against the wall.

When I got the chance, I asked him if I could go get our Sergeant. The major said, 'Yeah, you go get him so I can tell him what kind of outfit he runs.'

I ran over the where our Master Sergeant Emory Noland was, told him my story, and we got back over there quick.

It turns out that this major was fresh in from the States, had never seen combat, and had just arrived on station. We were the only troops he could find in the middle of the night because we were still working. Now the major starts yelling at Emory too. Emory is one of the most laid-back men you'll ever meet. Nothing upset him, and we all thought the world of him. Emory listens to the major for a little while, and he starts to wind down. Em asks the major if he is through, and the major says, 'Yes, now get out of my sight.'

Sergeant Noland heads straight over to Sergeant Wayne Stocks' quarters and tells him the story. I guess you know, but Wayne was crew chief for the Colonel – Joe Mason. Colonel Mason was not an easy man to work for or around, and we stayed pretty clear of him...which was easy where we worked, but Wayne goes over and wakes up Colonel Mason and tells him what happened.

That major from the States never saw the sun at Bodney.

Joe Mason had him on a plane and headed back home before dawn. The colonel could be tough on us, but by golly no major was going to abuse his men, and he also knew some powerful generals.

It also shows you what Mason thought of his crew chief that Wayne got us help instead of getting himself in trouble. Waking up the colonel in the middle of the night could be hard on your stripes."

This story was later verified by Em Noland, who typically took it in stride. When Wayne Stock was initially approached about this story, he laughed and modestly allowed, "Oh that can't be right. I never had any pull with Mason. I was just a nobody to the boss." When confronted with the full story verified by Noland, Wayne sheepishly agreed, "Well, maybe I did help a little now that you mention it."

Even the toughest colonel knew better than to cross the man who maintained his engine.

Belly Tanks and Messerschmitts

By Marc L. Hamel and Stanley Miles

Pilot Stanley "Stan" Miles of the 486th Fighter Squadron had quite an eventful day on May 13th, 1944. The 352nd was were detailed for a bomber escort mission, but what looked like an easy mission turned interesting quickly as the Luftwaffe showed up with a huge mass of fighters to contest the skies in the Tribsees area.

"I remember that mission well," Miles relates. "It was May 13th, towards the end of my tour (which ended June 7th, 1944). It was a beautiful day and we flew up over the Baltic escorting the bombers. You could see Denmark and Sweden and Norway. It was just gorgeous up there. It seemed like you could see forever. We flew over this big German fleet, and as we went over, they started throwing a smokescreen to obscure themselves.

"We had been with the bombers for maybe an hour with nothing unusual happening when we got a radio transmission that there was a big gaggle of over one hundred bandits headed our way. As we got in over land a little ways, we got a transmission from Woodrow 'Woody' Anderson saying he had spotted the bandits at one o'clock high. Sure enough, here they came."

Stan Miles in RAF boots displays the his Bolt's artwork of the lady Joy. (Stan Miles)

Flight Leader Frank Greene's encounter report details what happened:

"The enemy formation made an attack on the wing of bombers to our rear. They came in from 'down sun' making a quarter frontal attack on the bomber formation. I called my flight to break after them as they climbed to the "up sun" side in preparation for a second attack. Just as we caught up with them, we were joined by our relief. Half of the enemy aircraft broke for the deck, and half stayed up."

Miles' encounter report detailed it further:

"We went right through their formation and they dropped their belly tanks. One of these tanks tumbled down and hit my left horizontal stabilizer. It jammed something so that I could not climb no matter how hard I pulled the stick back. I could do anything else, but not climb.

"I looked out ahead, maybe 300 yards or more, and this Me-109 popped up in front of me. Just flying along level away from me. He was a long way out there, so I gave it the throttle and closed in on him. I fired and saw a few hits on him. So I closed in a little more, fired, and saw more tracers hit. I continued my attack on the Me-109 and fired a burst from 30 degrees and

350 yards, and saw hits along the fuselage. The E/A took evasive action, but I closed to 200 yards astern and gave him another short burst and got more hits.

"About that time the German pilot bailed out. I had my gun-camera running, so I got some good shots of my strikes on his plane and the pilot jumping out. My wingman was still with me so I eased around, came back and got some nice film footage of the German pilot in his chute.

"I recall that there was some talk at the time of whether or not it was right to shoot up a German pilot in his chute. We would sit around at Bodney and discuss this. One school of thought was that if you did not shoot the guy, he would land and be right back up fighting you the next day. I know that one pilot was extremely bitter about the fact that his brother had been killed by the Germans earlier in the war. He undoubtedly would have shot this guy. I could not do it however, and just took the film footage of him.

"About this time my wingman, Charles 'Griff' Griffiths, slid up next to me and called, 'Your tail is all beat up.' So we headed for home. Since I could not climb, we set a course for home at about 12,000 feet. We took a lot of flak coming out at that altitude over the Zuider Zee and the Channel coast. However, we made it back to Bodney and I set up for a landing. As I couldn't get the stick back, I made a bumpy 'wheel' landing, and that was that . . ." —just another "day at the office" for Stan and his Mustang PZ-S. Stan logged 5:40 hours and expended 261 rounds in downing the enemy aircraft.

Of course, he relates all of this with the casual grace of an experienced veteran pilot—which he certainly was by May of 1944. However, Carleton "Bud" Fuhrman, a self-described "greenhorn" at the time, saw the proceedings from a different perspective:

"I flew my third combat mission on May 13th, 1944, and the records I have indicate the target was Prenzlau, Germany. I was assigned the position of wingman to Colonel Joe L. Mason, the Group Commander leading the mission. Now, 56 years later, it seems they would have put a more experienced pilot in that position.

"The thing I most remember about the mission was when a large group of Me-109s made a dive at some bombers we were escorting. There were estimates that the attacking German fighters numbered as high as 150 planes. The attack, as I saw it, came down from the right heading for the bombers, which were on our left. I have stated before that this attack looked like a cartoon of a swarm of bees attacking someone. It was just a large group in front followed by a long string of stragglers. Col Mason headed for them and radioed 'Let's dive through them to see if we can scatter them,' as there was little chance of making a decent attack on that many.

"There must have two others in our flight besides Col. Mason and me, but I cannot remember who they were. We dove through this mess of German fighters, and it was more like driving at high speed across a busy highway intersection. It was just luck that got us through safely. I can remember seeing a '109 pass in front and above me with his guns firing. I don't know who he was shooting at, but maybe he was just as excited as I was and was firing by mistake.

"In an instant I was through and out the other side of the enemy 'swarm' and looking around I found myself all alone, the '109's just specks disappearing down to the left. I couldn't find anyone to join up with and there was chatter on the radio about heading back to base. As I was alone, I climbed to about thirty thousand feet (just under contrail level) and headed home. From the radio calls there seemed to be a lot of us flying alone. Since the target area was northeast of Berlin, I would say this event took place north of Berlin. It was a long ride back. As I was rather new to Bodney, and the field was not easy to spot from the air, I had to land elsewhere and refuel to get back to the base. My logbook shows '7 hours flown and 2 landings,' reflecting that I had to land for fuel before making it back to Bodney."

The mission was quite a success for the 352nd Fighter Group, with 16 aerial and one strafing victory. Joe Mason, Group C.O., claimed two 109's destroyed and one damaged, plus an Fw-190 destroyed. There was great controversy at the time over his claim because the two 109's actually collided while "breaking" to avoid the 352nd's attack. Mason's claim was based on the premise that, since he was leading the attack that caused them to "break" and collide, he should be credited for those two enemy aircraft destroyed and he was given credit for them. Having watched this action, ace Ed "Hell-Er-Bust" Heller's comments about this are unprintable.

The 328th Squadron's John Coleman and Francis Horne each scored doubles, over 109's and 190's respectively. In the 486th, "Woody" Anderson, Frank Green, and Don Higgins all added a 109 to Stan's victory. J.C. Meyer's 487th pilots added four more enemy aircraft. These were credited to Marion Nutter, George Preddy, and the fourth was a shared victory over a Ju-88 by Carl Luksic and Glen Moran.

Sadly, this tally was counted against the loss of the "hell-for-leather" pilot, Frank "Killer" Cutler, a 9.5 victory ace. In his zeal to achieve additional victories, Cutler was seen by the 487th's Nutter to overrun an Me-109 he was bouncing from behind. Both aircraft dissolved into a fireball, with neither pilot escaping.

It is reliably reported that several Stateside girls preferred Bill Halton (at right) to Rhett Butler, erm...Clark Gable... when they saw this photo.

Impressing Clark Gable by Reese Lewis

Reese Lewis, who worked in our Group Intelligence operation during WWII, relates this story of a visit to Bodney by Clark Gable.

"I don't know if you remember our Group briefing room (no 352nd pilot could ever forget it), but it was quite beautifully done and when Clark Gable entered it on his visit to Bodney, he whistled in surprise and remarked, 'If only CB could see this!' He was obviously referring to the famous movie producer Cecil B. DeMille.

"That room was quite a setting thanks to Major Sid Marks and Harry Henry, who had it constructed and added all the frills. It was Sid Marks who coined the phrase, the 'Battling Bastards of Bodney,' and erected the sign over the entrance which paraphrased Flo Ziegfield: 'Through these portals pass the bravest pilots in the world.'

"You may not remember the large flak map at the rear of the briefing room, either because you were in a hurry or possibly preferring not to look at it out of fear. It showed solid red to indicate the flak concentrations.

"Gable was interested in it while I was updating it. I explained to him what all of the red indicated—all the flak you pilots had to pass through to get to your targets. Again, he whistled. Not being a pilot, it was probably the first time he realized the intensity of flak you faced in addition to the Focke-Wulfs and Messerschmidts."

"Fastest Guns in the West"
by Bob Powell

Bodney, unlike most of the other fighter bases, was a big grass meadow with a rise diagonally across the middle of the field which sometimes caused problems in crosswind takeoffs and landings, but enabled the Group to get into the air and assemble quickly to head out for missions over Europe.

At our mission briefings, we were given our "start engines," "takeoff," and approximate return compass course from wherever the mission was to take us. Most of us wrote these down on the back of one of our hands for quick reference.

Since our Squadron Operations officer had posted our names and flight positions for the mission, we went to our planes, checked them with the crew chiefs, took that "last minute pee," and got into the cockpits to wait for the "start engine" time. We started our engines at the prescribed time and watched our assigned flight leader taxi out, and falling into our correct positions to taxi, sometimes checked our mags as we zig-zagged along the the perimeter track to the takeoff area. Occasionally, taxiing downwind too slowly caused our radiators to overheat and "popoff," resulting in the pilot having to move out and abort. Hence, we tried to move along rapidly to our takeoff positions when going downwind.

Once there, we lined up four abreast just 15 or 20 feet apart and slightly echeloned from our flight leader left to right, the other three flights in the squadron doing the same behind us—four flights (16 planes plus a couple of spares to fill in if anyone was forced to abort before reaching the Continent).

As "takeoff time" neared, we glued our eyes on the plane to the left, waiting for the flight leader's signal. Meanwhile, he was watching the tower for the takeoff signal, a green light. When the flight leader moved his head forward, it meant his throttle was going forward at that precise second, and he started rolling. Each member of the flight pushed his throttle forward, too, staying in the same echeloned position as the flight gathered takeoff speed. As soon as the tails of the first flight started lifting, the second flight started rolling, and then the third and fourth flights, while the next squadron to take off moved into position behind them.

As the first flight became airborne, it would proceed straight ahead for a few seconds and then start a gradual 180-degree turn to the left. Each of the following three flights would make a slightly shorter turn so that when the full 180-degree turn was completed, the squadron was already in formation in fours on course for their rendezvous with the bombers which were already approaching the Continent.

The other two squadrons, following the same procedure, would now be off to the left and right of the lead squadron, which was headed by the Group leader.

With this procedure the 352nd was able to get into the air faster than any of the other ETO Groups, setting a record of 48 airplanes in the air in fewer than four minutes. It always reminded me of the legendary "gunslingers" who claimed to be the "fastest guns in the west."

A Blue D-Day for the Bluenosers

By Marc Hamel

We have all experienced or heard about the sacrifices made by Allied forces in the crusade to free Europe. However, D-Day has a significant meaning to the men who served with the 352nd Fighter Group. An incident occurred that pre-dawn morning which lives deep in the memory of all of those witnessing it.

The 352nd FG (under 8th Fighter Command) arrived in England approximately 11 months prior to D-Day and became operational September 9, 1943. In the ensuing months, the Group made quite a name for itself under the leadership of such officers as Col. Joe Mason, Lt. Col. Ev Stewart, Lt. Col. Luther Richmond, and Lt. Col. J.C. Meyer.

The scene that greeted the 352nd on the morning of June 6th – the "new" tower. (Sheldon Berlow)

First flying escort missions in the rugged and well-armed P-47 Thunderbolt, the Group began transitioning to the P-51B/C Mustang when the 486th received the first seven on March 1, 1944. Properly exploiting the ship's strong suits, the 352nd began tallying big scores over the Continent. By early June, names such as Stephen "Andy" Andrew, Don Bryan, Frank Cutler, Ed Heller, Carl Luksic, J.C. Meyer, Virgil Meroney, Glenn Moran, George Preddy, and Jack Thornell were on the list of 8th FC "aces." Against this background, frantic preparations for D-Day operations began the night of June 5, 1944.

Carleton L. "Bud" Fuhrman, who flew through Frascotti's takeoff fireball D-Day morn. (Bud Fuhrman)

486th Squadron pilot Carleton 'Bud' Fuhrman recalls, "We all knew it was coming, but just when we were not sure. The first indication we had was when we returned from our last mission prior to June 6th. As soon as we had parked our planes, the crews were busily painting the black and white stripes that would identify Allied aircraft. These were later to become known as 'Invasion Stripes.' The crew chiefs were quick to inform us that the next time we flew these planes, it would be D-Day.

"About 7:00 in the evening of June 5th, some of the pilots were standing around the bar in the Officer's Club. Colonel Joe Mason came in and said, 'Boys, you had better get to bed, because you will need your sleep before the night is over.' We left immediately, went to our quarters, and then to bed. I had barely dropped off to sleep when an orderly came around and announced that all pilots were to report for breakfast at 2200 hours, with briefing at midnight.

"At the briefing, we found out that the long awaited invasion of the Continent was beginning, and that we would be taking off at 0230 in the morning. I went to my airplane to make sure everything was as I wanted. The instrument panel lights were adjusted to the proper level, the altimeter was set, and all switches were placed in the proper position. After that there was little to do but stand around in the dark so our eyes would have maximum night vision. When we first went to our planes, the moon was trying to shine, but by takeoff time it had clouded over completely. It was very dark. Even a drop or two of rain fell to add to the gloom.

"At the prescribed time we all started our engines, and the Mustangs started to move out to takeoff position. It was your job to know which airplane your Flight Leader would be flying, and join in on the rest of the flight as they came taxiing by. This could only be accomplished by knowing the exact place your Flight Leader's set of navigation lights would be coming from, as in the dark nothing else could be distinguished.

"As our airfield was a sod field and had no runways, a string of lights had been laid across the field to guide our takeoff path. We were taking off in our normal formation of four airplanes abreast, were scheduled to be the second flight off, and were using the Southwest to Northeast takeoff direction.

"The first flight taking off hit the string of bulbs and put them out. Now it was our turn. I was flying Number Four position, which is the far right of our four plane formation. The next plane over to the left was piloted by Lt. Robert 'Bob' Frascotti, who was my element leader and flew the Number Three position. Donald 'Red' Whinnem was to Frascotti's left, with our Flight Leader Corcoran on the far left."

Unknown to the rest of the Corcoran's flight that morning, Charles "Griff" Griffiths in his Mustang "MARJ" had inadvertently lined up to Fuhrman's right in the blackness. Griff relates, "The fighter planes resembled fireflies of varying colors and size as I taxied in to my assigned takeoff position. The field was quite inadequately lighted for takeoffs at night on a sod surface."

Fuhrman continues, "As the string of lights were out, my Flight Leader, Captain Martin 'Corky' Corcoran lined up on what he thought was the correct heading. Almost as one, we all advanced our throttles to full power and began our takeoff run. As we were approaching flying speed, my plane hit a bump and bounced into the air. At first I was going to let it back down on the ground a little longer to gain more speed. Then I heard a voice in my mind that almost shouted, 'Keep it up! Don't let it down! Keep it up!' For some reason I responded to this voice, and was flying along watching the lights of Frascotti's airplane just to my lower left. Suddenly this enormous fireball appeared almost in front of me. I saw two streaks of fire spread out and I thought someone had dropped his wing tanks and they had exploded (one hundred and fifty gallons of high-octane would make a big explosion). There was no time to dodge or do anything but fly through the flames. Thank Heaven, that good old Rolls-Royce Merlin engine never missed a beat. When I came out from the other side of the blaze, I could see nothing but darkness, momentarily blinded. I pulled back on the control stick as hard as I dared to gain altitude. The plane shuddered to warn of a stall, and I hoped I could see my instruments before I got too far out of control. It seemed like a long time, but in fact it could only have been 5 or 6 seconds, and I started to make out my instruments and gained control of my flight path."

Donald "Red" Whinnem in his A-2 emblazoned "HMS" referring to his bird "HMS Hellion," code PZ-Wbar. (Al Wallace)

"Red" Whinnem (best friends with Bob Frascotti since Flight School) adds, "I lined up and began my takeoff with Corky. My friend Bob was off to my right, but was a little behind me as we lifted off. I saw a blur go by to my right but didn't know what it was. I didn't actually see the fireball that Bud did as I was beyond it at the time, but thought that maybe the Germans were bombing us. As it turns out, that blur I saw was the tower that I just barely slipped over with my wing. I continued on with the mission not really knowing what had happened. I didn't know the full story until I returned."

Griff, on the far right, had a similar experience. "I advanced my throttle and propeller settings to their full-power position while at the same time jockeying to maintain an acceptable position in the flight. The plane was heavy with fuel and ammunition, and seemed to want to cling to the earth. One more second passed, and then the sky, earth, and my tiny position amongst them was illuminated as though it were midday. I applied war-emergency boost to the throttle, and the fighter shuddered while still on the ground. I felt the fighter leave the ground, settle back and quiver again before finally staggering into the air just as it struck the posts of the volley ball court."

Fuhrman continues, "By the time I spotted the lights of the rest of my flight they were just disappearing into the low clouds and I could not join them. I turned off course a few degrees and climbed through the overcast. I was quickly on top of the clouds and the sight I saw was almost overwhelming; it was as if I had come on a vast highway in the sky, all the lights of airplanes headed for the invasion area. I turned and headed off to our patrol area and did not see any more of our airplanes. Later another set of lights appeared and we chased each other around in circles like a couple of fireflies. A voice said over the R/T, 'Is that you, Fuhrman?' I recognized it as Lt. Gremaux, and answered yes to his question. I don't know how he guessed it was me in the dark. We stayed together in our patrol area as long as we had fuel and then returned to our base at about 08:00 in the morning."

A nice casual photo of Bob Frascotti in his Class A's (Bud Fuhrman)

In the gloom and pre-dawn darkness, Robert "Bob" Frascotti had collided with Bodney's "new" control tower, then under construction. The first flight off had lined up correctly as the string of lights still shone across the grass turf. Having no reference point, the second flight had erred to the right, directly into obstructions. As Griffiths recalls, "I did not realize that the leader of my flight was aligning the fighters thirty degrees to the right of the assigned takeoff direction." In visiting the field at Bodney today, it is easy to realize how a pilot could make this mistake while turning through some 130 degrees in the dark."

The flames from Frascotti's collision served as a beacon to guide the remaining 352nd FG aircraft into the air safely that morning.

Leonard "Jim" Gremaux took off in the first flight that morning, and saw this beacon from an aerial perspective. "We were to fly top cover over the Cherbourg Peninsula. We were the first flight away from the field, and it was dark and overcast. When Frascotti hit the tower, the flash from the field blinded me momentarily. I continued circling left like we were taught, and here comes this other group circling right, which scared the hell out of me. So I put the nose into the clouds, and ended up separated from my flight. I went on to the area we were supposed to patrol. After a while I see this plane come in and he breaks into me, so I broke into him. It turned out to be Fuhrman, so we joined up as a two-man team. We flew out over that huge fleet that day, and I never saw so many boats in my life. Later, on the way North towards home we flew under the overcast over London between several wires. Back at base when I told somebody this, they laughed and said the barrage balloons were up in the clouds. We had flown between the wires, but didn't hit any."

Fuhrman elaborates, "Our return flight took us up along the west side of the Cherbourg peninsula, and we could see the immense fleet of the invasion force now that it was daylight. It was a sight to see. After we got to England we started to descend but there were layers of clouds down there. We went through this layer and came out over the edge of London. There were barrage balloons everywhere with their cables hanging down. We spent a few minutes doing some fancy dodging to miss everything. We got through it okay, and found our way back to Bodney. It sure was a wild morning."

"Punchy" Powell relates his experience over in the 328th Squadron. "As the 486th Squadron lined up to takeoff, the 328th pilots were preparing for our own first mission in our Ready Room, located on the rise just a few hundred feet behind the new tower. The atmosphere was deadly serious as we were being briefed for our takeoff scheduled a short time later. Our Flight Surgeon, Dr. Kenneth Lemon, had just finished saying, 'I have orders to keep you flying until the invasion beaches are secure even if it means giving you stimulants to keep you going.' It promised to be a long day, and it was.

"Suddenly there was a tremendous explosion and a lightning-like flash quickly followed by the sounds of machine-gun fire.

Bob Lyons and Punchy Powell give their all for the morale of the girls back home. (Punchy Powell)

Our immediate reaction went from panic to realization. Horrible questions pierced our minds along with the sounds. Several pilots hit the floor as they recognized the sound of machine-gun fire. Some few ran out the door. Some went through the windows to get outside. Momentarily, not knowing what was happening, near panic resulted. One of the young pilots foolishly grabbed a fire extinguisher and ran out the door to fight the blaze which could now be seen through the open doorway. With the .50 caliber ammo from the burning fighter exploding, he only put himself in greater harm's way. There was nothing he could do to quench the roaring fire."

Ground officer Richard "Dick" DeBruin writes, "We went to the 328th FS operations area about midnight to observe preparations on the flight line. As the first takeoff time approached, we stood on top of an air raid shelter embankment. We were close to the new tower (which was nearing completion) and the volleyball net. Soon we heard the first flight of four aircraft roaring down the field. The second flight of four then followed. To our surprise we could see a plane coming towards us. Sensing danger, we dove off the embankment and hugged the ground, as a P-51 flew directly into the control tower and exploded. Another Mustang passed right over us and flew through the volley ball net. With ammunition exploding from the destroyed plane, we ran for shelter behind the 328th FS Operations building."

"Red" Whinnem recently allowed, "When I returned to Bodney and they told me what had happened to Bob, I went over to the Parachute Room in one of our Nissens and had a cry for myself. Willie O. (Major Willie Otto Jackson, 486th FS C.O.) came and found me and told me that he understood what I was going through. He said that I was posted for the next mission that day, but that he would scratch me so I would not have to fly. I told him no, that I wanted to get back and fly and keep going or I probably wouldn't again. I think I flew three missions that day, and there were lots of missions for that whole week. Everyone was really good to me."

On return to Bodney after his mission, Fuhrman continues, "It was then that I learned that the fireball was caused when Bob Frascotti had flown into the new control tower being built and had exploded. We were almost 30 degrees too far to the right of the desired takeoff course. We also found out that in the confusion of getting into takeoff position, five airplanes instead of four had taken off in one formation. The fifth airplane was piloted by Lt. Griffiths.

There was a volleyball court near and to one side of the control tower, and one of the volleyball court posts had been knocked down by the passage of one of our airplanes. On examination of our airplanes we discovered it had been Lt. Griffiths' airplane which had hit the post as there were fragments of wood in the rim of one of the wheels of his airplane. My airplane showed scorch marks from my passage through the flames. The closeness of the volleyball court to the control tower meant that I too had been very close to hitting the tower along with Frascotti. I still believe to this day that the voice saved my life, and that I passed just above the tower."

Griffiths adds, "The field was clearly lit after Frascotti's collision, as the fire maintained a brilliant glow. The remaining fighters took off with ample lighting and rendezvoused before proceeding to Normandy."

Powell sums up Frascotti's sacrifice with, "Sometime after the crash, as the 328th pilots took to the air and their pre-assigned positions in the wall of aircraft screening the invasion beaches from

Frascotti's PZ-Fbar "Umbriago" with Carmen Mascola. The artwork lettering changed from a light shade to a dark shade in the spring of '44, possibly for more visibility against the very light blue nose color. (Earl Martin)

the Luftwaffe, their P-51s roared into the night by the still-burning flames of Frascotti's 'Umbriago.' It was D-Day, and thousands of brave, young foot-soldiers were wading ashore on the beaches. Our mission was to insure that the Luftwaffe would not reach the invaders. The virtual wall of fighters south of the beachheads indeed provided an impossible barrier for the German planes. Many of the 352nd FG pilots flew three missions totaling some sixteen hours. Dr. Lemon was right. It was a long day!"

Powell, then a young Lieutenant, adds, "On my way to escape through a window in the panic after Frascotti's plane exploded, I stepped on the neck of Major Harold "Hal" Lund who had wisely sought the safety of the floor. Gingerly rubbing his neck the following day, Major Lund asked who that S.O.B. was that stepped on him. No one spoke up. Some forty years later however, when I was searching for 352nd veterans and found the retired Major Lund living in California, I finally admitted that it was I who had bruised the Major's neck in the panic of that fateful night that Bob Frascotti became our first D-Day casualty. We were extremely fortunate the other planes avoided crashing.

Charles Frascotti remembers his brother. "Bob was a very outgoing person who loved people and life. He was fun, and enjoyed clowning around and joking with everyone. He grew up in our very close knit and patriotic family (Dad served in France in WWI where he was gassed). Bob was also very popular with his peers, having been elected to President of his high school class for four years running, and

PZ-Ddot (reason for dot unknown)
"Flying Scot II/Vicious Virgie j.g." of
Scotty Cunningham and Bud Fuhrman, 42-106472.

Griff Griffith's P-51B Marj that he was flying on D-Day morning. (Harold Stanfield)

was a superior athlete. I would say that he was he probably was the typical fighter pilot."

"Red" Whinnem recalls, "Bob was the nicest guy you would ever want to know, and a great athlete as well. He could stir up laughter anywhere he went, and was great fun to be with. Bob would sing a song or tell a story and cheer everyone up. You couldn't ask for a better friend, and he was closer than a brother to me."

Every year after the war until his death, "Red" traveled on Memorial Day to Bob's grave in Milford, Massachusetts to pay his respects and remember.

Chaplain George
by Bob Powell

One of the most loved veterans of the 352nd Fighter Group was Chaplain George Cameron. George was not your typical military chaplain and the stories of how he carried out his difficult duties are many. Only a few are recorded here.

When the 352nd completed their training and shipped out for England late in June of 1943, many of our newly commissioned pilots had married after being commissioned and went overseas leaving their wives pregnant. Unfortunately, seven of these pilots never lived to see their sons or daughters. Aware of this sad situation, George Cameron decided to raise enough money to send a $1,000 War Bond to each of these children we called our "War Bond" babies. Taking into account the value of $1,000 in those days, this was to be no easy task. But George never looked for the easy way.

On paydays, he would start poker games, usually putting a good-sized amount in the "pot" to get it rolling. However, he also required the winner of every hand to contribute a small percentage of his winnings to his War Bond Baby Fund. Some of these "pots" got up into the hundreds of dollars.

Since he played taps on his trumpet every night, he put out the word that he would, before playing our "lights out" call, accept requests for other tunes to be played, but each request cost one British Pound note, equivalent to about $4. And, if there were some tunes you did *not* want played, that required another pound note. Four dollars would buy a lot of beer, fish and chips and other favors in those days, but these British notes went into George's fund. It wasn't long before George had obtained the amount needed to accomplish his goal—a $1,000 War Bond for each of the now fatherless babies, which were presented to the seven widowed mothers by an Air Corps Chaplain back in the U.S. wherever they lived.

One of my fondest memories of Chaplain George was our most unusual "sunrise service" on Easter Sunday, 1944. This was about the time we were switching over from Thunderbolts to Mustangs. George arranged to use the long, 108 gallon papier-

Chaplain George Cameron playing the bugle.

Chaplain George Cameron conducts Easter services at Bodney, England, 1944

mache belly tanks for the service. He had them arranged tip to tip in rows as pews for this special service, four or five to a row in two deep sections to seat the troops.

He set up his rostrum in front of the two sections and the backdrop for the service was a P-47 Thunderbolt and a P-51 Mustang angled slightly with their noses near each other. The bellytank pews were hard, but most significant. Just a few hours after hearing his special sermon and his blessing on us, we took off on another mission over Europe and I can assure you that there were no atheists in those cockpits.

My most personal relationship with Chaplain George had to do with D-Day. When we realized late on June 5th that D-Day was there, and Colonel Joe advised us to get to bed early as the next day would be a long one, I took a few minutes to write a letter to my folks. When I finished it I took it to Chaplain Cameron and said, "George, if by some chance I don't make it back through these upcoming missions, would you please mail this to my folks? And, if I do make it through my remaining missions, please return it to me." He told me he would.

A few months and a number of missions later I finished my combat tours and went back to George for my letter. He said, "Punchy, I mailed that letter to your folks."

When I asked him why, he said, "I read your letter, and I decided that anyone who could write such a wonderful letter should have it read by those to whom it was addressed. I hope you will forgive me for doing so." I still have that letter in my collection of letters I wrote and received from home during World War II. I never faulted George for mailing it.

Although he never talked about his prior service, we understood that George Cameron had served a hitch with the North China Marines prior to WWII, returned home and became a minister, later being assigned to the 352nd Fighter Group training up East. What a blessing he was for our Group!

After the war, George became the first Chaplain of the Air Force Academy and was one of the key people who contributed to the design of that beautiful building now serving as a religious home to our Air Force Academy cadets.

The Old Man We Called "Pop"
By Bob Powell

Although most of our pilots were youngsters fresh out of high school or college, one of the exceptions was Lt. Col. Eugene L. "Pop" Clark, born in 1911. When he arrived with the 352nd in Bodney as Group Executive Officer he was at the ripe old age of 32, about ten years over our average age of 21, so he was "Pop," from the steel mills of Pittsburgh, Pennsylvania.

"Pop" had entered the Army Air Corps after one year at the University of Pittsburgh and three years as a pre-med student at Creighton University. He got his flight training at Randolph and Kelly Fields in Texas and was commissioned on February 2, 1939.

Colonel Joe L. Mason, Group Commander, and the other pilots in Headquarters like Colonel Clark didn't fly missions as frequently as the pilots of the three squadrons, but before he was shot down on May 21, 1944, "Pop" Clark had established himself as one of the more aggressive group leaders and was credited with two aerial victories and two others destroyed in strafing attacks plus a "probable" aerial victory shared with Lieutenants Bill Hendrian and Dave Zimms. The "old man" could hold his own with the best of them.

But "Pop" is most fondly remembered for taking an American Red Cross nurse for a ride in his single-seater P-47. Katie Cullen, assigned to Bodney by the ARC, persistently pestered the 352nd pilots to take her up but none would. Although the Thunderbolt cockpit was fairly large, it wasn't designed for a pilot and passenger, and she continued to be denied a flight until "Pop" Clark finally told her he would. But first, he explained to her that he would have to sit on her lap to be able to control the airplane.

"Pop" was a big man of some 200 pounds and Katie was probably no more than 120 pounds, but she was so determined to get that flight that she quickly agreed to that procedure. Takeoff in the 2,000 horsepower fighter exerted a bunch of "G's" when the pilot moved the throttle to full takeoff power, quickly multiplying the Colonel's 200 pounds of pressure on his female passenger, but Katie got her ride. We have no way of knowing what aerobatics "Pop" might have performed for Katie, but when she climbed out of the cockpit she said, "It was great, but he nearly crushed me." We also aren't sure that "Colonel Joe" didn't have something to say to "Pop" after that stunt.

Like most wartime pilots "Pop" didn't talk about the war much to his family, but his widow, Mary Clark, and his son, Mike, have provided us some things they remember his telling them.

Flying his "Panama Pacito," he was hit by ground fire on a strafing run, taking a bullet in his radiator that forced him to crash land in Germany. He said he tried to light a book of matches to torch his airplane so the Germans couldn't get it, but he was unable to do so. In crash-landing his plane his head smashed into the gun sight and the blood from his head wound got on the matches and made them impossible to light. He was captured soon after landing and one of the first German soldiers to reach him took his jacket. He was later taken to a hospital and treated for a skull fracture and sent to Stalag Luft 3.

Arriving at camp the German guards wanted to shoot him because they heard an American pilot had strafed some civilians, and they thought it was him. However, the other prisoners gathered around him prevented them from doing so. Later, after the Germans had processed and reviewed his gun-camera film they determined that was not true.

In the beginning they treated him rather badly—for the first couple of weeks during intensive interrogations, even beating him up, but quit when they realized any information he would have was old and would be of no use to them.

He said that when he was first in prison he wasn't able to communicate with his family and one night he had a strange dream in which he saw the Virgin Mary. After his release, back with the family in Pittsburgh, his Mom told him that during the time when nobody knew his whereabouts and she was very worried about him, she had the same dream and after that she quit worrying about him, knowing that he was all right.

In the German prison camps, the American and British prisoners were separated in different barracks and both made plans to escape. However, a few nights before the American plan was to be executed, "Pop" had an accident. Bending down over a fire they had, a rock in the fire shattered from the heat and a piece struck him in the eyebrow. Since they didn't want to leave anyone behind, knowing he would likely be tortured for information about the escape, they cancelled their plans. By the time he had healed and they were ready again, the English prisoners had made their escape on which the movie, "The Great Escape," was based. This caused the Germans to carefully search all the other barracks and they discovered the tunnels the Americans had made.

During his stay in the German prison camp Colonel Clark lost 60 pounds—down from 210 to 150. He said the food was terrible and they would eat almost anything they could get their hands on. He said they were given one loaf of hard black bread a week for each five of them and they disregarded the bugs in their soup because they provided protein. And, that the Red Cross parcels were wonderful.

The prisoners concocted a still and used things in the Red Cross Parcels to ferment some 'hooch' and celebrated the Fourth of July marching around the compound, some simulating the dress of the three men in the famous Revolutionary War painting, one with a bandage on his head. The German guards kept telling them to settle down but they kept marching and singing until the guards trained their machine guns on them.

It was during this period that he gave himself up to the Lord. He said that after doing so he felt like he was being watched over and helped by God, keeping his spirits up and this is why he never skipped church after returning home. He said it was his epiphany and he always believed after that.

The prisoners of Stalag Luft 3 made a forced march to Nuremburg in the cold winter of 1944-45 with snow on the ground. They slept in barns and were in a railway depot in Nuremburg when it was bombed by the Allies. "Pop" was liberated April 29, 1945 by the 7th Army.

Mike also relates a couple of incidents from his father's cadet days.

Once, while in flight training at Kelly Field he got into an inverted spin and couldn't recover. He rolled back the canopy and tried to bail-out but he hadn't rolled it back far enough and he got stuck halfway out. When he was thrown back into the cockpit by the spinning plane,

Pop Clark in "Windy" Parlee's Mustang.
(Clark Family via Powell)

the aerodynamics of the plane somehow changed and it went into a normal spin—which he managed to get under control after dropping from 12,000 feet to 2,000 feet. It just ***happened*** that when he recovered from the spin he found himself in the landing pattern and proceeded to land the plane. He was accused by his fellow pilots of showing off.

Checking him out afterwards, the base doctor told him to lie down but he said he felt fine. The doctor repeated his order and made him lie down. Just minutes later, he started shaking so badly that the bed was shaking, too. The doctor had anticipated this uncontrollable reaction caused by a delayed adrenalin rush. After that he was fine.

In another incident when he observed a P-26 crash at the end of the runway on takeoff, he ran up to see if the pilot was OK and found him unhurt, but the plane a total wreck. He asked the pilot, "Do you mind if I cut the star out (insignia) out of the wing?"

The pilot said, "You can take the whole damn plane for all I care." So Pop pulled out his knife and cut out the star, took it back to the barracks and got all the cadets in the squadron to sign it. That star is Mike's proud possession, along with photos and names of all the guys who signed it.

Those of us who flew with the 352nd early in the war had great respect for the "old man" who led our squadrons and group on many missions over Europe. Unlike some of the higher-ranking pilots, "Pop" never hesitated to attack ground targets because of the high risks in these low-level attacks. And, of course, it was on one of these that he met his Waterloo in combat.

The name of Colonel Clark's plane at Bodney, "Panama Papacito," was based on his service in Panama between September, 1939 and September, 1942. Ironically, after the war, he would be reassigned for duty there again.

Focke-Wulf 190 as viewed from below shortly before it was attacked.

Rough Day in June

Drisko and the 487th On D-Day

The traditional "first day" at the office is usually a trying time, spent trying to fit in and not make obvious goofs. If you ever thought you had it rough, try Bill Drisko's first day on for size. Bill was assigned to the 487th Fighter Squadron's "Meyer's Maulers" of the 352nd FG. This squadron achieved the most victories in the 8th Army Air Force, and is in the top four all-time squadrons in scoring for the Air Force in all wars.

Bill begins, "I was in Class 44-A and we graduated from Cadets on January 7th of 1944 at Foster Field near Victoria, Texas. I had gone into the cadet program nine months earlier, leaving high school early on March 25th of 1943. Of course the war had been going on since December 7th of '41. The English had been at it even before that, so we had been very, very busy over these months trying to learn to fly and win our wings. We were quite proud that we had won these wings, and were now Officer Pilots

Bill Drisko ready for action in his P-51B. (Bill Drisko)

487th FS area looking west at D-Day, with "Meyer's Maulers" sign being supported by crewman

ready to go to a Replacement Training Unit (RTU) to learn to fly a fighter. We were sent to Bartow, Florida for P-51 training, and in May were ready to go somewhere overseas. After all, the Army Air Force had bases in England, Burma, Italy, Attu, Alaska, and all over the South Sea islands going towards Japan. So we knew the war was going full blast. Production was going full blast. We had been in the war for two or three years and were still in it. It wasn't over, but we were going to be sure we won this war.

"We were finally assigned orders sending us to New York, and about thirty guys (wondering where we were going to end up) arrived in the City. We decided that this meant we must be going to Europe—probably England. We were right. Europe it was. They put us in a hotel in downtown, and that was the only time I've been in downtown New York City. We were there three days, getting up and reporting in by phone. The ranking man, a First Lieutenant, would be in charge and take all of the calls. We would go out and see the city, hit the nightclubs and bars, and had a grand old time...we thought. The Army Air Force would call every once in a while, and two or three of the guys would go out to La Guardia, and one-by-one we were taken out of there. When I got on a C-54 transport I was with thirty other Lieutenants and Flight Officers headed for England. We flew through Gander and Goose Bay (Labrador) to Prestwick, Scotland (a huge base). That is where we stayed for a few days.

"In a few days I had orders with seven other guys to get on a train, go to London and then on to our base. We still didn't know where our base was—no idea. This was getting along into the early part of June, the second or third day. We were finally traveling to go into combat, which we had been training for over a year. Back on March 15th I had turned 19 years old, and I was pretty proud of the fact that I had checked out on the P-51 Mustang

Aerial view of Bodney looking northwest. Officer's quarters in lower right corner. 328th communal area in lower center. Worn path of main runway SW to NE visible in photo center. (Paul Grabb)

The 487th ground crewmen's Nissens up on the North Site at Bodney. Bomb shelter mound at center.

on the March 14th in RTU training. I now had 200 hours in the Mustang and figured I pretty much knew how to fly that airplane. Anyway, we took a train into London and they had given us four or five of these little rations the size of a Cracker Jacks box. They were named 'Breakfast,' 'Lunch,' and 'Supper,' a cold bunch of stuff with crackers and funny chocolate bars. The Air Force didn't bother with what meals you got, just so long you had the right number of boxes. So we had our meals, got into London, and received orders to go to Bodney Airdrome, near a train station called Brandon. I had heard of the 4th Group 'Red Noses' with Blakeslee, and the 56th with Zemke, but I hadn't heard of the 352nd Fighter Group. So I said, "Good. We'll go to the 352nd." So we took the train to Brandon. I didn't realize at the time we had gone from Scotland all the way clear down to London, and were now heading back up north towards Norwich, England. I was taking a train to my combat base was all I knew.

"We arrived that evening at a little train siding with a small building that couldn't really be called a station. It did have a phone, so the 1st Lt. in charge of we seven said, 'Well, we will just call.' Apparently our orders must have had enough on them to know where to call. We all waited around for a couple of hours, and it was getting later on in the evening, not long 'til dark. In a couple of hours a G.I. truck came along and the driver said, 'Get in. We are going to the base.' We got in and rode through the trees and woods on the left side of the road, bumping along.

"We arrived at a gate and looked out. It was getting dark and we couldn't see a runway, but we could see a grass field with lights. They looked like flare pots but I was told they were a string of lights. I thought to myself that we only flew at night once or twice in practice, and then we had landing lights and all on a runway. Now this was a grass field, at night, and they were going to take off. It all seemed a little scary to me, but it looked like everyone was doing it. All through cadets you did what you were supposed to do and were told

487th Armorers are bombing up Mustangs for ground support duty during the Invasion.

to do like everyone else did. Everybody did it, and it worked. I said, 'I guess if they fly at night I will too.' Also, the airplanes looked kind of funny as they had put funny black and white stripes on the side and wings. I had never seen any paint scheme like that in my whole life. I thought 'This 352nd is really something else!' All these stripes on 'em, and no idea what it meant.

"We had a hard time getting them to let us on the base. We had been standing around at the gate looking at the planes and the field, and they told us, 'You can't come in. This base has been closed down.' I thought that was kind of silly, but there was a war on, so what did I know? I was over where they flew combat, so maybe this was normal. After about 30 or 45 minutes they said, 'Well, since you are replacement pilots, you can come in.' So they let us in and sent us to Headquarters where we signed in. We were told to hang around a while. Pretty soon then this Lieutenant Colonel comes in and to this day I don't remember who he was. He said 'Well boys, there are seven of you and three squadrons. I am going to assign two to one squadron, two to another, and three to

Mission briefing in Maycrete hut in HQ's area. Note flight path in ribbon on the map.

the other. You go here, and you go here, and you go here.' He may have known who needed what, but we were green beginners anyway. He then told us, 'We have good airplanes here. It will run with the '109, climb with it, and it'll dive with it too. Good hunting!' I really didn't feel too much like a hunter at that time. I'd been constipated for a week due to travel, and I felt more like the Hunted. We all saluted and he said, 'Okay good. Now we are going to have a big briefing at midnight and you all had better be there. Go check out the base and get some bedding.' It was about seven or eight o'clock in the evening by then.

"Now we hadn't the slightest idea why they'd have a briefing at midnight. I'd never heard of such a thing. Of course, I wasn't in combat before, so what would I know anyway? They said 'Be at briefing' so we said, 'Okay.' They gave us something to eat at a mess hall (wherever it was), then someone took us to draw bedding. I was assigned to a barracks and got to a bed. There was somebody's laundry on it and his pictures of him and his family off to one side. I said, 'Whose stuff is this?' A guy offhand told me, 'His name was Hall. He got killed today. Just throw those off to one side and someone will come and get them.' Now that sort of surprised me. Someone got killed today and that was that? I thought, 'Wait a minute. Hold it. This must be a kind of a serious situation here.' But it happened that he had been killed. There were not that many killed that often, but I didn't know that at the time and I was shook up a little bit. It brought home the fact that we were in a war. I went ahead and tried to get some rest before the midnight briefing.

William "Flaps" Fowler poses with his bird "Stardust." (Flaps Fowler)

"Near midnight we got over to the briefing room with all of the pilots standing around talking and gabbing. They all had their .45's in their shoulder holsters, a couple or three I saw had great big mustaches, and they were wearing their boots and flying gear. Some were talking about all the boats they saw that day. They said they had never seen that many boats in their life as what they saw out there that day in the English Channel. Again, I hadn't the slightest what that meant either. Well, there were a lot of boats in the Channel.

"At midnight they called us into the Briefing Room and somebody barked 'Atten-Hut!' and everybody popped-to real quick, standing at attention. This big tall red-head Bird Colonel strode down the aisle, and it sounded like the movies to me. He said, 'Men, this is it.' I thought that was just the way John Wayne would do it in the movies. He started telling us about it, and what happened was they were having something called 'D-Day.' He said, 'This is the big invasion that we have been talking about. It is going to happen right now and tomorrow, and we are going to fly five or six hours. You'll get your assignments where to fly.' By the way, I wasn't going to fly—it would be another three or four days before I would go up. The pilots were going to take off at 2 am and fly five or six hours in a pattern in a sector. They were directed to go down southwest of England and cross the Channel there. Not straight south. The only ones over the Channel straight south were to be P-38 Lightnings because they could be recognized easily and so wouldn't be shot at by our troops. The colonel said all our planes in the theater had the black and white stripes so they could be recognized. It was then that I really understood it was D-Day.

"I was talking to Bill 'Flaps' Fowler at the reunion, and he guesses that Group C.O. Colonel Mason somehow must have been spooked by that D-Day situation. He had been talking about it and worrying about it for a long time. Bill told me 'We didn't pay a lot of attention to Mason's worrying, but that night I saw you replacement guys standing around in there. I thought, my God man, maybe Mason is right. This is the big D-Day and we are all going to get killed and will be replaced by you new guys.' He began to wonder what he had gotten himself into. That didn't happen to be the case, they just need some replacements, and it was time for us to be there.

"Those guys planned to take off and I planned to get myself squared away in my quarters. I got all squared away, and I guess it was about one or two o'clock or so, I was laying there listening. It seems to me it had started a light drizzle and out group was ready to take off. I heard them running up and taking off, and all of a sudden 'BOOM!' There was a great big loud explosion and I saw a light from a great fire that was spreading all over everything out there. I couldn't imagine what it was but I went on to bed. The next morning I found out one of our guys named Frascotti had been in a flight that drifted the wrong direction in the dark and he ran into the New Tower building. That is when I began to think, 'War is Hell.'

Sgt. Gillenwater, Drisko's Assistant Crew Chief, shows the fatigue of D-Day Ops on the ground crewmen. (Al Giesting)

"They did take off and run two or three missions that day. Early that morning, around four or five o'clock I heard this droning. A heavy droning of airplanes for a long time. 'What are they doing flying at three, four, or five o'clock in the morning?' I found out they were transporting the paratroopers and they had started out early for D-Day.

"By the next day we knew what D-Day was all about. In our squadron we kept a map, and our intelligence officer Lt. Seymour Joseph kept a little running commentary of where our troops were on the beach. It looked like a very small area we had taken, but eventually over the next few days it widened out, we had a beach-head, and held on. So that was my experience with the unit, the night before D-Day."

What a stressful first day! Bill was first told he could not enter the base, then was given the bunk of a man who just died. He was then faced with a unit that made night takeoffs into rainy skies from a sod field with jury-rigged lighting, all after midnight briefings. It would make any man wonder with what unit he joined up.

Lt. Malcolm C. Pickering

For more of the full story of the first flights of D-Day morning and Bob Frascotti's takeoff accident, see "Blue D-Day for the Bluenosers," Air Classics, July 1998, Volume 34, Number 7, reprinted in part on page 103 herein.

Bill Drisko finishes, "The day after D-Day I found that I had no more 'medical problems' that bothered me previously. Probably war was the answer to that. I went ahead over the next few weeks and flew six missions. Then on the seventh mission (on June 25th, flying George Arnold's P-51B HO-A, serial number 43-6902) a '109 got behind me. We were at about 21,000 feet and I was working hard to stay in formation since I was still a little inexperienced at altitude formation flying. I heard a funny noise, saw some 'puffs of cotton,' I wondered what was going on, and about that time someone starts yelling. Then my airplane went into a violent spin after something had exploded out there on the wing. There wasn't any wing left! I was in one helluva position, and was suddenly wishing I was back in Bartlesville, Oklahoma where I belonged. I knew I was over France, but didn't know exactly where. I later found this was over Soissons-Epernay. After what seemed like an eternal struggle to rid myself of all of the harness, etc., I was finally able to free myself from the spinning plane. Finally the chute popped open and I relaxed and floated to earth.

"I spent most of the summer in France with the French Underground and they took good care of me, and I had some interesting experiences. The first week of September I got liberated. We happened to be in Paris when it was liberated and I got in on all of that, still wearing my French clothes. I came on back a week or so later."

The full story of Bill's escape and evasion may be found under "Drisko's Unexpected Vacation in France" in the "Bluenosed Bastards Of Bodney" unit history by Robert "Punchy" Powell, Tom Ivie, and Sam Sox, Jr.

The author wishes to thank the 352nd FG Association members, including Bill and Rick Drisko, Dick DeBruin, and Punchy Powell, as well as Sam Sox and Tom Ivie. As always, thanks go to my family of Patrice, Chelsea, and Andrew for their support.

Bill Drisko PR shot.

Chickens Don't Fly! by Iggy Marinello

Frank Tenuto and I went into the village one day on our bikes, those British bikes with the brakes on the handlebars, and noticed there was a sort of farmers' market auction in progress. Tenuto said we ought to bid on some of those chickens being auctioned and suggested we get a local person to do our bidding, figuring the locals could get a better price for us.

We headed back to the air base on our bikes with three chickens in each hand and did fine until we started down a hill, started going too fast, and had to put on the brakes to make a turn. Up went the chickens. We had to stop and chase them down, but the story has a good ending—we enjoyed a good chicken dinner that night.

Iggy relaxing on "The Kelly Kid." (Iggy Marinello)

Later, when in Belgium, Frank was assigned a jeep since he was in charge of the refueling units. He and I asked the other guys for donations so we could go barter candy, soap and cigarettes for goodies from the Belgians. Since I could speak a little French, we usually came back with wine, eggs, bread and other things for a nice meal.

A Night on the Town by Ray Mitchell

After we had been confined on a troop ship crossing the Atlantic for several days and immediately finding ourselves under quarantine at a pilot replacement depot in the town of Stone near Hull, England for orientation before being assigned to a combat unit, Lt. Bobby Dodd and I got bored and took things into our own hands. Let's face it—2nd Lieutenants aren't very smart. They just aren't aware of it.

Bobby was a fun-loving maverick and for some unknown reason, I went along with his shenanigans. Noting that security was rather lax, we decided to slip off the base and go into town one evening. After enjoying our "officers night out," we hailed a taxi to take us the short distance back to the military base Everything was going fine until we noticed that our cab driver was having difficulty staying on the typically narrow English road with its many sharp turns.

We hadn't gone far when he failed to make one of those turns and over we went. Although we were not hurt, that's when we discovered that our driver was very drunk. Too late! So, we set the car upright again and took off walking the remaining distance back, leaving the driver to figure out his next move.

Knowing that it would not be wise to call the local authorities since we were officially AWOL for the evening and would be in deep trouble, we proceeded to walk the remaining short distance back to the base and found an unpatrolled location and slipped back onto the base—much wiser for our experience.

"It's Supermouse/Sweet Sue" of Boddy Dodd, 44 11626.

"My Goodness! My Guinness!"

This story will probably never make the *Guinness Book of Records*, but it has all the qualifications for that recognition. Who else but Lt. Bobby Dodd, a pilot in the 328th Squadron, ever received a promotion in rank in an English pub? It is possible, of course, that some may have promoted ***themselves*** after a few pints.

Bobby Dodd graduated from pilot training at Aloe Field in Victoria, Texas in the Spring of 1945, not as as a Second Lieutenant, but as a Flight Officer. He was told at the time that the rank was given to pilots considered too immature to be an officer. Although this kind of a selective judgment could have been made by someone in command at that particular flight school, the real reason was more likely that only so many officer slots were available at that time, requiring a very small percentage of each graduating class to be graduated as Flight Officers, a rank equivalent to a Warrant Officer in the Army Air Corps at that time.

Bobby arrived in the ETO in July of 1944 and was assigned to the 352nd FG. He remained as a Flight Officer until October when he was promoted to the rank of 2nd Lieutenant. Just 23 days later, however, he was promoted again, to the rank of 1st Lieutenant by his squadron commander, Major George E. Preddy.

It seems that Major Preddy didn't care much for the Flight Officer rank and felt that all those who had qualified as combat pilots should be commissioned officers and he upped Bobby's rank to make up for what he considered unfairness.

Not to take anything away from Bobby, who obviously deserved the new rank, but it is highly possible that a few pints of the famous "Guiness Stout" might have been a factor in this quick promotion.

Submitted by Dennis Sparks

There's a Time to Hold 'Em... and a Time to Fold 'Em

by Richard Brookins as told to Bob Powell

"On this particular mission my 'Jug' was out of commission, so I was assigned to fly P-47D PE-E as its pilot was not scheduled for the mission. We were to give long range fighter escort and cover for several boxes of the B-17s and B-24s. We took off from Bodney flying four abreast as usual, and formed up in tight formation. Our flight leader followed the flight in front of us, while we three kept it tucked up in tight formation.

"There was considerable cloud cover that day, so our No. 1 was either following the flight in front of him or was navigating through the clouds. As we proceeded on our assigned climb out course, going through cloud layer after cloud layer, we were due to be at around 16,000 feet by the time we departed the English coast.

Lt. Brookins after a session with Duchess I

"Suddenly I heard an explosion in my engine and I couldn't remain in formation for lack of power. I called my flight leader and reported my dilemma. My oil pressure had dropped to zero. In spite of having on my oxygen mask, I could smell something burning. The engine had frozen and I could see a stationary prop blade in front of me. It seemed so unreal at the time. Oil was pouring out of the left side of the engine compartment and rapidly covering the left side of the aircraft and my canopy. I could see flames coming from the cowl flap area and I knew I'd have to bail out.

"As hundreds of fighters and bombers were climbing out on course to Germany, I turned north to get as far from the fighter/bomber stream as possible. It would have been nice to descend to a much lower altitude, but the burning aircraft gave me little choice but to bail out as soon as possible. I released the canopy, rolled over, and climbed out of the burning plane and never saw it again. As I was in free fall, I found it difficult to orient myself as there was considerable cloud coverage and the ground was not visible at all. I pulled the D-ring and nothing happened. In a panic I clawed at my seat pack and thankfully the chute opened.

"What an eerie feeling! I had no idea what my altitude was but I could hear the roar of the many engines in the fighter/bomber stream. It sounded like they were all around me. I prayed that one of them wouldn't plow into me. It seemed that I floated earthward for 10 to 15 minutes before I popped through the last bit of cloud cover and could see the ground 1000 to 2000 feet below me. I landed uneventfully in an uninhabited farming area and was picked up and returned to my base at Bodney."

Operation Bulbasket
By Paul McCue

On 6 June 1944 (D-Day) two members of the SAS (Britain's elite special forces group) were dropped by parachute behind the lines in enemy-occupied France, soon after to be followed by others to a force of fifty men. Their task was to work with the maquis (the French Underground organization) to disrupt in any way possible the movement of German troops north to repel the invasion troops landing in Normandy. But it was not until the early nineties, with the release of previously classified documents and the unrelenting research of British author Paul McCue that their story came to light. Only that part of this highly dangerous operation is told here, that which reveals the strange events that involved a 352nd Fighter Group pilot.

The 486th's Lincoln Bundy, posing in a P-51B shortly before his death. (Sheldon Berlow)

They moved frequently to avoid the danger of their camps being discovered by the enemy using their radio direction-finding equipment or by informers. Water was a big factor as that region suffered one of its longest droughts in years. Their camp near Verrieres held a number of attractions. It was close to the DZ at la Font d'Usson and the forest of young trees had a pleasant stream of fresh water running through it. But their commander, John Tonkin, was beginning to feel uneasy there as two of their number had possibly been captured and made to reveal the location of their camp. And, worse than that, they learned that local villagers had learned of their whereabouts and a couple of young women, anxious to seek a hint of adventure, had sought out the camp in the forest to try to get to know the brave young Englishmen who had come to fight for them.

Angrily, they mounted a maquis guard and forbade any visitors. Undeterred, however, two or three of the troopers had taken advantage of Tonkin's absence and strolled into the village.for a drink at the bar, hoping to pursue friendships with the local girls. It was now obvious that the Germans would soon know their location.

Capt John Tonkin

Other events added to Tonkin's unease. First a man from a nearby hamlet warned the maquis of a suspect car seen several times touring the forest and the village. It was thought that he might be a direction-finding vehicle, although no mention was made of antennae or aerials. And, secondly, a downed American pilot was brought into the camp by the maquis on 1 July. Introducing himself as Lt. Lincoln Bundy of the 486th Squadron of the 352nd Fighter Group, the flyer had a ready story of how his Mustang fighter had been shot down by flak. He had then been picked up by the Resistance and passed along from maquis to maquis until he had reached Verrieres. But the SAS were suspicious of the circumstances and, fearing a German "plant," immediately radioed to London to check the American's identity.

To guard against what was now deemed a probability the Germans knew something of the activity in the forest, a decision was made to move out and another base was set up just a few kilometers to the south in the Bois des Cartes. The reason for not moving further was that Tonkin was expecting a critical re-supply drop in the DZ on the night of 3 or 4 July, including another Eureka set promised in a radio message from England on the 30th.

But even the new camp did little to make the SAS feel safer. On the 2nd a German plane dropped leaflets urging the maquis to give themselves up to the German authorities in Poitiers with a promise of no harm to them. While the maquis knew better than to accept this promise and laughed off the invitation, Tonkin and Dieudonne, the maquis commander, were nevertheless concerned that the Germans were showing so much interest in the area. Worse was the all-consuming issue of water supply.

Taking into account that it had been two days since their two men had disappeared and there had been no German reaction. It was a risk, but they decided to return to Verrieres for its water source until they could find another. But even as they moved back to the forest, German troops were mobilizing with the express purpose of dealing with "Bulbasket."

The SS Security Police, under command of SS-Obersturmfuhrer Hoffmann, already had information that the SAS camp was somewhere in the area and on 1 July had sent their agents to the area to pinpoint the camp's location. John Tonkin went out looking for a suitable site but soon after he returned. In early morning, disaster struck. Hoffmann's forces had stealthily crept forward under cover of darkness to ring the area and, either by fortune, or the Germans cannily permitted it, Tonkin had probably passed right through the enemy's cordon without detecting it.

Mosquito FB VI MM403 of 464 Squadron, RAAF

In the quiet of the forest camp, forty of the SAS party, the American flyer Bundy and nine of the maquisards slept unsuspectingly. In the early hours of the morning Dieudonne had been briefly awakened by the sound of vehicle engines, but thinking it was a replenishment party returning, and being so tired, he fell asleep again. But some of his men were soon to find the significance of the noises.

Maurice Salmoni, known as "Pierrot le Corse," left the camp about six o'clock to return a wheelbarrow he had borrowed from a nearby village and it was thought that he was the first to be captured by the Germans, taken by surprise and without a shot fired. It was also later surmised that Salmoni was forced to confirm the camp's precise location. About an hour later, two maquisards sprang the trap.

They had been part of a group which had gone into Verrieres the previous evening to obtain supplies but they were friendly with a couple of local girls and were given the opportunity to spend the night in the village. They readily accepted and warned their returning comrades to expect their return to camp at around seven o'clock the following morning, 3 July 1944, a morning not to be forgotten by Marcel Weber:

"Towards 7 o'clock coming from Verrieres with my comrade, Pierre Lecellier, we arrived within sight of the village close to our camp. We were in civilian clothes . . . suddenly, forty metres ahead I noticed something strange. The bushes seemed to move. I stopped and at that moment a German soldier, camouflaged with foliage, stood up and shouted to us, in German, 'Surrender, you are surrounded.' Quickly and together we drew our weapons, but at the same moment behind us six other Germans showed themselves. They had let us go by them and they shouted at us to stop or they would open fire.

"Pierre ran towards the village and in among the houses there. From then on I didn't see him again. I turned back and, with the Germans firing at me, went around the village. There were Germans in the house and they fired on me from the windows. I ran down a street and into the middle of a group of them. One grabbed me by the neck but I struggled free by elbowing him in the stomach. He let go and with some difficulty I ran off through the shrubbery opposite. The Germans continued to shoot at me and now that I was discovered I made towards the camp, zig-zagging as I ran. I felt a strong burning sensation in my right thigh and soon felt blood flowing—having been 'winged' by a bullet. I continued despite the pain and reached the camp where I found my comrades standing ready with their weapons, having been alerted by the German gunfire. I ran through the middle of them and shouted, 'Look out! The Germans are coming! I've just lost Pierre and I'm wounded.' I went on and lay down among a group of the SAS. A lieutenant, with the aid of the medical orderly, made me a wound dressing."

Lt Richard Crisp

The SAS had been caught completely by surprise. John Tonkin was asleep in his sleeping bag when the first shells and mortars landed. Moving to the northern edge of the forest he saw a considerable number of German troops working their way toward the camp in an attempt to trap the SAS group. He knew it was pointless to try to fight their way out as a group, so they followed their planned procedure of small groups scattering into the forest to the west in ones and twos concealing themselves, as their best bet.

Sleeping just a few metres from the SAS party, the maquis group was likewise caught unprepared though some of them, awakened by the gunfire, were up. Almost immediately, however, mortar bombs crashed down into the forest as the Germans attacked before their element of surprise was lost.

Grabbing their weapons, the maquis crouched in the undergrowth waiting and when they saw the forest growth begin to move about 15 metres away they opened fire on the dozen

Left to right, Tomkins, Stephens, and Crisp

or so German troops, who immediately went to ground returning a deadly fusillade in the general direction of the French. The latter slipped away and regrouped with their leader, Dieudonne, in time to confront another party of Germans, this time throwing grenades before melting away into the woods towards the stream.

The battle raged for some time, but ultimately, the Germans were successful.

News of the disaster spread rapidly throughout the towns and villages, inevitably becoming distorted to the extent that all the SAS were reported killed along with the maquisards, including some who had not been there.

Thirty SAS men, three of whom had been badly wounded, were captured along with the medical orderly, Corporal Allen, and the USAAF pilot, Lieutenant Bundy. Only four had escaped. Flushed with their success, the Germans were eager to glean as much information as possible from their prisoners, but ultimately were disappointed. No details were given regarding the home airdrome, the flight from England, types of aircraft, the organization and so on. There was no information regarding the interrogation of Lt. Bundy.

There was quite a squabble among the various units involved with regard to who would have responsibility for handling the prisoners. However, the Fuhrer himself, Adolf Hitler, had issued an edict on 18 October 1942 that would determine the fate of the prisoners. It was called Kommandobefehl, "Command Order," which was to bring shame on the name of the German armed forces. Infuriated by the growing number of Commando-type operations against the occupied continent, and particularly the channel island of Sark, Hitler had retaliated with this order:

> *In the future, Germany, in the face of these sabotage troops of the British and their accomplices, will resort to the same procedure, that is, they will be ruthlessly mowed down by the German troops in combat, wherever they may appear.*
>
> *From now on all enemies on so-called Commando missions in Europe or Africa, challenged by German troops, even if they are to all appearances soldiers in uniform or demolition troops, whether armed or unarmed, in battle or in flight, are to be slaughtered to the last man. It does not make any difference whether they are dropped by parachute. Even if these individuals, when found, should apparently be prepared to give themselves up, no pardon is to be granted them on principle.*
>
> *. . . If individual members of such Commandos, such as agents, saboteurs, etc., fall into the hands of the military forces by some other means, through the police in occupied territories, for instance, they are to be handed over immediately to the SD. Any imprisonment under military guard, in Prisoner of War stockades, for instance, etc., is strictly prohibited, even if this is only intended for a short time.*
>
> *. . . I will hold responsible, under Military Law, for failing to carry out this order, all commanders and officers who either have neglected their duty of instructing the troops about this order, or asked against this order when it was to be executed.*

The order claimed that Commando troops fell outside the protection of the Geneva Convention and there was little doubt that any German officer refusing to implement the order could expect similarly brutal treatment to that designed for the

Commandos. Therefore, the Commander had little choice but to turn the prisoners over to the SD. Later, trying to find a way out, the SD commander, Koestlin, tried to have the prisoners turned over to the Luftwaffe. However, neither the German Air Force nor other commands were willing to take over the prisoners with a death sentence already hanging over them.

Peter Banks

Finally, having exhausted all options Koestlin called for Oberleutenant Vogt, a prewar Protestant clergyman hardly suited for this task. He then personally ordered Vogt to execute the prisoners. The next day Vogt was to select the place of execution and dig the required graves. The day after that, the 7th, he was to carry out the execution.

The execution itself was to be performed by a firing squad of Vogt's men. Hauptmann Schonig was to be present and an interpreter was assigned to assist with the proceedings. Although appeals were made to the Chief of Staff to not go through with the execution, he felt that he was bound by the Kommandobefehl and could not therefore allow his mind to be changed.

One final initiative by Schonig casts some doubt on Koestlin's expressed claim to have been bound by orders. The Intelligence Officer did not feel that Lieutenant Bundy, the American pilot who had been shot down and captured with the SAS, should fall under the Kommandobefehl and Schonig therefore presented his point to Koestlin. Where here he might have exercised a lenient stance with little fear of later challenge, Koestlin instead replied that the American officer had acted in a common cause with the SAS and the maquis. It was therefore the Chief of Staff's opinion that Bundy did come under the Kommandobefehl and thus the fate of all the prisoners was firmly sealed.

Oberleutnant Dr. Egon Deter, assistant to Schonig in 80th Corps Intelligence, was ordered to accompany Schonig along with Sonderfuhrer Hoenig. The three officers arrived in time to watch the SAS prisoners and Lieutenant Bundy ordered into five trucks.

In their car they followed the convoy out of Pontiers. After some thirty kilometers the vehicles turned off the long straight Nro road and finally to the village of Saint-Sauvant, stopping in a forest ride.

First to dismount was the firing party, which immediately took up positions with two riflemen to each prisoner. The prisoners themselves were then ordered out of the trucks and ushered to the opposite side of the ride from the firing squad. The three wounded troopers, Pascoe, Ogg and Williams, had been left in Pontiers, but the remainder of those captured in Verrieres had been joined by two others, Sergeant Eccles and Corporal Bateman. The two NCOs had been held by the Germans since their capture at St. Benoit on 28/29 June and had now been permitted to join their comrades only to share the same fate. Deter noticed a conversation between Honigschmidt and one of the prisoners and was later told that the exchange had to do with sparing one of the prisoners but it could not be determined if it had to do with Lieutenant Bundy and that will never be known.

As dawn broke on Friday, 7 July, Schonig called for Lieutenant Richard Crisp, just 20 years old but the senior ranking prisoner. Limping from a leg wound from when he was captured, Crisp might have escaped at Verrieres had he not remained to look after one of the seriously wounded men, a gesture typical of the young officer. Now, the condemmed troopers could have no better leader for their final moments. Turning to Honigschmidt, Schonig asked the interpreter to translate the order of execution to Crisp and to also convey his own sense of personal shame, as a German officer, in having to carry out the Kommandoberfeld. Honigschmidt did so as Schonig struggled to look the young Englishman in the eyes.

Returning to the men, Crisp conveyed the confirmation of their death sentence, a fate no doubt anticipated following the unexplained journey through the night to the forest. The prisoners were allowed to smoke a final cigarette and take their leave of one another. As final handshakes and embraces were exchanged, Vogt called his firing squad to order.

It was difficult to even begin to imagine the thoughts of the thirty British and one American as they then turned to face their executioners. It was reported that some chose to throw their watches, photographs and other personal effects to soldiers of the firing squad for safe-keeping. Not wishing to watch the proceedings any further, Deter left Schonig to witness the final act and walked back to the car. As he reached it, a salvo of firing shattered the early morning air, followed shortly by a number of single pistol shots as a coup de grace were swiftly administered.

Schonig spent a few minutes grimly collecting identity discs from the bodies before he returned to his car and, with Deter and Honigschmidt, returned to Pontiers. Behind them, Vogt's men were left to bury the bodies in three shallow graves which had previously been prepared just inside the trees at the edge of the ride out of sight of the place of execution. Reporting to Koestlin back in Poitiers, Schonig, who had stood about ten metres to one side, described the prisoner's last moments:

"The execution was accomplished militarily and with dignity. The parachutists died in an exemplary, brave and calm manner after the decision of the execution had been made known to them in the English language by an interpreter ... the prisoners—they were not chained—stood in a line. They linked arms...The fire order was given by Oberleutnant Vogt."

The three wounded prisoners in the hospital were later executed by lethal injection. But the maquis sought their own revenge on those who had betrayed them at Pontier, taking them into custody, including the woman who had fraternized with the Germans, and executing them. War can be a ruthless business.

The British, too, sought revenge. In July Tonkin learned of enemy forces massing for an attack on the maquis near Pontier and another at St. Germain. He promptly got a radio call off to England suggesting these two concentrations as targets for the RAF. Tonkin's reports indicate that an RAF Mosquito attack was mounted on the Poitiers barracks target he had identified on 24 July. Witnesses said the Mosquito's bombing was right on target. The surviving SAS troops of the Pontiers massacre later were assigned to another SAS unit and continued to harass the German troops and their transportation and communications systems until relieved and returned to England—but that is not the end of this story.

As Christmas of 1944 approached, the Poitiers area bore few lasting reminders of the enemies occupation, although the war raged on the borders of Germany. The French countryside had gradually settled back into its traditional way of life, including a wild boar hunt. The huntsmen, however, came upon an area of broken branches and found three patches of displaced earth and a cautious exploratory dig soon revealed the reason. Not far below the surface were decaying, but still clothed, remains of a number of bodies.

This traumatic find blighted the hunt, but the discovery of bodies in shallow graves was not an uncommon event in France in 1944. Not only had the Germans often disposed of maquisards

Memorial of La Couarde at Verrieres

killed in action or executed in this way, but the maquis had quietly buried many bodies of Germans or condemned collaborators. Furthermore, it was known that a pitched battle had taken place in that area between the Germans and a maquis group. It was therefore not an immediate priority to alert the police. However, a letter was dispatched to the Gendarmerie in the town of Melle.

On 18 December police officials came to investigate and made careful note of all they found in the three ditches. Carefully excavating one grave they found five bodies dressed in uniform, leaving little doubt the bodies were of allied soldiers. The next day saw a host of officials descend on the site, excavating the three graves fully.

The first grave was 3 metres long, 1.8 metres across and 1.2 metrs deep. It contained eleven bodies. Of these eleven, ten were dressed in blouses with flap pockets and trousers with two front pockets. The eleventh wore the same type of trousers but had no blouse, which was found in the grave.

John Fielding pays tribute to the executed at Saint Sauvant.

The second grave measured 3 metres long by 2 metres across and 1.2 metres deep. It contained nine bodies, all wearing the same type of clothing with one exception, which wore a pullover with a zip-fastener.

The third grave, which measured 5 metres by 1.8 and 1.3 deep, contained eleven bodies. Ten were dressed in a manner similar to their comrades but the eleventh wore drill trousers, a leather waistcoat (jacket) and laced shoes.

The autopsy report of Dr. Maupetit was grim reading. With two riflemen to each prisoner, the Germans had conducted the execution with ruthless efficiency. Positive identification was made difficult as it became clear that most of the soldiers' "dog tags" had been removed. However, various items of clothing helped I.D. the troopers, but the body dressed in French civilian clothing gave rise to a number of theories but was eventually realized to be that of Lieutenant Bundy, USAAF.

Since the pilot had been hidden for a while by the French, he had undoubtedly been fitted out with appropriate civilian clothing. And it was concluded that the thirty-one bodies were deemed to be Lieutenant Bundy, twenty-eight of the thirty-one SAS taken prisoner at Verrieres, and Sergeant Eccles and Corporal Bateman, captured previously at St. Benoit.

The bodies were then placed in coffins on 21 December in the presence of the Mayor of Rom. Each body was wrapped in a shroud and a plate was nailed at the head and outside each coffin, bearing a number 1-31 in Roman numerals. On each body a corked glass tube, sealed with wax, was placed. This showed the same number as on the coffin and contained a description of the victim and the result of the autopsy. Finally, a number of strands were taken from the hair of each victim and placed in numbered packets together with all personal belongings found on the bodies. The packets were then placed in a locked box in the care of the Gendarmerie for later forwarding to British authorities.

On the evening of 21 December the coffins were taken to Rom and were tended by the FFI until the morning of the 23rd. They were then taken to the Town Hall where they lay in state until the afternoon. At 3 o'clock, watched by a huge crowd of local people who came to pay their respects, the thirty-one bodies were re-buried with full military honors in a corner of the village graveyard in Rom.

The only documentation released by the SHAEF Inquiry Court was a summary of the Court's findings which concluded that "thirty-one Allied military personnel met their deaths at the hands of the German armed forces in the Bois de Guron, near Rom, Deux-Sevres, France, on or about Friday 7th July 1944. Of the 31, the bodies of only 17 had been positively identified and these included Lincoln D. Bundy. Although the evidence was inconclusive, the bodies of the remaining 17 found were in all probability identified."

The Court also confirmed that the 31 men were prisoners of war at the time of their deaths, having been in the custody of the Germans for three or more days, and should have been entitled to the privileges and immunities accorded to POWs by the laws and usages of war and the terms of the Geneva Convention of 1929. There was no evidence to suggest that the prisoners had done anything to deprive themselves of the protection of the Geneva Convention. Without doubt, the Allies were convinced that a war crime had been committed.

In subsequent attempts to determine how the SAS location was betrayed, three possibilities were considered. The first was the SAS suspicion at the time of the arrival of the American pilot, Lincoln Bundy. On his arrival they immediately radioed to London to check his identity. No answer was received prior to the Verrieres disaster, nor indeed afterwards. This in itself led to post-war speculation that Bundy had been a German imposter and had been able to report on the camp's location. In reality, however, Bundy's identity was confirmed. He was executed with the SAS at Saint-Sauvant and is buried with them at Rom. The third was the reported radio-direction vehicle in the vicinity reported to the SAS.

Intriguingly, though there is little or no doubt that Bundy is buried at Rom, official American records still leave room for conjecture. The headstone on Bundy's grave at Rom only states 'believed to be buried here' while in England, his name appears on a USAAF Memorial to those 'missing in action' with no known grave at the American Military Cemetery near Cambridge.

In 1995, Paul McCue phoned me from England, (since I was/am the 352nd Editor/Historian) inquiring about Lieutenant Lincoln Bundy and I told him there was no evidence of what had happened to Lieutenant Bundy. McCue then told me he had discovered and visited the Lieutenant's grave in Rom, France while researching a story of the SAS operations in France in 1944 and had learned what happened to the American pilot.

After our telcon during which he told me the story of Bundy's fate, I called the Sheriff in the small Utah town where the Bundy family had lived and asked him if he knew any Bundys, telling him about McCue's surprising discovery. He said, "I live next door to his sister." A short time later I was talking to members of the Bundy family with the news Paul McCue had given me, the first knowledge they had of the true fate of their missing family member. It was a touching moment for them and for me.

Afterwards, I gave Paul McCue their names, addresses and telephone numbers and he sent them all the information he had learned about Lieutenant Bundy including photos of the gravesite in Rom, thus giving them closure on the sad loss of their son and brother. The Bundy family in Utah numbers in the hundreds and this story was the highlight of their annual family reunion. Later, when McCue's book, SAS – Operation Bulbasket, was published, they obtained copies for the family. Although this excerpted account includes all of the material about Lieutenant Bundy which was in the book, the account of the SAS actions are much more detailed. – Bob Powell

Lincoln Bundy's grave, set apart as an American airman, with the graves of the others executed during Operation Bulbasket.

D-Day on the Flightline
by Al Giesting, Al Barrows and Art Weatherwalks

AL GIESTING, Crew Chief

"D-Day started for the 487TH Squadron on the afternoon of June 5th. We had our early morning mission, and were getting ready for a second in the afternoon, but the second mission was scrubbed and the field was suddenly closed for anyone coming or leaving.

Al Giesting

"Flight Chiefs were called to a meeting while buckets of black and white paint were being distributed to each plane along with some brushes. When the meeting ended and the Flight Chiefs came back to the lines, they instructed the ground crews on just how the large black and white stripes were to be applied to every aircraft in preparation for what was obviously D-Day.

"Starting early in the evening everyone got busy as the painting had to be finished by eight or nine o'clock. Sgt. Gillenwater (my Assistant Crew Chief) and I got busy on 'Gracie,' Lt. Pickering's plane. We took chalk, made lines at the proper spacing, and started to paint. He painted the black and I painted the white ones. We got finished in plenty of time, and sort of touched up our masterpiece. The armorers, meantime, checked their ammo and guns, and everything was ready to go.

"Sometime around 1:00 am the first flight of the 486th took off. There were some men standing in front of the 487th Operations watching the 486th takeoff. Earlier, there had been some kind of small lights strung out marking the runway part of the way from the southwest corner of the field, but they were shorted out when one of the planes taxied over them, leaving the field pitch-black dark. The first flight got off okay. The second lined up too far to the south, but got off okay too. The third flight was the one that hit the tower. I know all do not agree with this, but it is as I saw and remember it. The rest of the flights got off okay due to the fire burning at the tower. I think one plane came back and landed. *(The airplane that to abort had clipped the volleyball court near the 328th Ready Room where their pilots were being briefed for their part of the mission)*. The 487th and 328th took off some time later and got off okay too.

"Everyone was busy from then on. and the Crew Chiefs and Assistants didn't get off the line for about five days. The cooks had a weapons carrier they used to distribute sandwiches, food, and coffee during the day. Some caught naps in the cockpits of the planes, on the wings, or at A-Flight, on covers thrown in the old straw stack. The weather the first day wasn't too good but we were used to it. We kept right on working, if not on our own plane, helping another crew with theirs. .

"I remember my plane coming back with a blown tail wheel tire. Some rookie pilot, I forgot who, didn't unlock the tail wheel after landing. (Several new pilots had reported into Bodney just prior to D-Day). Tires were a little short in supply, but Tech Supply promised me one in a couple of hours. Sgts Richard Germain and Lister, the instrument men, came to me to remove my flight indicator to put in another plane to get it going. I told them the tail wheel would be in soon, and they would have to remove the flight indicator over my dead body. They left without it and my plane was in the air in two hours. If you ever let them take parts, you soon had a 'hangar queen.'

"Lt. Pickering, my pilot as of D-Day, flew a lot of hours. Most of the action was shooting up supply trains and such. Every time they would come back they'd talk about the ships in the water and the landing. Sanford Moats had to land on the beach, out of gas, and they gave him just enough to get back. He talked about the bodies piled up like cordwood on the beach.

"Finally, they released the men on the flightline and said we could leave when our plane was ready to go. I remember getting to the barracks and laying down for about an hours nap, but didn't wake up until the next morning, still wearing my cap and shoes. Slept right through the night and missed supper. I got up and went to the latrine, showered and shaved, changed clothes, and went to breakfast. Then I went up to the line and started all over again. Several days after D-Day, our pilots started escorting bombers again."

AL BARROWS, Armorer

"On June 5, 1944, all leaves were cancelled and no one (including civilian workers) was allowed to leave the base. Double guards were posted all around the base perimeter. I was assigned guard duty most of the day until 8:00 pm (armorers routinely got guard duty and KP. Crews were exempt, and the complement of communications men was quite small). I believe we were actually issued live ammunition and guards were placed around the entire base within sight of each other. Dick Wykoff was close enough that we were able talk to each other.

"From my position in the woods overlooking the flightline area I could see the crews painting our P-51's with wide black and white stripes like zebra's. Our planes had previously been painted with black and white markings, but the orders came down to paint these stripes, which literally covered much of the plane, around the fuselage and around both wings. Watching from my post it was an awesome sight.

"At 8:00 pm I was relieved from guard duty, had supper, and reported down on the line. We started stockpiling 250 and 500 lb. bombs and ammo in preparation for "D-DAY" (by now we were certain that this was it). At about 2:30 am the first flights took off. I had returned to my barracks for a catnap, and did not observe the resulting tragedy. Lt. Bob Frascotti, during take-off, crashed into the new tower being constructed, but not yet in use. My squadron (the 487th) wasn't scheduled to take off until last, and I was just leaving my barracks some distance from the field when this accident occurred. Although I didn't see it happen, I could see the sky aglow over a patch of trees between me and the field. My route down to the 487th line area took me right past the scene of the accident, which was burning furiously as I went by.

"The planes flew continuously, each flight alternatively carrying wing tanks, 250 lb. or 500 lb. bombs. Little me (135 lbs) had quite a struggle dragging those 500 pounders on a hydraulic cart. This was over the rough ground and up an incline to my plane, and then attaching them to the bomb shackles. After working the rest of the night I got a couple of hours sleep in the morning. That afternoon half of the squadron came down with food poisoning from eating contaminated rice pudding. That meant we lucky ones had to work that much harder. The rest of the day was more of the same, with the sick returning to work the next day and our routine returning to normal.

Al Barrows, armorer, under the 487th Armament sign.

"I neglected to mention that my pilot at this time was Lt. Robert Berkshire, who was later shot down over France, ended up in a Paris hospital and returned for a visit after Paris was liberated but still on crutches with a broken leg."

ART WEATHERWALKS, 328th Squadron

Late afternoon on June 5, 1944 I checked out the Squadron Bulletin Board next to the Orderly Room. Nice! There was my name on the list to pull AA (Anti-Aircraft) duty that night. I knew of only two AA pits with .50 calibre machine guns. One was where the flight line crossed the east-west road through the base and the other was near the new tower under construction.

I had drawn duty at both of them one time or another but I could never understand this duty since we weren't allowed to fire these guns. My assignment was to the pit near the new tower then under construction, and it was the midnight to 4:00 am time slot. My pit partner, a Pfc from one of the other squadrons, was unknown to me since I had never met him.

ArtWeatherwalks waits for his pilot's return.

When pulling this duty it was common for one to nap while the other kept watch although there was nothing to watch but the dark sky. Around 1:00 am I heard aircraft engines starting up and wondered what was going on. We had never flown night missions from Bodney.

Then I saw red, white and green marker lights moving past the old tower which was still being used toward the Guardhouse in the southwest-northeast runway area. The planes seemed to all be gathering at that point. It was a beautiful sight so I woke my partner up so he could see it, too. We just sat there and listened to the first flight takeoff.

Having been on duty since midnight we kept wondering what was happening. Then the second flight gunned their engines and started moving. You could tell by the sounds of them that they were moving closer and gathering speed for takeoff. From our position in the gun pit we couldn't actually see the planes since the field had a rise across it that obscured your view of the takeoff point and until they topped this rise.

Suddenly there was a tremendously loud roar over our heads followed by two thud-thud noises. One of the planes had crashed into the new tower in the darkness. The first thud was the plane hitting the tower and the second was the fuel tanks in the plane blowing.

We immediately got out of the pit and started running. The field phone started ringing in the pit and after about a dozen rings I crawled back to the pit to answer it. You could feel the heat from the burning plane. The Sergeant of the Guard warning all the guard points and AA crews not to leave their posts. Who did he think he was kidding? Had we stayed there we would have been broiled on one side and crisp on the other since the pit was only 50 or 60 feet from the tower. We left! We got the hell out of there!

We never heard if he did anything about our leaving our post. Apparently not! Anyway, I went to the barracks and picked up a clean pair of undershorts and went to take a shower. After I got all cleaned up I went up to the AA pit area. All I could do was just hang around until my shift ended at 4:00 am, and that's how D-Day started for me. I went to my plane and did what all good crew chiefs and armorers were doing at the time—keeping those planes flying throughout D-Day.

A Pilot's Best Friend

Invariably when tales of aerial battles are written, they center on pilots who jousted the enemy high over embattled foreign lands. This is a story of the "other half" of the equation, the ground crews. Walk up to a pilot and ask him about his crew. Like a pitcher and a catcher, or a quarterback and a receiver, a pilot will wax eloquent about how his crew was essential to his success.

On the grassy field at Bodney, England, the fighter aircraft nestled in their revetments had a beehive of activity around them. It took a strong team of men and good logistics to insure these birds were ready to fly when needed. In general, each plane had four men maintaining its health; a Crew Chief (CC), an Assistant Crew Chief (ACC), an Armorer (Arm), and a Radio Man (Radio). The Crew Chief and his Assistant were responsible for the readiness of everything to do with the plane except the guns, bombs, and the radio gear. The latter were handled logically by the Armorer and the Radio Man. The Radio Man (and occasionally the Armorers) normally had several aircraft to look after, and would often work in groups to get the job done.

Lest anyone guess that these were cushy jobs, imagine working on a complex aircraft, in any weather, outside with no shelter, at any time of day or night. Yet these uncomplaining ground men did what it took to insure that the 352nd Fighter Group could fulfill its escort role for the bombers of the 'Mighty Eighth.' Through sleepless nights these men skinned frozen knuckles and changed impossible-to-reach spark plugs in rain showers to guarantee their pilot had every chance of succeeding and returning home. While a mission was being flown, the ground crewman might catch a little sleep or maybe have time to wash their coveralls in gasoline to try to remove some of the accumulated oil and grime. When missions were coming fast and furious, the crews caught naps in wooden drop tank crates so they wouldn't have to leave the flight line. Such was their lot.

Here's to them.

Bluenoser armorers hang and fuse a bomb

Standing left to right: Unknown, Dick Linn, "Huey" Long. Kneeling: Luman Morey, Stan Swenszkowski, Cy Hall, Art Nellen, unknown. Aircraft is PZ-Y, "Miss Lace" of Mac McKibben. (Earl Martin)

The "Luck of the Irish" Gang

Pilot Chester V. "Chet" Harker formed a strong bond with his crew in over two tours flown with the 486th Squadron. Their numerous aircraft were all named after Chet's wife "Cile." They also carried the wording "Luck Of The Irish" with appropriate nose-art.

Chet recalls, "I got the nickname of 'Irish' prior to joining the USAAF because of a number of hi-jinks that I managed to scrape through during college years and shortly after. When our P-47's first started arriving in England, S/Sgt. Charles Agee (CC), Sgt. William "Rebel" Harris (ACC), Paul Magula (Armorer) and John Aurellio (radioman, part time with more than one plane to service) were given the task of keeping the 'Cile' aircraft airborne.. This crew stayed together until sometime shortly after we received P-51's. Since the P-51 was somewhat easier to keep flying, some crews were split. Charles Agee became crew chief on Ed Heller's plane and Rebel was promoted to crew chief on mine (The Heller/Agee bond was extremely strong as well, with the two even teaming up postwar in the Air National Guard).

"Rebel was with me from the time we hit England until after the war ended, keeping my aircraft flying through the 109 combat missions plus many other test, training, and other flights. I just noticed I made one hell of a mistake—it wasn't my aircraft! Rebel loaned it to me for short periods to make sure that his work was top notch. The line crews had their own "druthers" as to who they wanted flying their machines. You can believe me when I say that he was proud of that plane and hated to see anyone else fly it *(this is strongly verified by other squadron pilots – author)*. Once we got in the air on a combat mission, or even a training mission, Rebel wouldn't leave the flight line until his 'baby' was back on the ground and safely tucked in for the night.

The early P-47's we received at Bodney had problems with oil leaks around the rocker boxes covering the valves, as well as with the 'O' rings of the pushrod housings. This caused a loss of visibility, especially on landing, when oil from the top 5 cylinders landed on the windscreen. As a result, it was SOP (standard operating procedure) to pull the cowling shortly after landing so the crew could go to work as soon as the engine cooled off. Until neoprene 'O' rings arrived from the States, the rubber ones (which couldn't take the heat) had to be replaced every second or third mission, as well as the gaskets under the rocker box covers. Charlie Agee and Rebel had me write home to 'Cile and ask her to send some 'dress box' cardboard with her next care package. Security restrictions wouldn't let me tell her that the cardboard was to be cut up and used as gaskets. Rank has no privileges when it came to keeping the plane in the air, so several times the guys put me to work (the menial job) smoothing the cover sealing surfaces with fine emory cloth stretched over a piece of glass. Ingenuity was the 'word of the day' when it came to flight crews.

"Here is an event was concerns the day I had flak burst under my P-47D 'Cile' in the vicinity of the Ruhr Valley. It took out the bottom five cylinders (mostly sheared spark plug leads) of that R-2800 engine. Since we were on our way home from the mission, I just kept rumbling back at reduced power. Upon landing, I taxied up to the revetment but the engine seized from lack of oil before I could get it parked properly. Once again, the crew worked all night installing a new engine and bottom cowling. The plane was buttoned up and ready for a test-hop by daylight. I'll bet industry would love to have some dedicated workers like the crew members we had on our fighter aircraft."

Pilot Chet Harker (left), crew chief Charlie Agee (next), and Rebel Harris (with screwdriver at right) inspect a detail of their plane "Cile, Luck of the Irish." Chet Harker)

At times things could go wrong that defied all attempts by the crew to repair them. On one of the bomber escort missions into Germany, we had been relieved from escorting the bombers and were on the way home. At about 25,000 ft. in the neighborhood of the Zuider Zee, good old 'Cile's engine quit. I lost about 4,000 ft. before I could get it chugging again. While trying to climb back up to join the gang it quit again. I radioed to my wingman to stay with me until we reached the North Sea and sent the rest of the flight on home. The engine quit a total of seven times before we got our 'feet wet,' and another six times crossing the water. At one time I was down to about 700 ft. but preferred to stay with the plane since the survival time in the North Sea at that time of the year was about 8 minutes. Finally crossing into England, I called an emergency at the first airfield I saw. 'Cile' and I cut out several B-24's in the pattern that were returning from their missions, landed, and the engine quit for the last time while still rolling. I coasted off into the grass and checked into Base Operations. I asked if they had some way that I could get back to home base so I could pick up Rebel and return to fix the plane. I was asked 'Can you fly an Oxford?' I didn't even know what one looked like but soon found out. It had 2 engines that were started by cranking up a spring in each engine while standing on the wing of the plane. The aircraft engineer and I, using a road map, zig-zagged back to Bodney. While I was being debriefed by intelligence about the combat part of the mission, Rebel was loading everything but the kitchen sink into the Oxford. Back at the bomber base, I believe he changed the fuel pump and made some other adjustments. After a ground run-up, I lifted off, stayed at a low altitude, and made it back to Bodney. Back there, they practically rebuilt parts of the engine *(in the open of course – author)* and the next day I took it to altitude on a test flight. It quit again so the plane was grounded until a new engine could be fitted. The old engine was sent back to the manufacturer for failure analysis. Heard through the grapevine that it had a cracked cylinder which caused the engine to flood out.

The 352nd Meets the Future Queen

When 352nd FG Public Information Officer Sheldon Berlow determined to build a better relationship with the Brits, he decided to name the next new Mustang after the *then* Princess Elizabeth (now Queen Elizabeth II). Ace William Whisner was tagged to get the aircraft. A special ceremony was arranged for the Princess to personally dedicate the new plane, and on that day, several Mustangs were lined up in front of the Headquarters and Control Tower with the new P-51 emblazoned with the name *Princess Elizabeth* parked front and center.

When Colonel Mason came out to see that all was in order he almost went into shock. Parked next the next to the new *Princess Elizabeth* was Lt. Cy Doleac's Mustang named "EX-LAX—Shht 'n' Git." As might be expected, Colonel "Joe" quickly ordered one of the G.I.s to get that "G.D." airplane out of there NOW! This maneuver was completed in doubletime—another Mustang was moved into place just a few minutes before the young princess arrived on the scene.

It is to Rebel's credit (as crew chief) and the rest of the crew that we never had an abort. In only one case, while he was acting as Assistant on the P-47, did I abort a flight. I knew the left mag was a little weak when I checked mags before takeoff but usually the engine would smooth out after full power was applied on takeoff. Didn't happen! I stayed within sight of the field and landed after the rest of the squadron took off.

In spite of military protocol, when working or just plain goofing off, Rebel and I were more like brothers than officer and enlisted man. I recall one mission when returning from escorting bombers, we were cleared to look for targets of opportunity. I got too close to one of the locomotives I was strafing and the boiler blew up, throwing a black layer of coal soot all over the plane. By flying through a cloud, the windshield cleared enough so that I could return to Bodney. When Rebel saw the plane, he chewed me out royally. I thought for a while he was going to make me clean and wax it. To this day, we are friends, see each other whenever possible, and keep Ma Bell happy with frequent phone calls. I have a lot of respect for this man who put in many hours keeping his plane and me in the air. Rebel has often told 'Cile that his job was to get me back home to her safely.

"One last memory of the enlisted men regards when I had the job of Operations Officer for the 486th Squadron. It was my good fortune to have an enlisted staff that really knew what they were doing. The Sergeant in charge was a whiz at making coffee, and when I was riding desk instead of flying, he made sure I always had a full cup in front of me. I didn't want to insult or criticize his good intentions, so I did my best to sip until I dropped. I didn't find out until much later that the enlisted personnel had a 'pool' going as to how many cups of coffee I drank on a non-flying day. They picked a nice rainy day to do the counting. The lucky number turned out to be 32 cups. I'm not sure who won the bet, but I know I went 'cold turkey' and didn't touch another cup of coffee for five years. Now, 55 years later, I still restrict my coffee intake to one mug daily. Thanks to a fun-loving, conscientious bunch of guys, I have never been bothered by ulcers.

Powell and Company

Robert "Punchy" Powell flew in the thick of the fighting with the 328th Squadron. The plane he shared with his crew was named for Punchy's home state: "The West 'by gawd' Virginian." Punchy allows, "My first crew was on the P-47D (coded PE-K) I shared briefly with Jamie Laing. The Crew Chief was S/Sgt Jim Loughrey of South Plainfield, NJ. Jim shared the same hometown as later my P-51 Crew Chief, Bob Lyons. The other members of the crew were: Cpl. Frank Ruvo and Cpl Clinton Haynes.

"Jim was a really easy-going, unflappable, stocky guy who had sort of a fixed grin on his chubby face. In contrast, he seemed to be quite serious most of the time, particularly about his job of keeping that plane in tip-top shape. We didn't carry on many lengthy conversations, but this was probably my fault. I was a cocky bastard (I guess I hadn't had my gold bars long) and there isn't anyone more stupid to a S/Sgt with some time in the service than a new, shiny-bar, shavetail Second Lieutenant. Maybe that's what caused that permanent grin on Jim's face.

"Being such a new second John, I was still suffering from my indoctrination as an 'officer and a gentleman' as the saying went, and hadn't relaxed enough in the service to really know how to talk to enlisted men—maybe thinking I wasn't supposed to be too informal with them. However, Jim had been around awhile and he probably recognized me for what I was—a dumb-John shavetail with wings. We never did develop quite the close relationship that I later had with Bob Lyons, but I am sure that Jim respected me as a pilot and one who didn't get too reckless with his airplane, and believe me, that was HIS airplane, not ours. It just had our names on it. Jim was an excellent mechanic and when he said the plane was ready to go I never questioned it. I had great confidence in him. Unfortunately, Jim died a number of years before we started getting together at reunions and I never had an opportunity to thank him for being the outstanding person he was and for the care he gave me (and his airplane).

Robert "Punchy" Powell with his crew of PE-K "Jamie Boy/The West By Gawd Virginian" and his crew. L-R: Jim Loughrey, Punchy Powell, Frank Ruvo, Clint Haynes

"**Frank Ruvo** was a quiet little guy who did his job well, didn't have a lot to say,, but worked well with Jim Loughrey. He was part of the team and I can't say that during that early period of my combat days we ever really just talked other than his answering a question I might ask or his asking me something.

"**Clint Haynes** (The Chief) was my armorer on both the P-47 and the P-51, and it was impossible not to like this big guy. I had always held Indians in high regard, even though I shot a lot of them playing cowboy and Indians as a boy—emulating the movie scenes I watched on Saturdays when I would go to our county seat and spend all afternoon at the theatre for ten cents. Clint was Indian, the son of a Chief in his tribe. He was a big, jolly man nobody had the guts to buck, but he was easy to know and easy to like. I liked him. More importantly, he was good at keeping my guns working and I could ask little more than that.

"When we went to the Mustangs my ground crew consisted of S/Sgt Robert W. Lyons from South Plainfield, NJ, Assistant Crew Chief Walter C. Hughes and Cpl E. L. Kingsbury. Except for Bob Lyons, I can't say that I really had a close relationship with this crew either. Maybe I was still being 'the officer,' I don't know. However, this was a great crew, too, and I know how hard they worked to keep me safe in the air. They sometimes worked through the night to make sure that my plane was ready for the next day's mission. The various crews had their own little group of friends and they would help each other make engine changes when necessary and keep those planes on ready.

"Although these 56 years have dimmed my memory of some things, I recall that it was Bob Lyons who made an engine change on my ship. Bob came to me one early morning after they had worked most of the night, and told me my plane would ready as soon as it had been flight-tested. Having had the lessons of courtesy taught to me by my Mother as a boy, I am sure that I thanked my crew many times during those days. However, I never really had the opportunity to show them my real appreciation until our reunion days when

Bob Lyons and Clint Haynes started gathering with us annually. Now, every year, unashamed and untethered by the long-gone officialdom of rank differences, I hug these two guys with all the love and affection in my body. I thank them and my God that they cared enough to insure that my airplane was ready for any combat, and for their patience with their young lieutenant pilot during those years of our youth.

"Today, and everyday of my life now, I pray that they will remain in good health and be Blessed with joy in their remaining years. I know that without their skills, I would not have been blessed with long life, good health and much happiness. Through them, I try to thank all of those who got me through the war for only God knows that I did not do it on my own. I only regret that the others are not here to share these days with us."

Starck Mad and Even Stevens

Walter "Wally" Starck piloted "Starck Mad" during the war with the 487th squadron, and downed enough German aircraft to be noted as an ace. He expands here with, "Completing flight training at primary, basic and advanced schools gained me my silver pilot wings and a commission as a 2nd Lieutenant. And, as luck would have it, my first assignment was to a Fighter Squadron, to fly the Republic Aviation's P-47 Thunderbolt Fighter. I was finally there.

My respect for the enlisted man at this time was not very great. There was no association between Officers and enlisted personnel. We were billeted separately, we ate in separate facilities, we wore uniforms of better quality, and Officers were admitted before enlisted men at every facility and function. During flying training while in the states, pilots did not have an assigned aircraft or crew, so there was no rapport built between the pilot and crew. The crew's responsibility was to keep the aircraft in flying shape, serviced and clean and ready for a pilot to use. That was my feeling about the aircraft crew duties. Respect for their performance at their job was reduced by an instance as follows. During a training mission, the aircraft I flew developed an oil leak in the engine. Oil splattered over windscreen so heavily that I could barely see to fly or to land the aircraft. Upon return to the flight line, I entered the oil problem in the A/C Form-1 as 'Oil on Windshield.' Upon arrival at squadron operations a few minutes later, I noted that the aircraft I had just flown was called in as 'In commission,' meaning, ready to fly again. I raced out to the aircraft and checked the entry in the Form-1 citing corrective action to the problem. There it was. 'Oil wiped off windshield.' Needless to say, that action did little to enhance renewed respect for an enlisted's performance. As I look back to that time and example I have to chuckle. That crew chief did take appropriate action on the problem entered on the form.

Upon arrival overseas at Bodney Air Base, near Watton, England, things changed. Pilots were assigned an aircraft and that aircraft had a Crew Chief, an Assistant Crew Chief and an Armorer. What a far cry from the day when I had to report my aircraft out of commission with the complaint, "oil on windshield," and find it is back in commission only minutes later.

My crew was tops. They cared for that aircraft in great detail with spit and polish. Every instrument functioned with precision. The servicing of the craft was complete. The machine guns were oiled and polished. The ammunition bays were clean and filled with rounds secure in their attaching links. Prior to departure on a mission, the aircraft engine was run and fine tuned for operation. Upon my arrival at the aircraft, the crew assisted me in a walk-around and check of items he deemed necessary. They helped me enter the aircraft, fastening harnesses, oxygen hose, radio cords. They were the helping hands that made getting ready to fly a pleasure.

Upon return from a mission, they were at hand to check on the aircraft performance, the functioning of the guns, the radio. They had a deep interest in everything. After I departed for debriefing, the crew set about correcting whatever malfunctions were identified. Engine vibrations or rough operation called for a checking and changing of plugs. A machine gun misfire or stoppage required removal and complete stripping of the firing mechanism to determine the cause and take corrective action. Failure of the ammo belts to feed into machine gun resulted in removal of all the rounds from the clips and then loading each round into clips manually. Time consuming and tedious work. But the entire crew worked at that task. There were no stoppages from faulty links after that."

It must be noted that gun stoppages were epidemic in the 8th Fighter Command when the new P-51B's and C's arrived in late 1943 and early 1944. The B/C's thin wing necessitated the guns be installed at an angle, making a sharp break-over curve for ammo belts entering the .50 calibre gun breaches. This was cured with the thicker-winged D and K models and their upright gun installations. Many early Mustang units fitted electric booster motors to insure feeding. However, it was discovered in the 352nd FG that if the rounds were all perfectly loaded into their clips making up the belts, there would be no jams. This therefore meant that each round had to be hand-loaded and inspected, rather than the easier machine-loaded method. This is the finger-bleeding work of the armorers to which Wally Starck is referring.

"There were several aborts due to radio failure. Some were shortly after take off while others while well along into a mission. Major effort was expended to find the reason. Radios were replaced. Antennas were replaced. The fault never showed up...but the radio failed. (Of interest, some 45 years later, I heard that my Crew Chief was not awarded a medal because his aircraft did not complete ten consecutive combat missions. Not only did he not get rewarded for his diligent efforts but I was not always granted credit for a combat sortie. We both lost out, but I think he felt his loss more deeply than I.)

D-Day was just over the horizon. We all felt its imminence. There was a stand-down and the crews were busy painting invasion stripes on the wings and fuselage of the aircraft. Extra efforts were made to ensure the readiness of the aircraft. Leaves and passes were cancelled. Pilots were busy assisting their crews in checking over everything possible with their aircraft. In my case, I feel sure I was more of a hindrance than a help. But I did my part, making sure things were in good shape. Then it happened. Awakened at 11 PM, 5 June, and told there was a briefing at Group Headquarters at 11:30. There, we were told D-Day had arrived and our 1st take-off was scheduled for 02:30 AM. Upon arrival at our squadron dispersal area, we found our crew was already there. They had pre-flighted the aircraft, ensured the main and belly tanks were full of fuel, guns checked and re-checked and the aircraft was in a GO condition. I don't know how long they had been on that flight line in the darkness, but it must have been as long as we had been up. Again, they were there, making sure all possible was done to insure the superb functioning of that aircraft for a successful mission and the safe return of their pilot; me.

"One wonders what my crew's thoughts were while we were off giving top cover to that gigantic invasion fleet. The unknown can conjure up any number of anxieties, and I am sure they each had theirs. My flight was four hours and thirty-five minutes long. As I completed the landing roll and taxied to the dispersal area, there was my crew, standing by with all smiles, happy for my safe return. After discussing the performance of the mission, and the condition of my aircraft, my crew chief advised me to go directly to 487th Operations to see what was happening. He said his instructions were to immediately service the aircraft and ready it for another mission. That was correct, and I was scheduled to take the second mission after I had eaten lunch. That mission was three hours and fifteen minutes long and I landed about 9 pm. Again, the crew was waiting for my return and were deeply concerned about the results of that mission. I was grateful for their presence, since I was weak from sitting strapped in that seat for so long on my second mission of the day. The Squadron Doc had given me a Benzedrine tablet with instructions to swallow it just before take-off....not before. It would stave off fatigue, but for only five hours, and then all the pent-up fatigue would hit me at once. My crew chief, Sgt. Stevens, had a cup of water for me to swallow the tablet before I taxied out for takeoff. It was he who assisted my exit from the cockpit upon return from the second mission. Fatigue was already setting in and I felt weak and spent. As I left the flight line, my crew, even though the hour was late, began the task of readying the aircraft for tomorrow's mission. My crew. Let me briefly introduce you.

"Keith F. 'Steve' Stevens, Crew Chief—He was small of stature but hard working; dawn to dusk and beyond. Steve knew his aircraft and would

Walter "Wally" Starck and his crew of HO-X "Starck Mad."
Left to right: Keith Stevens, Wally, Robert McKinney, Bill McCurtain. June 1944.
(Sheldon Berlow)

leave no stone unturned to find the answer to a problem.

"Robert L. 'Bob' McKinney, Assistant Crew Chief—Bob was a fine young man who was the helping hand of the Crew Chief. Always on the job, doing his thing.

"William R. 'Bill' McCurtain, Armorer - Bill is a happy, congenial and outgoing person who knows his armaments. Ever ready to explain the cause of a malfunction in the machine guns or the .50 calibre links. When the aircraft came back from a mission and the guns had been fired, he would quip, 'I see you have messed up my guns again.'

When Major General William E Kepner referred to the 352nd Fighter Group as "Second To None," I can't help but feel he was also referring to my aircraft crew. Postscript: I wonder if my crew ever really knew how I felt about them.

The "Boss"

The 486th Squadron's Commanding Officer until April of 1944 was Lt. Colonel Luther H. Richmond. Luther led from the front, and commanded great respect with his troops. Though his relationship with the enlisted men was a bit different than the other pilots quoted here, he is fondly remembered by them today. "My relations with my crew were probably unlike that of some of the junior officers. After all, I was the commander and was supposed to set the example. I did not have an 'old buddy' relationship with my crew like some of the more junior pilots did. But, they knew that I was behind them and depended on them. Our relationship was properly distant, but also close. I respected them and depended on them and they knew it and respected me. Remember, I was of the 'old school' where officers and enlisted did not mix fraternally. I think it all worked out for the best anyway."

"Little One"

Over in the 328th Squadron, the team of aggressive Don Bryan and his crew rolled up a score of 13.33 aerial victories. Don recalls, "We started getting our aircraft at Bodney in Summer of 1943. I was smart enough to go to the 'C-Flight' Line Chief, M/Sgt. Harold Tall Chief [his real name] to ask him who he thought was the best Crew Chief in my flight. He said, 'Staff Sergeant Kirk Noyes.' I said, 'Okay, I'll take him.' He was then crewing one of the first aircraft that we got in the 328th Squadron; a P47 D-2. I said to Tall Chief, 'When we get the next '47, put Noyes on it and it'll be my plane.' Chief came back with, 'No way. Once I give a crew an aircraft, it's theirs from then on. Take it or leave it.' After stewing on it for a while, I made the most important decision of my life [I didn't know it at the time]—I took Kirk and his crew Joseph A Bleckinger and Glen Van Andel. They are why I'm here to write this now.

It's kind of funny how the crew really owned the plane, and they just let us borrow it for our missions. Once we were having trouble with the artificial horizon. Time and time again, Kirk would fix it and it would flub up after an hour or so. Finally, as I was getting ready for a mission, Kirk told me he thought he had it beat. He warned me that if it didn't work, I had to abort. I said 'Okay.' The damn thing flubbed up about the time I got airborne. The weather was CAVU (clear air, visibility unlimited) so I continued the mission. When I landed, Kirk hopped up on the wing pleased as punch as he thought the instrument worked okay. When I told him it never worked, he chewed my fanny out like I have never had it done before or after. The fact that it would spoil his perfect record of no aborts, that it was CAVU, and that I was good at needle and ball just didn't matter. Kirk always wanted his aircraft to be perfect.

"The thing that saved my fanny in one-feld-swoop happened early in 1944. We were starting to get water injection for the Thunderbolt's Pratt and Whitney. It gave the P-47 five or six more inches of mercury boost, and added about two hundred more horsepower. I asked Kirk, 'When are we going to get water injection in [Little One]?" (as I had named the plane for my wife Frances) He told me, 'Never.' 'Little One' was a P47 D-2 and we were not going to get water injection. Talk about a jolt. I was already flying the oldest aircraft in the Squadron, and now I was going to be the flight leader with the slowest plane. Kirk asked me, 'What would you prefer? Have the engine blow or get shot down?' I told him that was a stupid question, so he said, 'All right, I'll pull the stops in the turbo supercharger, and you can have all the boost it can put out. But just remember, the books say the engine will blow in about 30 seconds.' I told him, 'Do it.' A bit later I found myself on the deck with a gaggle of Fw-190's above me. No way could a '47 out-turn or out-climb '190's on the deck. I just threw

Bryan's "Little One III" displays its scoreboard. (Don Bryan)

everything open and prepared to bail if the engine blew. I was in a right-hand turn, and the '190's looked like they were coming from everywhere. Luckily, they were poor shots and were all firing at me. Firing guns will really slow you up. The Germans each stalled out, and I had 'Little One' hanging on the prop. I saw I couldn't make the clouds so I split-essed down through them again. The same thing happened, but this time I had a lot more speed. I made it into the clouds and pulled off the power. I am sure that I had full power on for at least 4 minutes. I later checked the books—with no stops in the turbo supercharger at low altitude, it would put out between 90 and 100 inches of mercury boost. If five or six inches gives an additional 200 horses, what does 35 inches do? When I got back from this mission I was late. Kirk hopped up on one wing and Blec on the other. They didn't say anything, but for the first and only time they unstrapped me and walked me into 328th Operations. Kirk checked the aircraft [Hell, he took it apart], but couldn't find anything wrong with it. I tested it and he checked it again. I eventually flew 'Little One' for fourteen more missions before we got a P-51B.

"Have you ever seen a 'ticked' Crew Chief? When I was flying and it was cold or I got a bit scared I always had to use the relief tube *(this was rubber tube with a funnel that a pilot could urinate into and it would be carried outside the aircraft—author)*. Trouble was, in really cold weather it would freeze up after the first use. I would have to get down to warm air for a thaw if I had to use it again. 'Little One III' was especially bad about tube freeze-ups, and on one mission I had to use it a second time. I was at about 25,000 feet. Well, the tube froze up and I ended up holding a full tube in my hand. That was okay, as I was leading and I planned to drop down a few thousand feet and thaw it out. Just then some SOB yelled, 'Break!' I started flying again real fast. The next day we didn't have a mission, though it was a bright, clear, cold day. We were all sitting out side the Operations shack in the sun. I saw Kirk get into the cockpit and close the canopy. All of a sudden Kirk rolled the canopy back and came boiling out. He stomped up to me and asked what the G— D— H— I did in his plane? Only then did I remember about the relief tube incident.

Here is a very funny thing about 'Little One III.' It was a P-51D-10. They were all fast, but 'Little One III' was so much faster than any other '51's in the outfit, D-10's or not. It had about 10 knots on any of the others. After what Kirk did with that P-47D-2...I wonder. No one will ever know, but I wouldn't put it past him to have done a little cheating."

Fahrenwald and German

Pilot Ted Fahrenwald's letters home reveal a lot about what he thought of his ground crew, and crews in general. Some of it is tongue-in-cheek, and some of it is embellished, but it gives a wartime glimpse of the comeraderie pilots often had with their crews.

September 19, 1943—"Comes bad weather like today and all day long I prance hither and yon, up and down the line of our planes were the mechanics toil and cuss, fix and smoke, spit and moan, polish and sometimes become baffled by their work.. From ship to ship I hop, and back to mine, shooting the breeze, swapping lies, bumming smokes, and lending a 'palsied' hand here and there. In general violating all of the non-fraternizing traditions of the Army. But I have the sweetest running engine and the highest polished wings, and the crew laughs with glee at my corniest jokes. Things running smoothly and we get a lot of fun out of life."

October 10, 1943—"Had a good day today. An amusing habit the ground crew has. Every mission calls for a new belly tank becuz we drop them off after having emptied them. So, the mechanics generally scribble various tender messages on 'em. Paint their thoughts of Nazzy's in big bold red letters. Bet the folks in occupied countries who find 'em get a bang out of them. Ugly caricatures with appropriate inscriptions painted alongside. Our tanks today all went in the drink however. If one I dropped ever floats back to England there will undoubtedly be a complaint registered by some shocked and outraged lady. If she registers no complaint, then she ain't no lady....."

December 26th, 1943—"A little before suppertime on Christmas Eve I walks out to my ship and met with nothing but profanity from my crew chief Robert German. I find that an order has come from Fighter Command that an immediate change must be made on all ships. New belly tank shackles or some such item. The work will force all of our ground crew to work all night long so the ships will be ready for a scheduled Christmas day mission. Something special the jerks at HQ have cooked up. Well, naturally the boys moan about having to work all night on this particular night. Especially since they have a party all lined up with great quantities of females and so on. It is sad, but c'est la guerre....I swipe four or five quarts of wine and wander on down to my ship. It's cold as the devil and raining in gusts. All around the field there are little lights blinking beneath or ships. When I find my ship there's a big Cletrac tug parked in front of it, giving power for a searchlight. My crew is working like mad and cussing away so I join 'em, and tell 'em, 'To hell with it for a while.' There we sit under the belly and wing of my ship pulling corks and passing the jugs around until way after midnight. Was a really nice party. Sat there swapping lies and working now and then, and after a couple corks had been pulled nobody minded the wind at all. When the colonel comes by to see how work is progressing, he sees the pilot and crew of one ship working like slaves. But had he shoved his arm up in the turbo-supercharger housing, he'd have found three or four bottles hidden there. So, about 4 am most of the ships are rigged and put to bed and their crews joined their pilots in the huts and got warm by the fireplace...

April 16, 1944—"This Sgt. German is a type alright. Best durned mechanic on the line and can diagnose the slightest aches and pains of my engine by just running 'er up. Listens to her tick with his homemade stethoscope—he sticks one end of a screwdriver into his ear and places the other end here and there on the machinery. I told him today when I landed that the engine was worth approximately seventy-five billion dollars...cash money! The ship shows his efforts too—runs like a Swiss watch, only faster."

April 23rd, 1944—"Me and my crew are thicker'n flies. Nacherly, I hate to have any other pilot fly my ship, as there are some knuckleheads in the outfit. Guys who habitually substitute brawn for brains, and are mighty rough on a precision machine like this ship. So, whenever I land and taxi back to my revetment, I announce officially that something's haywire with the ship, whereupon Sgt. German rips a couple of panels off'n the cowling to make it look good. So far, no one but me has flown 'er, and she's the best ship on the line. I've had a couple of very fine airplanes ruined by mishandling on the part of some meatheaded numbskulls."

The heroes on the ground. Whether they were informed of this or not, they were truly a pilot's best friend.

Mac McCarthy's "Patty IV" being serviced by 17th Service Sqdn.

Claude "Shorty" Thomas (right) gets some assistance with the Pratt and Whitney R-2800 in Miklajcyk's "The Syracusan." (Claude Thomas)

THE RUSSIA SHUTTLE — A "BLUENOSED" VIEW

by Marc L. Hamel

*An excellent contemporary account of the historic Shuttle Mission to Russia and Italy was penned by 486th Fighter Squadron (FS) pilot Lt. Donald W. "Mac" McKibben upon his return to base in July of 1944. He graciously gave his blessing to utilize the narrative for inclusion in this book. Likewise 486th pilots Carleton L. "Bud" Fuhrman, Leonard A. "Jim" Gremaux, Edwin L. "Ed" Heller, and the late Stephen W. "Andy" Andrew, Thomas W. Colby, Charles E. "Griff" Griffiths, George Hampson, Donald "Red" Whinnem, and Undo F. Rautio (crew chief) provided memories of this historic undertaking.**

The first 8th AAF three-way Shuttle Mission (code named "Frantic") to Russia, Italy, and back to England lifted off from English soil on June 21st, 1944. While the Shuttle's bomb groups and the 4th Fighter Group's involvement have been well covered, few have recognized the contributions of the 486th Fighter Squadron.

The basic concept behind any Shuttle Mission flown by the AAF in '44 was to confuse the defenders of the Reich by not exiting the target area in the same direction as the approach. A side benefit of this

Stephen "Andy" Andrew and Henry Miklajcyk stride from a briefing in 486th Operations.* (USAF Museum Collection)

Highly respected 4th Fighter Group CO Colonel Don Blakeslee led the fighter escort for the long-range penetration into Germany and on to Russia.

plan was the corresponding boost in morale for our visited Allies. More than 160 Third Division bombers were involved in this Shuttle, escorted by more than 60 P-51 Mustangs. By continuing westward into Russia after bombing Ruhrland, the Eighth's aircraft would theoretically endure the Luftwaffe's wrath only once. As was customary, escort duty would be divided among several units. The final long-range escort leg into Russia would be handled by the 4th FG and the 486th FS.

On June 10th, 1944, the 486th received confidential orders from AJAX headquarters that it would be attached to Col. Donald Blakeslee's three 4th FG squadrons for the mission to Russia. George Hampson, 486th FS Engineering officer, relates how the meeting at AJAX unfolded:

"The 352nd received orders that the 486th C.O. and Engineering Officer were to fly to Fighter Command HQ's near London for a secret meeting. Willie O. Jackson, who had replacing Luther Richmond as C.O. when Richmond was downed by flak in April, collared me saying that we'd be flying in the AT-6. I thought 'Oh God, here we go again' as I had been treated to a terrible 'joy-ride' by a hotshot pilot back in training. On the contrary, Willie O. flew straight and level at low altitude on a nice day.

"On arrival at AJAX, Willie O. knocked on a door to ask directions and a major in dress uniform opened it while drying his hands on a towel. It gave me the feeling that things were pretty cushy around there. Entering the mess hall, I was stunned to see a huge room full of nothing but high-ranking officers in dress uniforms seated at the tables. Willie O. spotted a few empty seats way at the end, so we sauntered over and sat down. We were politely chased off by a sergeant who told us it was the General Officers' table. We never did get to eat.

"General William Kepner was seated behind a huge desk with nothing on it but red, white, and blue telephones. Behind him was a wall-sized map of Europe. Facing him were Don Blakeslee, a 4th FG C.O. Jackson and me. Kepner explained the mission and then asked Blakeslee and the other officer if they could do it. With their raunchy hats, expensive boots, and cigars, they replied something like 'no sweat' and 'piece of cake.' I know that I wasn't asked anything, and Willie O. didn't say more than one or two words if any. You see, the 4th Group guys were members of the 'old boys' club, while Willie and I were the new kids on the block. It was a short meeting, and I wondered why they asked us to attend at all unless it was some way for the top brass to satisfy their egos."

Wrecked B-17's on June 22nd in Poltava, Russia after the German recon planes followed the US bombers into the Soviet airspace. A nighttime raid on the night of the 21st-22nd resulted in the damage seen here. (Jim Pietrzak)

Back at Bodney, twelve ground crew members (representing the various technical disciplines) were selected to go with the force to service the fighters in Russia, and they would fly as waist gunners in the Shuttle's B-17's. Tom Colby had this to say:

"It should be noted that all briefed pilots and enlisted men were sworn to extreme security measures. The men of the 328th and 487th Squadrons tried every trick in the book to gain knowledge of the hush-hush mission afoot. Lots of free drinks were offered and wild stories were passed out, including tales of missions to central Africa. However, no incidences of security breaks were ever reported."

The 486th prepared for the worst, and traveled a round trip to the 4th FG's home at Debden, England to collect 108-gallon paper-mache drop tanks on the 14th. Major Jackson, flying his PZ-J, led the squadron back to Debden the next morning for a scheduled liftoff on the 16th. The weather over Europe proved appalling for several days, so the mission was scrubbed. The eighteen pilots (including the two alternates) and twelve crewmen returned to Bodney on the 18th.

The 486th flew an escort mission to Magdeburg on the 20th, and upon return were notified that the Shuttle Mission was 'on' again for the next morning. The squadron packed once more and moved to Debden that evening.

The 486th FS aircraft that roosted overnight at Debden were a mixture of olive drab and natural aluminum P-51B and C models plus Jackson's new P-51D-5. They carried the sweeping blue cowl and spinner paint endemic to the 352nd FG, and full D-Day "Invasion Stripes" still encircled the fuselage and wings. The individual aircraft codes obscured by these stripes were boldly repainted on the cowling sides. Also, 108-gallon paper drop tanks were carried for maximum range as opposed to the normal 75-gallon metal tanks. Generally these aircraft had upwards of 250 hours each on their engines due to the intense grind of operations associated with the D-Day.

Liftoff

Don McKibben describes, "After taking a look at the foul weather that hung 200 feet over the field, hardly a man would have been eager to fly if this hadn't been the big show that we had sweated out for so long. However, the weather officer assured us that we would encounter nothing but sunny skies and pretty white cumulus clouds over the entire route, and it really turned out that England was the only place that was having bad weather. Forming above the overcast, we could hear Col. Blakeslee's 'Hello Dicton, Hello Dicton, Horseback, Horseback, Horseback,' and Dicton's reply, giving him his reciprocal course. When he got '90 degrees' we started out and settled back for a long haul." Lt. Joe Gerst flying his "Smoky Joe" (PZ-X) reluctantly had to abort due to a malfunction on takeoff, as did Martin "Corky" Corcoran in his "Button Nose" (PZ-L). Their places were eagerly filled by Lts. "Bud" Fuhrman and Ernest Bostrom.

"Colonel Blakeslee's navigation was faultless and at exactly the briefed minute, the red-nosed ships of the 4th Group and our blue-nosed planes swung over the bombers and began weaving out to the side and over the glistening B-17's. They had been escorted by other fighter groups to that point and had effectively bombed their target, Ruhland, and reformed into two compact boxes. At the waist guns of those big friends, our crews were keeping a nervous eye peeled for the enemy. We didn't intend to let the enemy get a crack at our ground crews."

Combat

Tom Colby, piloting "Lil' Evey," (PZ-T), recounts, "Warsaw was still under our left wing, and we were perhaps 50 to 75 miles south and just crossing the River Vistula when we were hit by perhaps twenty Messerschmitt Me-109 fighters. The engagement was short and sweet. I didn't see much as I had a 108 gallon tank hang up and couldn't immediately shake it off. I spent a lot of effort in violent maneuvers, and was finally successful at the end of the engagement." Major Jackson also had a "hung" tank, which later tumbled loose in formation to the consternation of those it narrowly missed.

The "sweet" part of the engagement was the 486th's score. Lt. Leo Northrop downed two Me-109's, Lts. Donald "Red" Whinnem and Edwin Heller each claimed one, and Yellow Leader Major Stephen "Andy" Andrew had a damaged claim.

Crew Chief Undo Rautio had a different view from the waist gun port of the B-17 in which he made the trip. "In due time, we arrived at and 'pasted' the target, and were taking our lumps from flak. Then the 109's came at us head on. I remember one passing so close that I could see the

An overconfident Lester Howell poses in his P-51B "Texas Knight" just prior to D-Day. Howell did not return from a solo bounce on multiple enemy aircraft. (Bud Fuhrman)

saw they were actually four 109's, three being camouflaged. I took a burst at one and zoomed up at full-throttle with all of them on my tail. I climbed in a spiral so steep I was just above stall. It seemed that when a 109 shot its guns, it stalled. Three of them stalled out, and the fourth got a burst in on me and hit my rudder. It almost caused my knee to hit my chin. He then stalled out. I took off and finally found the bomber string. I surmise that if I hadn't been so scared, I could have winged over and gotten that last 109 after he stalled out."

Lt. Northrop, flying his "Donna Dae V" (PZ-N), simultaneously took over

pilot's eyes behind his glasses, but his speed was so great that I couldn't swing the gun towards him. Then 'zap,' a 486th P-51 was on his tail. The last I saw of them were little specks diving towards earth. There was lots of gunfire from other locations, but my side remained clear. Then there was more flak, and I crouched down behind a very small bit of armor plate and sat on my flak jacket until things cooled off. Fortunately I didn't see any B-17's fatally hit, although one did trail smoke from the left outboard engine."

Ed Heller (flying "HELL-ER-BUST" (PZ-H), elaborated in his Encounter Report: "As Yellow Leader was closing in on the tail of a 109, I observed another Jerry queuing up on his tail. I immediately chased the Hun that was pursuing Yellow Leader and forced him to break for the deck. I followed him down to about 8,000 feet where I finished him off with a short burst from 200 yards dead astern. The Me- 109 poured smoke and I last saw the E/A when it went into the ground."

Heller expanded on this years later: "After I had shot down the 109, I started back to the bombers. At about Angels 20 (20,000 feet), I saw below me three 109's in a fight with one P-51, so I peeled over to help him. Well, when I got close I

the task of clearing Lt. Heller's tail, convincing an Me-109 pilot to hit the silk after a few very long range bursts. While firing at this Messerschmitt, Northrop was assailed from 11 o'clock low by two others, which he "broke" into twice in an effort to gain position. Northrop reported, "...as I turned into them again, this time we both fired all we had. He almost rammed me and I had to pull up my right wing in order to avoid him. By pulling up my right wing I succeeded in making a very tight left turn to check these two jokers. I saw the leader, the one I had shot at, going straight down with his whole engine and canopy in flames."

Lt. Whinnem downed another Me-109 while piloting his "HMS Hellion" (PZ-W). Red reported: "I saw this Me-109 milling around for a while. He was probably a green kid trying to 'build up points.' Anyway, he got up the nerve to make a pass at us. I was leading the flight, and as he swung down in front of us I kicked my rudder around and took a shot down at him. I am pretty sure I didn't hit him, but someone saw the guy jump before I lost him into the clouds. Anyway, I got credit for that one."

Luckily, the Luftwaffe chose to strike as the Mustang's drop tanks were nearly dry, thereby shortening their ultimate mission range. Bud Fuhrman piloting "The Flying Scot II" (PZ-D), recalls: "We used the 108-gallon paper tanks when we went to Russia, but I never liked them as they vibrated in the air stream when they were empty. We were told to try to hang onto these 108 gallon models as they would only have the usual 75 gallon ones in Russia. Of course, when we were bounced that was all forgotten."

Klaus J. Schiffler recently translated into English a section of the book "Jagdgeschwader 51 'Moelders': Eine Chronik" by Gebhard Aders and Werner Held. The translation covers this unit's viewpoint of the June 21st raid.

"On the 21st of June a American formation which had bombed Berlin flew across Warsaw toward the southeast. The I. and III./JG 51 which had been practicing air combat in the air when the ground control informed them of an enemy formation heading their way. Although their fuel tanks were almost dry, the Wing Commander assembled both of these Groups and ordered the Staff Squadron to take off. The German fighters overtook the B-17 formation and participated in a regulation attack from the front. Mustangs were above the bomber units, but the unit was aware of their position. The American escort fighters were unable to intervene before the German fighters had flown through the bomber formations. First Lt. Wever of the 3rd Staffel shot down a B-17. As the German fighters began to reassemble after the attack, it was raining with Mustangs. The I. Gruppe was scattered. Major Losigkeit held the III. Gruppe together and was able to ward off the Mustangs' attacks. Two P-51's were shot down, one of which crashed at the edge of the runway at Bobruisk."

Russia

Passing into eastern Poland and western Russia, evidence of the stark horrors of the Eastern Front were visible even from the air. Tom Colby relates, "We passed just to the south of Kiev, and part of it was on fire. We could see artillery and mortars shooting and falling." The scorched earth policy in Russia took on vivid meaning when the charred villages, shell holes, and tank tracks became visible from the air.

Rautio, in a B-17 relates, "We flew for a long time over dark green territory that seemed devoid of roads or human habitation, simply wilderness. All of the tension had thoroughly exhausted me and I fell asleep! I awoke just before we landed, and the Germans were already overhead photographing the whole area. We P-51 men were then flown by Russian DC-3 to Piryatin."

"Leaving the bombers, we altered our course southward and soon passed Russian P-39's as they flew towards a rendezvous with our bombers," McKibben continues. "We began to check our gas and a few of the boys began to sweat it out. After passing Kiev we all kept a watchful eye for our base. We were beginning to feel the effects of sitting on those hard dinghies for the last seven hours, and time seemed to drag before we sighted flares that marked Piryatin. You could sense the relief in Colonel Blakeslee's voice as he said, 'Well boys, here's the end of a perfect mission. Well, it wasn't quite perfect, and a 4th Group boy didn't help matters when he staged a minor accident on the lone steel mat runway and held the remainder of us in the air for 20 minutes."

Colby emphatically adds, "Believe me, this was serious as all of my tanks were dry, and when we finally landed I estimated only 15 to 20 minutes of fuel remaining. Blue Flight was very tight-lipped, tired and relieved!

"Speaking of 'relieved,' after more than eight hours in the air plus the extra 20 minutes holding, everyone's bladder was ready to explode. Our aircraft were dispersed in a helter-skelter fashion on the west side of the airstrip, and we immediately unbuttoned our drawers and 'let fly' in front of a huge audience. Later that day the elderly American base commander complained to Blakeslee about this unseemly conduct. Blakeslee just laughed it off."

A recently painted fine art print shows Col. Blakeslee having just landed in Piryatin. Fuhrman corrects, "In the painting of when we arrived in Russia, Blakeslee is pointing to his watch to show we arrived to the minute. He actually had little to do with that, as we were just following the bombers and their navigators that led us most of the way. Also it would be difficult to have previous knowledge on head or tail winds we might encounter during the flight. To bring it down to the exact minute was just luck. Pure luck." This is no reflection on Blakeslee's skill as a

Contrails of B-17s and B-24s on way to target.

pilot or popularity as a leader however, as he is well respected and thought of by those who flew on that mission.

After a truck ride to the briefing tent, an American Intelligence Officer interrogated the pilots while Russian girls provided refreshments. Fuhrman recalls having fresh white bread and jam. He adds, "Gee, that tasted good after the dark coarse bread we got in England. I often wondered where they managed to find the stuff. I'm sure the Russian people didn't have it."

Following a quick rest and evening chow, the pilots retired to their assigned tents with canvas cots. Colby relates, "I'd had the foresight to put two taped bottles of scotch in a suitable cranny of my right gun bay, and retrieved them in time for a twilight cocktail party. The group included Jackson, Higgins, McKibben, French, Heller, and Deacon Hively of the 4th Group among others. I'm sure someone thanked me for my generosity, but only momentarily." It seems the Luftwaffe had other plans for the Shuttle participants!

Germany Retaliates

"That night (the 21st), we had just gotten settled when all hell broke loose around us. We peered under the edge of the tent to see ack-ack breaking in the sky above us and searchlights groping through the night. We just laid there and took in the show. That is, until a couple of parachute flares broke into brilliance directly above our tent. There was a sudden rush of half dressed and undressed figures, making for the slit trenches at the end of our row of tents. Those sprinters among us made it in time to get the few remaining vacancies in the trenches. Raid-wise G.I.'s had just about filled the trenches with the first few bursts of ack-ack. The rest of us squatted in the shallow ditch and waited for the bombs to start falling. Meanwhile, tracers from the field defenses were trying to shoot out the Jerry flares. As the parachutes got lower to the ground, the Russians depressed their guns until tracers were coursing between our tents and directly above our heads. After the flares had been shot out, the ack-ack stopped firing and no bombs had been dropped, we went back to our sacks. We decided that Major Andrew and Lt. Northrop had broken some kind of dash record in their scramble for the slit trenches. Jerry didn't bother us again that night."

"It was a crazy deal all over when the Germans came after us in Russia," Red Whinnem adds. "We hit the slit trenches, and it felt like World War One all over again!"

Crew Chief Rautio pitches in, "That night the Germans blew up most of the B-17's at the bomber bases, ending the crew's mission, and stranding us!"

Colby continues, "During the raid that night I ran out with the group and nearly hung myself on a tent rope—as did others. I had a rope burn for a week. When I returned later, there was two inch tear in the cot at about chest level if I'd been lying on my back. I looked up to see a large tear in the tent, and felt under my cot and came away with a jagged 3-inch shell splinter!"

"We learned the next day that Poltava and Mirgorad, the two bomber bases, had suffered heavily in the raid," McKibben laments. "Only about half of the original force of bombers would be able to continue on to Italy. They then decided to evacuate us and our planes farther into the interior. That evening (June 22nd) our squadron went to Kharkov and the 4th Group went to Chugiev and Zaporoache."

Klaus J. Schiffler's translation of "Jagdgeschwader 51 'Moelders': Eine Chronik" continues with, "Two P-51's were shot down, one of which crashed at the edge of the runway at Bobruisk. In this aircraft was found a map with the flight layout pointing directly to Poltava. Capt. von Eichel-Streiber brought this map to Air Fleet HQ which confirmed a recon aircraft report which followed the bomber formations to Poltava on which a bombing attack was planned. That night eighty He 111 and Ju 88 bombers attacked this airfield and destroyed forty-seven B-17's. The only personnel loss that JG 51 sustained that day was Uffz. Robert Willms of the III. Gruppe, KIA over Domaczewo. He had no kills to his credit."

From S/Sgt Undo Rautio's point of view, "The Germans also tried for the '51's on the night of the 22nd, and we could do nothing but cringe in the cellar of a schoolhouse while the bombs shook us up. We could hear some ineffective AA gunfire in the distance. In the morning we found the Germans had missed our '51s, and we settled down with the Russians to await rescue by the ATC. The plan was to prepare the '51s for a long flight to Italy, then go by ATC through Iran, North Africa, Casablanca, and finally back to Bodney. That's the way it was, but in the meantime, we tented out, ate captured German food, and looked over the Stormoviks (crude, to say the least). One day a lone German Me-109 flew over at about 800 feet and never fired a shot. A Russian Airacobra attacked him from above, cannon banging, with shells ripping up our sod taxi-ways. The German 'poured on the coal' and roared off west streaming black exhaust smoke. The Airacobra was pretty, but not fast enough."

Kharkov, Russia

While Kharkov initially looked untouched, it was soon apparent that past fighting had gutted many buildings. The airfield was on the outskirts of town, and Colby recollects, "The runways were in a triangle, with hangars on the south side. Huge bomb craters, smashed steel hangars, and perhaps 30 to 40 wrecked Stormovik aircraft were scattered about. I'm sure the field had been shelled also, but the runways were made of huge concrete hexagons and they stood up. We landed and parked our aircraft in a line at the east end of the airfield."

Initial reception on June 22nd by the Russians was confused until an American interpreter arrived in a B-17. Transport was then arranged to get the pilots and enlisted men into Kharkov itself. The aircraft were guarded by Russian women soldiers who overnight picked wildflowers and left them on the planes. Colby recalls this, "Transport to and from the airfield was by open-top Studebaker G.I. trucks! The Russian truck drivers drove in a kamikaze fashion, and knew little English except 'stop' and 'go.' The drive from the airfield to the building where we stayed was about 2 miles and up a hill. The airfield itself was at the top of a hill and in spots had a view of the wrecked city. Kharkov, a pre-war city of about 250,000, was reduced to rubble with burned out brick and concrete buildings barely 10 feet high in most places—no roofs anywhere. Our building was perhaps 150 feet long by 80 feet wide, had at least two floors (we slept on the second), and a very broad staircase. It did have

A magnificent, moody view of Clermont Hall with a storm brewing. The pilots were quartered here before their Nissens were built on base. The house is just a short distance east of Bodney off the Watton Road. (Warren Jackman via Don McKibben)

Henry Miklajcyk, Stan Miles, Langhorn Gee, and Leo Northrop beside Northrop's "Donna Dae" P-47 named after his childhood friend and Hollywood starlet. Note the "Indian" insignia of the "Fighting 21st Pursuit" that became the 486th in Spring 1943. Taken early '43 at Republic Field, Farmingdale, Long Island (Stan Miles)

Lothar Fieg in his PE-F "Denali" 42-8456 on October 14th, 1943.
(Harry Miller)

Stan Miles on Northrop's "Donna Dae" P-47 at Republic Field, Farmingdale, Long Island (Stan Miles)

"Jean Louise" (PE-D) 42-8660 of Harry Miller with Crew Chief C. Dale on wing.
Note odd olive-drab color on 75 gallon tank, presumably a gas-stained grey painted version. (Harry Miller)

An unidentified B-24 buzzes the North Site communal area at Bodney.
(Warren Jackman via Don McKibben)

Full right view of Harry Miller's "Jean Louise" (PE-D) 42-8660. (Harry Miller)

Ray Cornick of the 328th. Note the visible stitching for the inside pockets added to his A-2 for escape kits and maps. (Harry Miller)

Inspiring Earl Abbott poses in his "Sandra Lee" PE-A 42-8425 in October of '44. Note brooms (fighter sweeps) and bombs mission markers. (Harry Miller)

328th FS ground officers. 328th Engineering Officer Frank Gustavson, Intelligence Officers Otto Ziebell and Dave Lee. Note Gustavson's A-2 jacket has an added shearling collar, presumably from a crew's B-6 jacket. PE-X "Blondie" 42-8419 of Bill Hendrian lurks in the background. (Harry Miller)

A true leader of men, Ev Stewart was C.O. of the 328th FS early in their career. He is shown at left with his crew of their PE-G "Sunny IV" 42-8437 in Winter 1943-44. (Warren Jackman via Don McKibben)

Earl Meyer in PZ-E "Bonnie Lee," 42-8459. Earl was most often wingman to Chet Harker. (Warren Jackman via Don McKibben)

PZ-A "Prairie Farmer/Spirit of Los Angeles City College" 42-8699 of Stephen "Andy" Andrew taxies on Bodney's field. (Roger Freeman)

Colonel Joe Mason and PZ-M "Gena, This Is It" 42-8466 in early December 1943. (Warren Jackman via Don McKibben)

Don Higgins clowns around in front of Willie O. Jackson's PZ-J. (Warren Jackman via Don McKibben)

Captain John H. Walker, Jr of 352nd HQs. (Warren Jackman via Don McKibben)

Visiting aircraft parked along Great Wood near Old Tower on July 20th, 1944 for Secretary of War Stimson's visit. (Warren Jackman via Don McKibben)

Fortress "DF-Q," 43-37890 of the 91st BG's 324th BS visits Bodney in July of 1944. Photo taken NE from Old Tower towards 486th Operations, visible at lower right. (Warren Jackman via Don McKibben)

Martin "Corky" Corcoran lounges on his PZ-L "Button Nose" 42-106439, July, 1944. Note Invasion Stripes and badly weathered blue nose paint in typical 486th light blue.

Corky (left) and 486th Intelligence Officer Warren Jackman with "Button Nose" July, 1944.

The photographs of the Russian Shuttle Mission pilots on the following pages were provided by Warren Jackman through Don McKibben.

Ernest Olaf Bostrom strikes a jaunty pose with his gear on July 5th, 1944 returning from the Shuttle.

Russian Shuttle Mission Pilots

(Photos Courtesy of Warren Jackman via Don McKibben)

Tom Colby on return from the Russia Shuttle had his well-worn "Lil' Evey" shear its rudder pivot on landing. Tom's grin reflects his relief at being home safely.

Charles "Griff" Griffiths returns
to the 486th area after the Shuttle mission.

David "Jerry" French with his gear upon returning from the Shuttle, July 5th, 1944.

Russian Shuttle Mission Pilots

(Photos Courtesy of Warren Jackman via Don McKibben)

Don Higgins grinning broadly on his return.

Russian Shuttle Mission Pilots

(Photos Courtesy of Warren Jackman via Don McKibben)

Ken Williams in front of 486th Operations after the Russian Shuttle. His jacket art reads "IZZY," the name of his earlier P-47.

Carleton "Bud" Fuhrman enters the 486th Parachute Room July 5, 1944 returning from the Russia Shuttle.

Leonard "Jim" Gremaux in his "suntans" uniform on return from the Russia Shuttle.

Louisiana native Willie O. Jackson flashes a grin after the Russian Shuttle Mission, July 5th, 1944. Plane is new PZ-J, 44-13398 acquired around June 15th. Nose paint was applied between June 15th and 21st, so it is still fresh and unfaded in this photo. Note definite light blue color compared to sky, and this is not due to weathering.

Russian Shuttle Mission Pilots

(Photos Courtesy of Warren Jackman via Don McKibben)

Squadron Leader Willie O. Jackson unpacks his gear from PZ-J, 44-13398 after the Russia Shuttle mission, July 5th, 1944. Note lighter, flat blue nose shade, not unusual in the 486th.

Donald "Red" Whinnen grins from his PZ-Wbar "HMS Hellion" in July 1944.

Willie O. Jackson (left) and Donald "Mac" McKibben check out the souvenir Russian Star on Mac's cap after the Shuttle Mission, July 6th, 1944. They are leaning on Joe Mason's PZ-M "This Is It."

Leo Northrup in his flight coveralls after the Shuttle Mission.

Warren "Buddy" Brasher in front of the 486 Operations Building in July 1944. Shuttle.

Blue, Blue, Blue, What Shade of Blue!
by Sam Sox, Jr

One of the most frustrating chores of an artist, modeler, or warbird restorer is trying to determine from black and white photographs what color a particular subject was. As is well known, very few color pictures or slides were taken in the ETO. What has appeared was via someone in the states sending color film to a loved one or as the Allies moved through France and Belgium, stumbling on film stock in deserted photography stores.

Brick from latrine at Bodney showing the late-war dark blue that closely matches the RAF's "Deep Sky Blue." This dark insignia blue shade has a very slight purple cast to it. (Bill Espie)

Another frustration for aviation historians has been in them trying to determine what the various shades of blue that were used by the squadrons of Mustangs in the 352nd FG. Anyone who has looked at a large quantity of pictures of the Groups Mustangs has noticed that there were * See foot note two to three different shades of gray recorded depicting the blue nacelles areas. Next was attempting determine if these varying tones of gray were different shades of blue or were the variances due to the photographer using different types of black and white film. There were two types of film being used by "official" photographers during WW II, each capturing color in different ways. As the researcher, you never had the privilege of knowing what type of film was used when a particular subject was photographed. There were basically 2 types of film used for still picture, Orthochromatic and Panchromatic. "Ortho" recorded Blues and yellows lighter - Reds were captured darker than the subject was in reality. Panchromatic film captured blues and yellows darker and the color red lighter. All this added to the confusion.

George Nunemacher's cast aluminum model of a P-51D made at Bodney in the summer of 1944. The blue on the nose is actual 352nd FG nose paint. Decals were added later. This correct mid-1944 shade is carried currently on Kermit Week's recreation of "Cripes A' Mighty 3rd." (George Nunemacher)

The 325nd was lucky in that 2 blue items which had been painted in 1944 survived and could be used as resources in tyeing down what the blues actually looked like. The first item was the name panel off Lt. Robert H "Punchy" Powell's first Mustang named "the West by Gawd West Virginian." Due to the fact that the Mustang had crashed on July 18, 1944 and portions burned, the panel bore smoke damage and could not be used for nailing down the exact shade of the most widely used shade of medium blue. The second artifact was a 1/72 scale ID aluminum model prepared by Sgt. George Nunemacher of the 486th FS.

In the mid-1980s while visiting with George at one of the Group Association

Reunions, I informed him of the problems I were having in identifying the various shades of blue. He told me that one of the Metalsmiths in the squadron had made in late Summer of 1944, a sand cast using a black 1/72 scale ID model. He melted down scrape aluminum and produced about 6 or so. When George completed polishing his model in September of 1944, he says he went over to the 486th FS Paint Shack and opened a can of blue and one of yellow and painted it. He told me that he still had it. He advised me that he had displayed it very little since the war always being careful to keep it away from any direct sun light. He volunteered to loan it to me. When I received it, I lightly polished the blue so as to remove oxidation that may have occurred and took it to a buddy who ran an auto repair shop. I had seen his extraordinary ability of matching almost any color. He completed the match and provided me with some sheets of the color from which I made color chips.

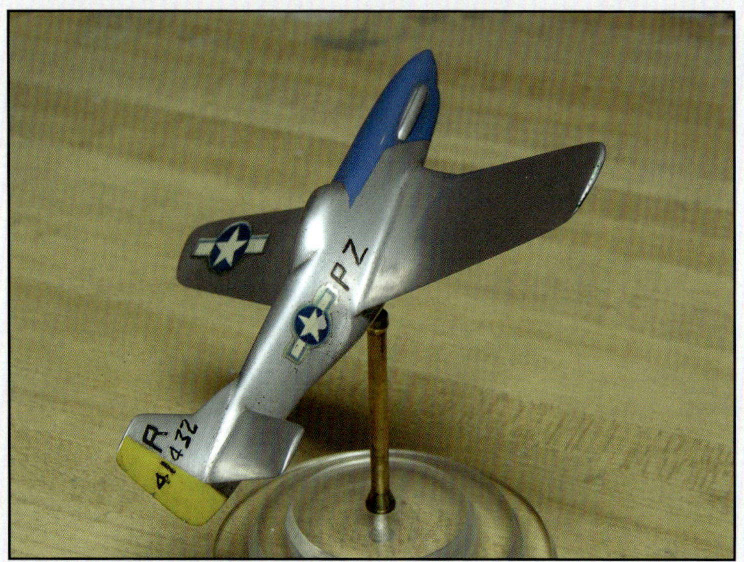

Another view of the wartime paint on George Nunemacher's model. (Nunemacher)

I noticed in reviewing the ever increasing number of pictures that had been taken in the Fall of 1944, the gray on the nacelle areas had changed significantly to a much darker gray indicating just more than a possible change in film type. The noted 8th AF author Roger Freeman stated in one of his many books that one of the stipulations of the USA/British Lend/Lease Agreements was the provision that the British would be responsible for filling US paint requirements. He indicated that the illusive blue was the British color Deep Sky Blue. Over the years, I had established a friendship with a fellow in England, Bill Espie, who as child, had as his playground the old air base at Bodney in East Anglia. He told me that he had previously been successful in locating from several dumps sites on the base some cans of red and OD paint. I asked him if would see if he could locate a sample of the paint from the old base. A few weeks later he called me and told me that he had visited the base on Oct 15, 1998. On one of the remaining walls in one of the Latrines in the 328th FS area, had found the remains of some artwork. On one of the bricks was the dark blue. He removed the brick and sent it to me. I had the paint computer matched and color chips prepared.

During a visit to England in May 1999, I visited the deHaviland Mosquito Museum and by chance mentioned to its curator Geoff Follett my problem and the story about how I had obtained a sample of a dark blue from the old base at Bodney. I needed to verify that what I had was indeed Deep Sky Blue. He said he would look into it for me. Shortly upon returning to the states, I received a letter. It not only contained a paint chip but the Pantone formula for paint. It was the same shade of blue as had been retrieved from the sample on the brick. I learned that it had also been applied to British Mosquito aircraft which flew Photo Reconnaissance missions.

At last after searching 15 years, the mystery had been uncovered and all the questions surrounding the varying shades of gray appearing on black and white photos and few bits of color photography answered. It should be noted that the first most prevalently applied shade medium blue has no accompanying Federal Standard 595 match (the US Military Color Standard).. Its source and ID is still a mystery.

*** Footnote**. A very light shade of blue was applied to a couple of the Olive Drab and Gray Mustangs in the 486th FS. It appears that several applications were being experimented with to determine what shade of blue showed up best and covered OD surfaces. This color too has never been officially identified.

Credits: George B Nunemacher, Bill Espie, and Geoff Follett.

John "Curly" Edwards' PE-E "Barbara M 4th" 44-13406 and its crew. (Paul Grabb)

Bodney Winter 1943-44

a patched but complete roof." These quarters were 'furnished' with two-foot high, room length raised platforms with straw mattresses that masqueraded as bunks. It resulted in twenty to thirty-man continuous beds.

As can be imagined, the Americans were a real curiosity for the Russian population, and small mixed groups gathered in the dirt square before the quarters attempting to be understood.

As described by Fuhrman and Colby, the restroom facilities were non-existent, resulting in a preference to "go" by the planes. Leo Northrop is shown here. (Bud Fuhrman)

"Everyone we met was pleasant and tried to make us feel at home, despite the obvious hardships they were under," relates McKibben. "The uniforms of the Russian men and women were rather grimy, but they themselves were clean and neat. The women were attractive, though not beautiful, and long after we left Russia those buxom belles were the source of great wonderment on the part of Americans who had only known English and American girls. Compared to those Russian beauties, our girls are flat-chested indeed!"

Colby adds, "We were warned by the B-17 commander who came in late in the day not to mess with any of the women or we could get shot! Any flirting was done in a group only, and in daylight." Fuhrman added that one of the Russian women soldiers encountered was a sniper by trade, with 23 German "kills" to her credit.

The latrine situation was quickly sussed out as Colby describes:

"Perhaps 25 yards from the main entrance was a wooden latrine, approximately a '100 holer' with human excrement built up around every hole and lime sprinkled generously atop. It smelled to high heaven! Russian troops and civilians used it constantly, and were in no way bashful. To urinate they would step away from the building and do their 'business' either standing or squatting and showed absolutely no concern. We did not watch, point, laugh, or in general comment because at least half of them (men and women) were armed."

Bud Fuhrman concurs, "There was a huge box about 2 feet high with holes cut around the edge. The place was so dirty that we didn't want to use it. A Russian soldier showed us how by leaping up, straddling a hole, pulling down his pants, and letting fly. This we probably could have managed, but with half of the city looking on, you had all the privacy of a goldfish bowl."

Colby elaborates, "We had another problem when it came time to urinate. We would walk as far away as we could get from the building, but at least in pairs as we were followed by very bold male children who would laugh, point, and shout to others probably insulting us in Russian. No one, and I repeat no one, would drop their pants and defecate. Man did this become a problem!"

The next day the men were informed that return to Piryatin would be delayed until the next morning (June 24th). The 486th spent the 23rd increasing their Russian vocabulary and looking over the area. Wary of mines, Colby and Lt. Warren Brashear made a three-mile jaunt through the desolation, only turning back when met by a huge, heavily armed, though well-intentioned Russian Captain.

Before takeoff for Piryatin, the men realized that something had to be done about the elimination problem. "It had been several days and most of us had headaches, dizziness, and nausea. I had a roll of toilet paper, stashed in the same gun bay as the scotch, so it was decided to go up to the airfield to be away from everyone. We rounded up a Russian truck driver, and after much yelling, he took us up to the field. Somewhere there is a photo of about nine of us squatting around PZ-T doing our utmost to eliminate all of those Vienna sausages, macaroni & cheese, and delicious white bread accumulated inside!"

"At daybreak on the 24th we took off for Piryatin, hoping eagerly that there would be a mission," Don McKibben continued. "There wasn't, so we spent the day sunning ourselves and taking advantage of the comparative luxury of outdoor showers. We learned that Jerry had tried to bomb the field the previous night, but missed by three miles."

All departed at daybreak on the 24th except Colby, who had a dead battery in his Mustang "Lil' Evey." I never felt so alone in my life," he said. After walking to the quarters over two miles away, he was able to borrow a C-10 generator and get started. Before leaving Colby was given a package by the C.O. of the B-17s to give to Capt. Hively, some thirty miles away at Chugiev. Colby was also ordered to report back to him with acknowledgment and answer! After circling over the sea-like landscape, he was able to locate Chugiev, deliver the package, and make it back to Piryatin about two hours later. The offending battery was duly recharged.

Chugiev Troubles

"We heard from the 4th Group boys of a big party that had been given for them in Chugiev, and Major Jackson decided to take us there that night. The dusk landing on the 24th at Chugiev was the beginning of our airplane troubles. Lt. Brashear taxied over an unseen abandoned slit trench and sank in, damaging his prop." Sheer Russian manpower lifted this 11,000 pound aircraft out and into a safe area. Fuhrman dryly describes the landing as "very interesting" with brush and shell holes flashing past his wingtips at dusk.

Colby adds, "There was no runway, just a large grass field. The grass was about 1-2 feet high and hid a lot of loosely filled abandoned slit trenches, holes, and ditches. The landings were hard and really beat up the aircraft. While we weren't taxiing fast, sinking into a trench was enough to shear the splines off of a prop hub. You didn't get very far after that." After a moderate celebration and another night in a community bed, the 486th rose to return to Piryatin on June 25th.

On takeoff that morning, problems again plagued the Bluenosers. Lt. Jim Gremaux's starter motor was damaged by water, putting that Mustang out of action. Gremaux would remain in Chugiev with Lt. Brashear while the remainder of the squadron (along with the 4th Group) lifted off, returned to Piryatin, and relaxed for a rumored mission on June 26th. Eventually transport was arranged for Lts. Brashear and Gremaux via Air Transport Command, and their damaged Mustangs were abandoned.

Chugiev also proved to be the bane of Ed Heller and his famous Mustang "HELL-ER-BUST." Having had his rudder repaired back in Piryatin, he optimistically lined up for takeoff with the rest of the unit on June 25th. On takeoff his engine failed. "I jammed on the brakes and stopped just short of a ditch," Heller recalls. "After checking the mags I again attempted to take off. This time the engine failed after the 'point of no return' and I ended up in that damned ditch with a busted prop blade."

Heller was determined to fly out in a Mustang, so he had the prop repaired with a bastard blade from a B-17. After these repairs were completed over a couple of days, the Mustang was again lined up for takeoff after ground runup. A few seconds after becoming airborne, a connecting rod came sailing through the cowling, and Heller jammed the P-51 back onto the ground and stopped without additional damage. He was forced to wait two weeks for a replacement engine, which was then manhandled into the Mustang. The rest of the Group was now long gone, so Ed decided to fly home via the southern Air Transport Command (ATC) route over the Middle East. After a short hop over to the large bomber base at Poltava for extra drop tanks, Heller lifted off and pointed his nose towards Teheran.

The Alert Flight

An alert flight of four ships was arranged the morning of June 25th at Piryatin. "Major Jackson came to me early and ordered me to take my four aircraft to the end of the runway and set up an 'alert flight' in case we needed to protect the takeoff for that day if we flew down to Italy," Colby details. "We were to ask no questions, just keep an eye on the tower and take off on a red flare and defend the entire airdrome!

"About this time an event took place that emphasized the warnings we had received about proper conduct in any and all fashion while in Russia. Trucks would occasionally use a road crossing the end of the runway to service Russian P-39's on the NE corner of the field. Though trucks were supposed to check for arriving or departing aircraft, one truck crossed just as a P-39 was landing and a collision resulted. The P-39 pilot was not using the metal runway, but was apparently trying to shorten his taxi time by approaching at an angle on the grass. I don't think the pilot was killed, but there was a big 'dust-up' and some shots were fired. We were told that as an example the truck driver had been shot— on the spot!

"After some 30 minutes on 'alert,' suddenly red flares were fired. Frankly I hesitated, then cranked up the engine and turned on the radio to hear a voice screaming to 'get airborne'! We lumbered down the runway fully loaded with fuel, including two 75 gallon drop tanks. We climbed at about 40" of manifold pressure and 2500 rpm, jettisoned drop tanks after clearing the field, and requested directions. Subsequently I was told there were Ju-88's in the area, but not where. I kept climbing and to the south as this was late morning and I wanted the sun behind or above me.

"We ultimately got to nearly 20,000 feet when a voice blared 'Here he comes!' and 'He's coming right at me!' In a scramble of hysterical girlish screams a series of contradictory instructions flowed, so in desperation I called for full power and dived at a steep angle for the center of the field. At a scream, 'He's going away!' I turned to starboard then another scream informed me I was wrong so I did a half roll and headed west. Not seeing a German plane, I dove to treetop level so we could see a shape on the horizon. Looking around, Blue Flight was in tight, so I radioed them to spread out and asked who saw anything. No answer, so at about 61" manifold pressure and 3000 rpms we flew on for what seemed like hours on 270 degrees west. After actually only a few minutes, I spotted an aircraft and we slowly but surely

Sgt. Harry Sanford poses with a lend-lease P-39 Airacobra in Russia. (Earl Martin)

closed in on it. I was directly astern and under him and closed to 25 yards. It was a P-39 at full bore also chasing our mysterious German! I called out not to fire, and pulled alongside left with my wingman, Bostrom, and our element leader and his wingman ('Bud' Fuhrman) on the right. The Russian pilot gaped at us as I don't think he knew us to be behind him!

We passed him by, continued on for a while and didn't see a thing ahead. I aborted the chase and returned to base, making very sure to call in our approach well in advance because of the itchy Russian gun batteries."

Bud Fuhrman had it even rougher on the 'Alert Flight,' a precursor to the tone of his entire day. "When I got the order to be on the 'Alert

Flight' I asked Jackson if I should take my plane and gear. He said 'No, use the plane and equipment there,' I guess figuring we would never be used. As it was, I was using Jerry French's airplane (formerly Chet Harker's "Cile II - Luck of the Irish") and the helmet and harness were too big. I had to use one hand on the earphone just to hear, and crossed my legs to keep from falling out. Nothing fit right! The yelling from the tower caused the radio to be so garbled we couldn't make out anything. Not a good way to fight in a Mustang!"

Colby points out that the aircraft had been run very hard for about 40 minutes, well exceeding the 5-minute time limit on maximum performance operation. On the ground a successful scramble was made to mount new drop tanks, connecting them with the scarce and invaluable glass breakaway tubes.

Colby finishes, "We reported to Jackson who questioned our maneuvers (rather sharply) as he was embarrassed that we hadn't downed the Ju-88 in full view of the airfield, bringing distinction to the 486th Squadron and America in general. It had been Hively who was screaming like an embattled virgin, and he was both embarrassed and angry. Col. Blakeslee was present, took it well, and told Jackson that we did fine considering the screaming and mixed instructions. It was rumored that the Ju- 88 was caught by P-39s. No way! He escaped and no one in my flight ever saw him at any time."

Fuhrman's "Milk Run"

That afternoon, June 26th, the 486th (minus Brashear, Gremaux, and Heller) took off to escort the bomber boys again. The target would be the oil refinery at Drohobvos, Poland, proceeding to a landing in Lucora, Italy. The mission was a "milk run" for all except Bud Fuhrman, who fills in the details for us.

"At our briefing we were informed that a ground crew, ready with spare wing tanks and gasoline, would be stationed along the side of the runway for anyone who might drop a tank on takeoff. This precaution was made because we were taking off from the steel mat runway which was anything but smooth, causing our fuel laden fighters to take some hard bounces. As the flight to Italy would be a long one, we needed all of the fuel we could carry. Our takeoff seemed normal, and the Squadron formed up and set course for the first check point.

"The pilot next to me signaled that I had lost one of my 75 gallon wing tanks. This had not become apparent to me as I was still using fuel from one of the internal tanks located in the wings. I pulled PZ*D out of formation and headed back to Piryatin. I quickly found the airfield, came around, and made a standard 360 degree overhead approach. When I swung onto my final approach to the field, there was a photo-recon P-38 landing ahead of me. He landed short and taxied slowly down the runway. Nothing the tower shouted at him seemed to speed him up, so I went around again and made another landing approach. This all cost me valuable time. When I landed and saw that my roll would carry me to the waiting fueling crew, I cut the mixture control causing the engine to cease firing. I rolled up to the crew and spun the aircraft around with the brakes. I will give the crew all of the praise, as almost before the propeller stopped windmilling, they had the new tank hung and filled with fuel. Others raced to get the sway braces adjusted and the fuel lines connected. Inside of five minutes I started up and was taxiing out for takeoff.

"I was quickly airborne and on my way again by myself. I had all of the compass headings on the back of my hand, which was standard practice to give us something to go by in case we got separated from the group. I climbed to 25,000 feet and set a very fast cruise to catch up with the squadron. I easily spotted where the bombers did their work, as there was a towering column of dense black smoke. I set off on the new course I knew our airplanes had taken.

"I always kept a sharp lookout for other aircraft, especially over enemy territory. This habit sure paid off, as when I glanced in my mirror, my heart jumped into my throat! Close and coming up fast were two Me-109's. I didn't need a second look to see their cannons sticking from the holes in their spinners. Just moments before I had glimpsed our airplanes ahead of me, though some miles away. Almost in one motion I broke left in a violent turn, reached down and flipped the fuel selector to an inboard wing tank, flipped the arming toggle to the wing tank release, and dropped the tanks. The

P-51 had an 85 gallon fuselage fuel tank behind the pilot, and being full it caused a very tail-heavy condition. My turn was so sharp that I thought the tail would pass me. As opposed to a lot of control stick back pressure in a normal steep turn, I had to push forward to keep control of the plane. This sudden turn caught the Me-109's by surprise as I was fast taking the initiative from them. The two E/A separated, one zoomed climbing right, the other swinging down to the left. I would have loved to have taken off after the one heading down, but I knew his buddy would be on me in a flash. No chance of catching the one climbing right.

"I called my squadron for help, and Jackson asked my position. I replied, 'Angels 25 and about 5 miles east of target area.' In a couple of minutes the squadron dove down to my aid. Their appearance caused the Me-109's to quickly lose all interest and disappear. Having dropped my wing tanks, I throttled back and flew a straight and steady course to one side of the bombers to conserve fuel. Thanks to good old PZ-D, I didn't have to visit the prearranged emergency fuel stop on the Island of Vis, and had enough fuel to reach Lucora with the rest of the squadron. PZ-D was very good for fuel economy and speed. All of the rest of the pilots thought it was an uneventful mission, but for me it was all of the excitement I wanted."

Italy

After crossing the Adriatic the aircraft came to roost in the talcum-like dust of Lucora Airfield (near Foggia). After an overnight stay in the base's tents, the 486th took off with the 4th Group for the short hop to Madna, Italy. The next six days (June 27th through July 1st) were spent resting up at Madna, and cavorting on the beaches working on suntans. Though missions were likely to be "on" at any time, poor weather over England prompted this extended rest period. Bud Fuhrman celebrated his birthday in Italy that year, turning 23 years on June 28th and enjoying a swim in the Adriatic.

McKibben describes the departure from Italy on July 2nd:

"We flew with the 15th AF on a mission to Budapest. Our Group swept the target area before the bombers and gave target support. Just as the bombers started their bomb run, a large formation of Me-109's were sighted and the

Russian/American camaraderie on the Shuttle. Standing, second from right is Ed Heller (tallest) and Tommy Colby, Red Whinnem, and Jerry French follow right to left. Kneeling front right is Warren Brashear. (Bud Fuhrman)

Group took out after them. Major Higgins, piloting Bill Reese's PZ-R/bar as his 'Bayou Baby' malfunctioned before leaving Bodney, got an Fw-190 and then shared in the destruction of an Me-109 with a 4th Group pilot. However, we wound up on the red side of the ledger on that mission by losing Major Andrew and Lt. Howell. Lt. Howell was last heard of when he went on a bounce." Howell called that he 'had a bunch surrounded' and for the flight to follow him, but unfortunately, they were already engaged in a fight. He was flying 'Texas Knight,' a P-51B-15 coded as PZ-V."

The late Charles Griffiths tells of the loss of Stephen "Andy" Andrew when the engine on his PZ-A failed. Andy's crew had installed a new engine in his plane just before leaving England. It had been "slow-timed" and test-hopped, but had flown no missions prior to the Shuttle Mission. The engine performed okay on the England/Russia/ Italy legs, but developed problems on the mission to Budapest.

"I was flying Andy's wing and Lt. Northrop was element leader. I don't remember who number four was because we lost him right after the Germans jumped us. It could have been Lt. Howell, who was lost on that mission. He left on his own going after a single bandit, and we heard him call for help later.

"Apparently, as the Jerries hit us Andy firewalled it to catch the Jerry he'd picked out. After a few seconds his engine froze, and he was staring at his prop and starting a steep descent. Northrop and I stayed with him, essing above him as he tried to restart his engine. At about 25,000 feet I suddenly saw tracers coming at me, and I think I set new records for evasive action while still trying to stay with Andrews. My head went full circle as they attacked again, then suddenly disappeared. I don't know why they left, unless another of our guys got on them. Meanwhile, back at the ranch, Andy was still going down and me with him. If anyone knows what it's like to deadstick a Mustang, you know that Andy was trying to find a suitable field. Then he radioed, 'Well, boys, I guess I've had it,' and asked for wind directions. It was fairly flat country, and he landed wheels-up in a plowed field. When we saw him wave and run away across a field, we strafed his plane in a couple of passes, left it burning, and left Andy to his fate."

Andrew evaded for a week, and covered 100 miles along his chosen escape route.before he stumbled into an SS Division's bivouac while evading at night near Baja, Hungary. He was captured and imprisoned at Stalag Luft III near Sagan and later moved (in January '45) to Stalag Luft VII in Bavaria where he remained until war's end. Liberated in April of 1945, he was able to visit Bodney briefly to renew old friendships before heading Stateside.

Colby added to the battle tale: "A single Me-109 plunged through the top cover, pulled out level at about 300 yards, and startled both me and McKibben. As we maneuvered to jump him, four Mustangs from another unit literally blew him apart. After a high power chase of two Me-109's towards Romania, I broke off, and it's a damn good thing I did. My engine was breaking down from the stress of some 275 hours and its hard use during the past week. I later signed off on the engine and flew it back to England rather than have it rebuilt in Italy and left with the 52nd Group."

Aviation author and editor Jon Guttman has kindly provided information on this day's events from the German and Hungarian viewpoint:

"In regard to the Axis side of the July 2, 1944 mission, the amount of aircraft they threw up does not make an easy sorting out of who got whom. All three of the Hungarian 101st Fighter Group's squadrons were up. Four Hungarian Me-109Gs were lost, resulting in the deaths of 2nd Lt. Erno Karnay and Corporal Sandor Beregszaszi of 101/1 Fighter Squadron, and Cpl. Pal Takats of 101/3, while Sgt. Pal Forro of 101/2 bailed out wounded. (Forro had also been among the three Hungarian and one German Me-109s downed by the 49th Fighter Squadron during the June 14 dogfight; shot down by 2nd Lt. Don Luttrell, he belly-landed in a grove of trees and emerged miraculously unhurt, although he said in his combat report that the only part of his Messerschmitt still in one piece was the engine).

('The Hungarian research book I have also noted that the following German units (operating from Vienna, Graz, Zeltweg, Wels and Seyring) also got involved in the air battle: I/ZG.1—Me-401s, II/JG 302 - Me-109G, II/ZG.I—Me-110sII/ JG.27 - Me-109Gs, I/ZG.76—Me-410s, III/JG.27 -

Me-109Gs, III/ZG.76—Me-410s, IV/JG.3 (Sturm) -w-190A8s and Me-109 G6s, II/JG.302—Me-109Gs, I/JG.10 - Me-109Gs, JG.108 (Jaegerschule), operating from Bratislava")

"I have only found partial German information thus far. JG.302 reported 10 of its pilots killed and one wounded who bailed out. I/ZG.76 claimed eight bombers, while III/ZG.76 claimed only one. Hungarian claims for July 2 included five B-24s and two confirmed P-51s credited to Sgt. Zoltan Rapos and Senior Cpl. Pal Szikora, 101/1—P-51, near Vereb, 2nd Lt. Gyogry Debrody, 101/3—P-51, near Pilisvorosvar, probably George L. Stanford, Jr. of the 335th Squadron, brought down as POW.

"Among the unconfirmed claims was a P-51 by Ensign Leo Kriszevszky of the 101/3 Squadron, which the Hungarians later traced to P-51B 43-6746 QP-X, the wreckage of which was found on the Yugoslav side of the border at Mostar. Its pilot was 1st Lt. Ralph Hofer, a 16 1/2-victory ace of the 4th Fighter Group's 335th Squadron. Stanford, whose Mustang had been disabled by a thrown rod and dove for the deck before finding his plane in flames and being forced to belly-land, stated that Hofer had been following him down, trying to guard his tail when he was jumped by an Me-109."

In addition to the 352nd Fighter Group's two losses, the 4th Group lost four P-51s and the 325th "Checkertail Clan" lost two (one downed over Budapest and one destroyed in a mid-air collision east of Budapest).

Mac McKibben concludes the details of the July 2nd melee with a dry, "The rest of the mission was uneventful. The following day (the 3rd) we ran a mission with our own 8th AF bombers to Arad, Rumania. Another milk run."

July 4th was uneventful except for the 486th's use of some flare pistols to create a little 'fireworks' in celebration. Some of the oat stubble surrounding the field was set smoldering by the flares, but as it was quickly doused with no harm done. That evening, the squadron was alerted with orders for a return mission to England.

Off For Home

"Target for the trip home was the marshaling yards at Bezier, near the south coast of France. The pilots of the 52nd Group at Madna were astounded when they heard the weather officer brief us for a 200 foot ceiling over England. In Italy they simply did not fly if there was any weather to contend with. Our route out to rendezvous took us directly over Rome and across Corsica. We rendezvoused with the Big Friends just south of the target. The only opposition was a lone Me-109, who was apparently rat-racing around the tail-

The 486th birds assemble at Debden on June 15th for the Russia Shuttle as the mission was originally "laid on." The weather dealt otherwise, so the actual mission was flown on June 21st. Foreground is Willie O. Jackson's spanking new PZ-J, 44-13398, with no blue nose applied as yet. Next, front to rear are Don Higgin's PZ-B "Bayou Baby" , Brashear's PZ-Bbar, Lester Howell's PZ-V "Texas Knight", Corky Corcoran's "Button Nose" PZ-L, Al Wallace's "Little Rebel" PZ-W, Ed Heller's famous "HELLER-BUST" PZ-Hbar, Leo Northrop's "Donna Dae V" PZ-N, Bud Fuhrman's "The Flying Scot II" PZ-Ddot, Mason's former "This Is It" PZ-M, Pappy Gignac's (KIA June 7th)"Gig's-Up II" PZ-G, and back row right to left Chet Harker's "Cile II, "Luck Of The Irish" PZ-H, Frank Greene's "Geronimo" PZ-I, Stan Miles' PZ-S, and Stephen Andrew's PZ-A. Joe Gerst's PZ-X "Smoky Joe" and McKibben's PZ-Y are not shown. Gerst and Corcoran aborted the actual mission, and Higgins' "Bayou Baby" ran rough, switching him to Bill Reese's PZ-Rbar for the Shuttle.

A wartime map of the ingress route to Poltava. The southern egress route shown is for pilots of aircraft damaged in Russia. They took the ATC route home.

end of the bomber formations. Major Higgins, who had taken over the squadron when Major Jackson had to abort, started after the Hun but he evaded him in a steep dive. We kept our wing tanks throughout the pass and were content to let the Jerry go. It was a long way from home and we needed the gas in our wing tanks.

"We left the bombers by flights of four as the gasoline supply began to dwindle and headed north toward England. Major Jackson and Lts. Williams, Whinnem, Northrop, and Bostrom all returned to Italy because of fuel shortages or mechanical difficulties, but soon effected repairs. The English coastline looked good, but what even looked better was the weather over England. There were plenty of holes in the overcast, and we thanked our lucky stars the weather officer had been wrong. After landing at Debden we made a dash for the bar where free beer was dispensed under a sign reading 'Welcome Home Col. Blakeslee and the Boys.'"

Colby adds, "No problems on the 20 minute flight back to Bodney from Debden, but for some reason I didn't join the others in buzzing and beating up the field...thank God. On landing I felt a severe jolt and a pull to the right. The next day the line chief came and showed me where the top rudder pin was sheared in two! " (This aircraft was later made the assigned mount of Jerry French as PZ-F when Colby received a new P-51D).

Ironically, on the truck ride from our planes to the mess hall and huts, we passed Gremaux and Brashear. They'd left several days before us on ATC from Chugiev, and we all arrived together!" The next day Jackson, Whinnem, Williams, Northrop, and Bostrom arrived safely back in England from Italy.

The "Wandering Boy" Heller Returns

Back in Russia, after laboriously fitting the new Merlin power plant; Heller finally lifted off from Poltava and pointed his nose towards Teheran (next stop on the southern ATC route). Predictably, the hydraulics blew after gear-up, leaving no flaps or brakes upon landing. After rocking the aircraft until the gear dropped and locked, Heller successfully ground looped to a stop at Teheran with no further damage. After another two weeks of waiting for parts (with a case of dysentery), repairs were made and the next leg of the return trip was completed to Cairo. After a bit of sightseeing, he was off again to Benghazi, flying low over the North African desert and viewing the battlefields. The next leg of the trip (to Casablanca) was undertaken crossing the Gulf of Sidra, Tunisia, and the Atlas Mountains. Upon landing and filing a flight plan for return to England, the resident General summarily grounded Heller and confiscated his Mustang for personal use, forcing Ed to complete his trip to England via ATC after all. As Heller notes, "What a trip!"

Colby concludes with this: "There was great speculation as to the value of these 'Shuttle Missions.' Much was made over the symbolism of the Allies being able to strike from three sides and the general mobility of the American Air Force. It wasn't for the German's sake! It was all designed to impress the Russians. At our level we all suspected this, but Col. Blakeslee informally made this clear to us."

While the overall Shuttle losses incurred by the bomber and fighter groups were high, the 486th FS acquitted itself well on the mission. Official credits show 5 ½ victories credited to the unit, versus one loss to enemy opposition (Howell). Operational accidents due to poor Russian field conditions claimed two more Mustangs – Heller's P-51 to the General in Casablanca, and the loss of Stephen Andrew to POW status because of engine failure.

The 486th FS Shuttle Mission participants were: pilots Majors Stephen W. Andrew and Willie O. Jackson, Captains Martin E. Corcoran (aborted) and Donald H. Higgins, Lieutenants Ernest O. Bostrom, Warren H. Brashear, Thomas W. Colby, David J. French, Carleton L. Fuhrman, Joseph L. Gerst (aborted), Leonard A. Gremaux, Charles E. Griffiths, Edwin L. Heller, Lester L. Howell, Donald W. McKibben, Leo W. Northrop, Donald Y. Whinnem, and Kenneth Williams. Ground crew participants were T/Sgt's Sheldon F. Ackerly and Clifford A. Craft, S/Sgt's Howard Jester, Alan L. Polworth, Undo F. Rautio, Edward F. Richards, Harry L. Sanford, Carl J. Scoffone (unable to participate due to illness), James R. Sturdevant John I. Sullivan, Walter J. Tyska and Harry L. VanDyk.

The Long Reach

by M/Gen. William E. Kepner

The daylight onslaught by steadily increasing fleets of American heavy bombers and long range fighters upon the vital targets deep in enemy territory and the concentrated fury of the Luftwaffe's stepped-up fighter forces in their effort to prevent it made inevitable the greatest head-on clash of air forces in history. The long, desperate battle which resulted is not yet over but it is now clear that the courage and the destructive fifties of our bomber crews and the skill, daring and teamwork of our escort fighter pilots are driving the defeated enemy to the ground. The tactics of these fighter pilots, unique because of their unprecedented mission, are the subject of this publication, and some fifty of the most outstanding of these... are its authors.

Without going into the development of the VIII Fighter Command in England from its first two groups to its present ability to escort a single mission with a thousand fighters, it is none the less necessary to pin the comments herein down to the particular phase of the European air war to which they apply. The fighter pilots quoted are taking for granted some knowledge of the new requirements, equipment and techniques which have kept pace with the rapid expansion of the opposing forces and the mounting ferocity of the battle.

The initial stage of fighter escort in this theatre presented nothing new in the basic escort principle as established by the RAF and our own forces in other theatres. It did represent the many unique and original ideas contributed by all as to the implementation of the old fundamental principles—particularly in finding a wary enemy, then developing teamwork by all the units and individuals comprising any unit down to a single "pair."

During this stage our P-47 Thunderbolts provided protection for the bombers on short penetrations of some 165 miles, roughly to Brussels, Cambrai, Paris, Argenton, etc., and part way protection to

L-R: General Kepner discusses strategy with Charley Bennett, Col. Joe Mason (leaning) and Col. Curly Edwards.

this limit of range as the bombers went out and again as they came home. The enemy, for the most part, simply waited to intercept until fuel limitation forced the Thunderbolts to leave. It was good fortune that the enemy's technique at that time was not well enough developed to stop the bomber fleets even though the major portion of their journey was unescorted. Of course, valuable experience was gained by our fighter plots during this period and certain signal victories gave promise of the fine performance of the P-47 fighter plane and the skill of our pilots which were to be demonstrated in the months immediately following.

About the first of August of 1943, two experiments began, one necessitated by the other, which by some vision and much trial and error, were to make possible the first really long range fighter escort. These were the use of belly tanks or wing tanks and the group by group relay system.

Putting tanks on the P-47s for longer range seems simple enough but the program was complicated by tactical considerations. Since our fighters were liable to attack or to be attacked the minute they crossed the enemy coast, there was hope only of saving the distance across the North Sea, and a tank of but 75 gallons capacity was used for this purpose. The hopes of high Air Force command were of course looking far beyond this modest achievement and these hopes received new impetus one day when the 4th Fighter Group took a chance and carried their tanks inland, holding them to the moment of combat, with the 56th and other Groups following suit. With this possibility demonstrated, larger and larger tanks were requested and became available. Farther and deeper went the Thunderbolts along with the Fortresses and Liberators, escorting them 450 miles now to any point within an arc range running through Kiel, Hanover, Karlsrube, Stuttgart, Vichy, Limoges and La Rochelle. The enemy tried early attacks upon our fighters to force them to drop tanks. However, it was too costly for him. He relapsed into his policy of avoiding fighter combat whenever possible in order to get at the bombers when, in the last reaches of their penetration, they were still unescorted. His opportunity to do this was now cut down to a much shorter period on the deepest missions and, on the middle distance missions, it did not exist at all. Long range escort was actually in being.

Since the faster speed of our fighters caused them to use up their range in a much shorter period than bombers covering the same distance, it became necessary to divide our fighter escort into relays. One Group would fly direct to a rendezvous point along the bomber route, stay with them until relieved by another Group or as long as possible, then return more or less directly to base. The long line of bombers was protected not only by the fighter group assigned for its protection for given periods but also by the constant stream of fighters en route to rendezvous and returning pretty much along the same route. Experience taught many lessons and this relay system became a large and permanent feature in the new aerial warfare science of long range escort.

The addition to our fighter force operations in October and December of 1943 of two groups of P-38 Lightnings created a new phase. With new built-in fuel capacity adding to their normal range, the P-38s could go to the bombers' target and provide protection during the most critical moments of bombing. Still, there were gaps in the escort. There were not enough P-38s to cover the bombers from the point where the Thunderbolts had to leave them and return and still cover the target area, too. Some operational difficulties experienced by the Lightnings flying at high altitudes in the winter season in this theatre added to the limitations, this handicap being overcome to some extent by a fine display of skill and courage by their pilots and leaders.

Complete long range fighter escort, round trip from England to Berlin, Munich and points east became possible with the addition to the operations of this Command in March, 1944 of a number of groups of P-51 Mustangs, by large measure longer in range than any other fighter on the battlefronts and superior in many other combat characteristics. The circle was now complete. Where the bombers go to bomb, the fighters could go to protect them. Together they wrought and are wreaking a vast destruction upon the enemy and together they have fought him out of the air to the point where he dares contest our passage only sporadically and at moments especially opportune for him.

This outline of the technical phases of long range fighter escort is not to lose sight of the fact that it has meant determined, courageous and skillful battle all the way, the greatest offensive fighter battle ever fought. The work of the P-38 Lightnings has already been mentioned. The brilliant, devastating accomplishments of the P-51 Mustang are our present enthusiasm. It is perhaps the moment to point out the following fact—by far the larger proportion of our escort fighters to date have been P-47 Thunderbolts and their share of the 2,321 enemy planes destroyed by this Command in combat, with 1,496 probably destroyed and damaged, is in ratio to their numbers. If it can be said that the P-38s struck the Luftwaffe in its vitals and the P-51s are delivering it the coup de grace, it was the Thunderbolt that broke its back. All working together under comprehensive plans by Fighter Command and its Wings cornered the German Air Force to the point where it had to fight.

The comments quoted in this publication were written during the most recent stages of the long range fighter war as outlined above. Their experience and opinions are accurate and authentic as of now. New fighter groups, groups about to become operational, fighter pilots in training, future students of tactics are warned, however of the rapidly changing techniques and must keep themselves abreast of them if they wish to have the best and most recent information available. The Command will do whatever is possible and practical by amendment, revision or re-publication to keep up to date.

It should also be pointed out and underlined that this material, although it may help in creating one, is not a manual of fighter tactics. It is a record of experience from the battle, in passing, of extreme value we believe to every fighter pilot and every fighter leader who will sit in a cockpit in any war theatre in the world. Many differences of opinion will be apparent. Be sure they are real differences, not just differences in expressing a similar idea. There are many comments, for example, as to how to avoid over-shooting. Yet, Captain Beeson says: "Over-shooting is a very good thing." But, he means, and explains, that speed of attack is a very good thing and he is seconded in this opinion by Major Mahurin, Colonel Zemke, Lt. Col. Schilling and other great fighter pilots. In general, fighter doctrine is confirmed and any basic principles are repeated so often that their soundness cannot be doubted.

Major General William E. Kepner
Commander, Eighth Fighter Command

Among the pilots contributing to this Fighter Command treatise were Capt. Virgil K. Meroney, first ace of the 352nd FG and Colonel John C. Meyer, C.O. of the 487th Squadron of the 352nd, who became the seventh ranked USAF fighter ace of all our U.S. wars. Their comments follow:

Capt. Virgil K. Meroney
487th FS, 352nd FG

After reading your letter and much thought, I have decided to write to you in a sort of outline form on several points I have learned in combat against the enemy.

When we get replacements (pilots), they know practically nothing of the many things that go towards making a good fighter pilot. Their training, before we get them, is a headache in itself; what with all the safety precautions and all that they have back in the States. So, I won't go into that. Here are some of my experiences:

A. Individual Defensive Combat

On several occasions I have been at a disadvantage when first seeing the enemy aircraft, but so far I have always been able to alter the situation by doing some quick thinking and acting. There is a lot of argument on whether the P-47 can outturn the Fw-90 and the Me-109. I feel that it all depends on the situation of the moment. If you are bounced from above you cannot outdive them since they had more speed to start with. Before you can gain any additional speed you are a "dead pigeon." Having learned that, I have always chopped my throttle and turned into them, turning in whatever direction I saw them coming from over my shoulder. I have always been able to outturn them in that manner. After that, you can, according to the situation, either press home your attack or get the hell out. When attacked by vastly superior numbers of enemy aircraft, I have always found that by turning into them and barrel-rolling into them, it always breaks their formation and I have been able to single one of them out and get him. It helps a helluva lot if your wingman is right with you. If, unfortunately, your flight finds itself alone, it works out better if the elements split. Still, it is advisable for them to keep each other in sight to offer mutual protection that old teamwork again.

B. Offensive Combat

When making an attack on an enemy aircraft, I like my wingman to stay back more than normal so that he can stay with me more easily. If he stays abreast of me his wing will very likely blank me out at times and he will lose me.

Usually, the e/a tries to hit the deck and kick the ship around while going down. It makes me happy when they do that because then all I have to do is close up and let them have it. But I have noticed that the smart boys try to turn into you and if you've come down from several thousand feet above them you cannot stay with them. I have always found it works out best if you pull back up and do a sort of wingover. In that manner, you are bound to end up on their tail.

Virgil Meroney looks pleased with his P-47 "Sweet Louise/ Josephine." (Sheldon Berlow)

To keep from overshooting, I have always started throttling way back as soon as I get in range and start firing. In that manner they are finished off before I overshoot them.

I think it is inadvisable to attack when you spot a single plane far below you even if the whole squadron is with you. If you do you are bound to be on the deck before you finish him off, the squadron drawn away from their job protecting the bombers, and a lot of time and altitude wasted.

When I first did any shooting, like a lot of other pilots, I opened up way out of range and I'm convinced that range estimation cannot be emphasized too strongly.

Group and Squadron Formation

Up until a few months ago I was an element leader, so I don't think there is much I can say concerning Group and Squadron formations and tactics. However, these are some of the ways our squadron operates:

Each flight can make a bounce; the others covering. We try to keep a section of two flights together at all times. In the flight, the element leader can make the attack while calling it in and the flight leader and his wingman will follow him. Insofar as the wingmen are concerned, they can lead the attack only if the flight or element leader cannot see the enemy aircraft. They are the ones who will tell the wingman to go ahead and they will follow. If the wingman peels off and leaves his flight he may get one or two e/a, but the chances are he won't get back to get his name in the Stars and Stripes—air discipline.

Position for a bounce is good if you have it, but usually, if you try to get it after seeing the e/a, he will get away. You have to go right after him the minute you see him.

Insofar as tactics are concerned, I think we should change them and experiment with different types of formations as much and as often as possible. If for no other reason than just to confuse the enemy. For example, send a Group on a penetration support so that they rendezvous head-on with the bombers, and tail-on, when giving withdrawal support. It doesn't take the Germans long to find out about new rules.

That's it, for what it's worth, if any. Now for that beer. I need it! I believe the main things are teamwork, confidence in your leaders, your aircraft and a fighting heart. KILL THE BASTARDS!

COMMAND COMMENT: "Capt. Meroney's remark about not waiting to position yourself for a bounce is a bit overeager. Every effort should be made to get into good attack position short of losing the run."

EDITOR'S NOTE: Capt. Virgil Meroney was the 352nd Fighter Group's first ace. He also flew combat in the Korean War and flew a mission with his son in the Vietnam War. His son was later killed in action.

Lt. Col. John C. Meyer, C.O.
487th Squadron, 352nd FG.

In every case when attacked by enemy aircraft I have turned into the attack. We have found that the turning characteristics of the P-47 as against the Me-109 and Fw-190 are very nearly equal. Since, when we are attacked the e/a has almost always come from above, he has excessive speed and turning inside him is a simple matter. If the e/a is sighted in time, it is often possible to turn into him for a frontal attack. On two occasions I was able to do this and the e/a was reluctant to trade a head on pass and broke for the deck. Thus I was able to turn a defensive situation into an offensive one.

The sun is a most effective offensive weapon and the Hun loves to use it. Whenever possible I try to make all turns into the sun and try never to fly with it at my back. Clouds are very effective for evasive action if there is 8/10's coverage or better. They are a good way to get home when you are alone.

When attacked by superior numbers I get the hell out of there using speed or clouds or, as a last resort, diving to the deck. An aggressive action in the initial phases of the attack will often give you a breather and a head start.

I had one experience which supports this last statement and also shows what teamwork can do. My wingman and I, attacking a pair of 109s, were in turn attacked by superior numbers of e/a. In spite of this (the e/a were still out of range for effective shooting) we continued our attack, each of us destroying one of the e/a, and then turned into our attackers. Our attackers broke off and regained the tactical advantage of altitude but during this brief interval we were able to effect our escape in the clouds. Showing a willingness to fight often discourages the Hun even when he outnumbers us, while on the other hand I have, by immediately breaking for the deck on other occasions, given the Hun a "shot in the arm," turning his half-hearted attack into an aggressive one.

487th CO J.C. Meyer is belted in by Sgt. Conkey, his crew chief.

I do not like the deck and this is especially true in the Pas de Calais area. I believe that it may be used effectively to avoid an area of numerically superior e/a because of the difficulty in seeing an aircraft on the deck from above. With all silver planes this excuse is even doubtful. The danger from small arms groundfire especially near the coast is great. I realize that I differ with some of my contemporaries in this respect, but two-thirds of our Squadron losses have been from enemy small arms fire. Just recently I led a 12-ship squadron on a fifty mile penetration of the Pas de Calais area on the deck. We were under fire along the entire route. We (the 487th) lost one pilot, three airplanes and three others damaged. I repeat. I don't like the deck and can see little advantage in being there. Caught on the deck by three 190's I was able to outrun them by using water injection.

Mainly it's my wingman's eyes that I want. One man cannot see enough. When attacked I want first for him to warn me, then for him to think. Every situation is different and the wingman must have initiative and agility to size up the situation properly and act accordingly. There is no rule of thumb for a wingman.

I attempt to attack out of the sun. If the e/a is surprised, he's duck soup, but time is an important factor and it should not be wasted in securing position. I like to attack quickly and at high speed. This gives the e/a less time to see you and less time to act. Also, speed can be converted to altitude on the breakaway. The wingman's primary duty is protection of his element leader. It takes the leader's entire attention to destroy an e/a. If he takes time to cover his own tail, he may find the enemy has flown the coop.

Effective gunnery takes maximum mental and physical concentration. The wingman flies directly in tail on the attack. This provides maneuverability and he is there to follow up the attack if his leader misses. Once, however, the wingman has cleared himself and is certain his element is not under attack he may move out and take one or the other e/a under attack more than one target is available. Good wingmen, smart wingmen, are an answer to a leader's prayers.

If surprise is not effected the e/a generally turn into the attack and down, thus causing the attacker to overshoot. When this happens I like to break off the attack and resume the tactical advantage of altitude. Often the e/a will pull out of his dive and attempt to climb back up. Then another attack can be made. A less experienced e/a pilot will often just break straight down. Then it is possible and often fairly easy to follow him. Usually on the way down he will kick, skid and roll his aircraft in violent evasive action for which the only answer is point blank range. Compressibility is a problem which must be taken into consideration when following an e/a in a dive.

The effect of superior numbers in a decision to attack is small. The tactical advantage of position—altitude, sun and direction of attack are the major factors. With these factors in my favor the number of e/a are irrelevant.

It is not wise to attack when the enemy has the advantage of altitude as long as he maintains it. If you're closing fast enough to overshoot, you're closing fast enough to get point blank range—and you can't miss.

I am not a good shot. Few of us are. To make up for this I hold my fire until I have a shot of less than 20 degrees deflection and until I'm within 300 yards. Good discipline on this score can make up for a great deal.

I like to attack at high speeds and break up into the sun, making the break hard just in case his friend is around. Then I like to get back that precious altitude.

Enroute to rendezvous we fly a formation which has for its basis mutual protection rather than flexibility or maneuverability. The Group is broken down into three squadrons of sixteen ships each. The squadrons are stepped up with the second squadron about 1,000 ft. higher into the sun and almost line abreast; and the third squadron 2,500 ft. higher than the lead squadron on the down sun side and line abreast. This alignment makes it impossible for any one squadron to be bounced out of the sun by an attacker who is not clear of the sun to one of the other squadrons. The flights fly almost line abreast (slight finger formation) providing mutual cross-cover between individual planes and flights within the squadron.

Upon rendezvous with the bombers the Group generally breaks down into eight ship sections of two flights, each operating independently and at various ranges from the main bomber force. One flight of this section remains in close support of the other on bounces. This method has been the most successful one tried by this organization. It has certain disadvantages in that they may run into superior forces in which case considerably more would have been achieved by keeping a larger portion of our forces intact. However, one main problem to date has been in seeking out the enemy, rather than his destruction once found. This deployment has been the best answer to that problem. These eight ships are under orders to remain within supporting distance of each other at all times. These sections operate above, below, around, ahead, behind and well out to the sides of the main bomber force. The extent of ranging is dependent upon many factors such as weather, number of friendly fighters in the vicinity and information on enemy disposition; the decision on this is left to the section leaders.

There is no rule of thumb limitation on who makes bounces. The primary job of the flight leaders is that of seeking out bounces while that of the others is flight protection. If any member of a flight sees a bounce and time permits, he notifies his flight leader and the flight leader leads the engagement. However, if the flight leader is unable to see the enemy, the one who spotted him takes over the lead or in some cases, of which there are many, when the time element is vital, the man who sees the enemy acts immediately, calling in the bounce as he goes.

Usually if the combat is of any size or duration, flights become separated. The element of two becoming separated, however, is a cardinal and costly sin. We find it almost impossible for elements to rejoin their squadrons or flights after any prolonged combat. However, there are generally friendly fighters in the vicinity all with the same intention and we join any of them. A friendly fighter is a friend indeed, no matter what outfit he's from.

Recently, we have tried imitating Hun formations but have not had any particular success with it.

On the defensive the eight ship section turns into them presenting a 64-gun array which the enemy is reluctant to face. If we are hopelessly outnumbered or low on fuel after the initial turn, the individual ships keep increasing their bank until in a vertical dive using the superior diving speed of the P-47 to escape. On one occasion when we were extremely low on gas one flight of our eight was bounced by three 190s. That flight broke for the deck with the 190s following. They soon broke off their attack and zoomed back up and we continued our dive and effected our escape.

We drop belly tanks when empty. Gas consumption is a primary tactical consideration in this theater and we don't like to use it to drag empty belly tanks around.

The number of aircraft to go down on a bounce is influenced purely by the number of enemy aircraft. In any case at least one flight stays several thousand feet above until the situation is carefully sized up at which time the leader of that flight makes the decision on whether to join the fray or stay aloft.

We pursue all attacks to conclusion if a favorable conclusion seems possible. In other words, if by continuing the pursuit it seems reasonable the enemy may be destroyed. There are exceptions to this, however. For instance, the Hun sometimes will send a single aircraft across our nose to draw us away from the bombers while their main force attacks the heavies. Our leaders must watch or this and make the decision.

Every effort is made to hit the enemy while he is forming for the bomber attack. Generally he forms ahead and well to the up sun side of the bomber force. A large part of our group force is deployed in that area.

Our Group was the first to attempt a penetration in force on the deck for a strafing mission. Out of this experiment I have these recommendations to make: That penetration to within ten miles of the coast be made on the deck, and then have the force zoom to eight or twelve thousand feet, navigate at this altitude, penetrate beyond the target and hit the deck again at some prominent point a short distance from the target and proceed to it. This, rather than penetration all the way on the deck where the enemy's small arm fire is intense and pin-point navigation impossible. When our aircraft are below 8,000 feet over enemy territory, they should be just as low as possible. Twenty feet above the ground is too high.

COMMAND COMMENT: The episode cited illustrating the value of initial aggressiveness seems a dangerously fine estimate of the situation, but only possible because the enemy was sighted well in advance. Otherwise, they might have been victims of a typical enemy sucker play.

EDITOR'S COMMENT: The on the deck mission mentioned by Colonel Meyer near the end of this report and earlier, too, was a mission over the Pas de Calais area on a day when air-to-ground visibility was minimal and we had difficulty finding targets of opportunity and many of our planes were damaged by ground fire resulting in losses by each squadron. A more detailed account of this low-level mission is provided in another chapter in this book ("Pas de Calais Malaise" by Marc Hamel).

Saved By The Germans: Jule Conard's Story

By Marc Hamel from material provided by Harold D. Simpson

The title of this chapter is something not often seen regarding USAAF pilots downed in the "Fatherland" of Germany. Understandably, the local citizens of the Reich were not too pleased with the RAF Bomber Command and Eighth Air Forces bomber crews laying waste to their cities. In many cases, downed aircrew members were summarily executed on the spot by irate civilians, unless a German soldier successfully intervened. So this story reads a bit different, and we'll see the good side of human nature in trying circumstances for a change.

Jule Conard poses with his "Spirit of Lufbery" (Sheldon Berlow)

1st Lt. Jule V. Conard flew with the 487th, and arrived at Bodney on May 1st, 1944 in the company of Cy Greer, Bud Fuhrman and Bernard Karl (the latter two joining the 486th). Jule was born in October of 1924, and originally joined the Air Corps as a mechanic in January of '41. He was accepted to and went through cadet training, completed flying school, and ended up flying single-engines, transitioning through Goxhill in the UK before arriving in East Anglia. Jule was, of course, under the command of J.C. Meyer, and often flew wingman to Major George Preddy.

On July 16th, 1944, Conard was flying wing to Preddy on an escort mission to Munich. The force consisted of 1,087 heavies and 712 fighter escorts. At about 10:30 gremlins appeared, and Preddy's Encounter Report details the bad news, "I was leading the squadron with Lt. Conard flying my wing. We were escorting B-17's on a penetration to Munich and ran into high solid overcast with tops at 29,000. We were at 30,000 in the vicinity of Aalen when Lt. Conard called that his engine was not functioning properly and that he would have to return to base. I called that I would go with him and made a turn to the right to pick him up. He had fallen far behind and lost altitude. He called again and said his engine had stopped completely. At this time he disappeared in the overcast. I told him to let me know what happened and he acknowledged the call. After waiting ten minutes I called him again but received no response. Nobody heard or saw him after this." Conard was listed as MIA, and some records listed his Mustang as downed by enemy fighter action — which we now know is false. In an interesting turn, Jule was flying Preddy's earlier P-51B mount "Cripes A' Mighty II," serial 42-106451, after George received his justly famous P-51D "Cripes A' Mighty 3rd."

While all this was happening, a 26-year-old German Essingress girl by the name of Frida Kummerle was working alone in the fields of nearby Hofgut Tachenweiler (a farm) near Aalen. She heard the distant crash of a plane, and noticed a parachute descending into the dense, remote forest of the nearby Schwabische Alps. This area is now a thin pine forest, but at the time was a near-impenetrable tree nursery. She immediately ran to her father Erich Truckenmuller, who gathered a search party and plunged into the Schmaler Hau forest. With Frida's directions, they soon found Conard in dense undergrowth with his

George Preddy's "Cripes A' Mighty II" P-51B serial number 42-106451 in which Conard was downed.

Carleton "Bud" Fuhrman (in Red Whinnem's "HMS Hellion" jacket) confers with his flight classmate and fellow 486th pilot Bernard "Barney" Karl (Bernard Karl)

leg broken. Erich quickly returned to the farm, hitched up the Brener Wagele (wagon), and headed back into the forest. They splinted Jule's leg with wood and strips of his parachute and took him back to the farm. There, the German Land Police ascertained his identity as Jule Vernon Conard, and preliminarily interrogated him.

Erwin Hafner was 12 years old at the time, and belonged to an illegal Catholic youth group (religious groups were forbidden by the Nazis). Simultaneously he participated in the Jungvolk HJ Youth Group, and was on a day trip to Bartholoma. The group had gathered at Langert, and there had already been an American air attack. Erwin allows, "We hiked up through the forest in the direction of Aalbaumle and then on in the direction of Tachenweiler. Suddenly we heard the noise of long-range bombers along with the repeating fire of automatic guns. Screaming airplane engines were heard. No question – an air battle was taking place above us. We took cover under some trees for a while. Due to the low cloud, we only briefly got a clear glimpse of an airplane as it broke through the clouds and disappeared behind a forest. Then a loud explosion and a dark cloud.

Excited, we continued in the direction of Tachenweiler. 'Up there, a parachute!,' was suddenly shouted from one of our group, and we all saw the pilot coming down to a landing. No question – It must have been the pilot of the airplane that had just crashed. When we finally arrived at the farm, a number of excited people were seen coming and going. Volkssturm (Peoples Army) with rifles slung over their shoulders were searching for the pilot who crashed. Suddenly, a girl yelled 'He's up there in that dense forest and is wounded.'

"A wagon loaded with straw was hooked up and together with the Landjager (Land Police), Karl Lang, departed to capture the pilot. We followed them into the wood, and after a time came upon the pilot. Since the pilot had broken his leg in the landing, a wooden splint was made and wrapped around his leg with pieces of his parachute. Fascinated, we stood around and saw for the first time a living U.S. pilot. He was literally shaking all over like leaves on a tree, understandable after the air battle and jump into foreign territory along with a serious injury. Regardless, the Land Police took on a very threatening attitude. One of them said something about hanging him from the next tree, saying 'These Amis (Americans) are destroying our cities and killing our innocent people.' With our limited school English, we sought to calm this very scared pilot, who sensed his life was at stake. Land Policeman Lang conducted himself properly. He did what police need to do. He recorded the pilot's particulars and made a first interrogation – 'Name? Age?' Our group leader George Gartner translated. Spontaneously the pilot pulled on a chain around his neck with his identification disc. We thereby saw something that we could never forget: the name Jule V. Conard. The next question, about flying his mission, was not answered by this brave young officer, because it was forbidden. Contrary to the crowd standing around, the Land Policeman accepted this answer, but not his statement that he was alone in the aircraft. Only after we explained to the policeman that it was not a bomber which crashed, but rather a single engine fighter, was he convinced.

Somewhat calmer, the pilot asked, 'Can I smoke?' As the policeman nodded, we saw what we would later learn in the Occupation to be the renowned American cigarette 'Camel.' Jule Conard opened his leather flight jacket and took out his pack. After our assuring words that nothing would happen to him in a prisoner of war camp, we said goodbye to the American prisoner and marched off in the direction of Bartholoma."

Erwin Hafner adds, "As the word got around the area, Erwin Schanzel and Marita Beyerler (both of Essingen) both recall that an entire procession had formed in the direction of Tachenweiler to see the American prisoner. He was at that time laying in the barn awaiting transport from the area. The crowd unanimously expressed that they were only able to find the pilot because Frida Kummerle was able to so accurately describe the exact landing place. This is clear to anyone acquainted with the forest which extends from Tachenweiler to Keonigbronn. Had Jule Conard landed with his parachute in this region unnoticed it is likely that he would not have been found for a long time…if at all. He was unable to get out of the forest by himself, and his shouts for help would have likewise gone unheard. The forest over the war years was unattended and deserted. In this respect, Frida Kummerle must rightfully be considered and recognized as the true rescuer of Jule V. Conard."

Cy Greer (center) with his crew of Lilli Marlene. (Cy Greer)

Conard was then transported to Luftwaffe Fliegerhorst (Air Station) at Crailsheim and was admitted to the Reserve Hospital for treatment. The Germans transmitted through the Red Cross a report named, "The Report On The Capture Of Members Of Enemy Air Forces For The American Prisoner: CONARD Jule."

It filled in a few details with, "The prisoner CONARD will be transmitted on 18 July 44 to Oberursel. Aircraft Type: Presumably Mustang. Plane has been destroyed so completely and burnt that recognition is impossible. The pilot has been captured on 16 July about 11:00 and has been admitted to the Reserve Hospital Crailsheim with fracture on right lower leg." Conard was then typically sent to the Dulag Luft interrogation center at Oberursel near Frankfurt. He was then imprisoned in an as-yet determined Stalag Luft for the duration of the war.

"We Three"
Ed Zellner: Pilot, Evader, Partisan
By Marc L. Hamel from interviews with Ed Zellner

Edmond Zellner accomplished many feats while flying with the 328th Squadron during World War II, not the least of which was destroying 3.25 aerial and 2 in ground attacks. As though this was not enough, this aggressive pilot successfully evaded capture and participated in partisan sabotage activities after being downed by flak in the summer of 1944. Today Zellner points with pride the fact that the Group Commander, Joe L. Mason usually designated Ed to fly his wing on missions. As can be seen from his victory tally, he also made the most of any opportunity in the "shooting" position.

Zellner pulled all of his fighter training together in the spring of 1944, starting with a four-way shared victory over an He-177 bomber (listed as a Focke-Wulf Fw-200 bomber in some documents) on March 22nd. His unit was still flying the rugged and well-armed P-47 Thunderbolt at the time. On May 19th he added a "probable" victory over a Messerschmitt 109 fighter and a confirmed shared victory over another Me-109.

Ed Zellner with "Needles" and "Ratsy," the spaniels. (Ed Zellner)

Of the former Zellner noted, "I chased him from 26,000 to 10,000 feet giving him five or six long bursts. There were many hits around the fuselage and cockpit, leading me to believe the pilot was killed." The plane went down in an uncontrolled spin, but confirmation unfortunately was not forthcoming. The day's second victory occurred after he joined a green-nosed (359th FG) Mustang upon being separated from his Squadron. He detailed, "As I was alone I joined him for mutual protection. The E/A (enemy aircraft) took evasive action and got in my line of flight so I fired upon him. There were many good hits about the wing root and fuselage and as my guns stopped the pilot bailed out." Ed had by this time transitioned to the longer-legged, nimble Mustang, and was flying his assigned P-51B-15 coded PE-U and named "WE THREE."

Two days later on the 21st he shared in the strafing destruction of several locomotives and burned two Junkers Ju-86's on the ground at a German airdrome. Following up, three days later Ed ripped an Me-109 from the air after pursuing it down to 2,000 feet. After receiving a telling burst from Zellner, the 109 flamed and went in vertically.

On May 27th Ed was again on the aerial scoreboard with an Fw-190 victory (shared with

Major Storch of the 357th FG), and added an Me-109 confirmed kill as well. The 190 at low-level was forced into the ground by the Zellner/Storch pairing and exploded. The 109 was one of a group of fifty attacking the bomber formations that day.

Ed Zellner in his P-51B "We Three" (Ed Zellner)

As Ed reported it, he "...got on the tail of an Me-109 and using short bursts fired at him until he went straight into the ground and exploded." On all three days his faithful 'WE THREE' carried him to victory.

In addition to escort missions, the 352nd FG's renowned blue-nosed Mustangs were involved in intensive support missions for the invasion of Europe during the early summer of '44. Ed received a new "WE THREE," a D-model with the bubble canopy, during this same period. Zellner narrates, "On the 30th of July, after the D-Day invasion, I was asked by Lt. McCarthy to fly for him. He wasn't feeling well or something. We were out over Germany and one burst of flak came up and shot the spinner off of the propeller, with one or two bursts on either side, but I got it home all right as nothing was otherwise really wrong with the aeroplane."

"WE THREE" was not repaired in time for of the next day's mission on July 31st, so Zellner flew another pilot's Mustang. He continues, "I went to about the same place in Germany. This was mission number 89 for me with only two more hours to go in my tour. We were escorting bombers to Munich. After breaking off from our escort duties, we went down to do a little strafing. I came across an Me-109 coming in for a landing, took a few shots at him, and he went right in. I never received credit for this one, as my gun camera was destroyed and I never turned in an encounter report on it.

"After we were strafing for a while, I finally climbed up a little bit to go home. Near Ath, Belgium the first burst of flak that came up hit my engine. I was at about 23,000 feet when hit. I just gave a little radio call to let someone know where I was and what was happening. The next thing I know the Mustang started burning and going down, so it was time to get out. I rolled the trim tab forward to a nose down position, rolled the plane over upside down, let things loose, and just popped out of the cockpit. (This was a procedure that many of the pilots practiced.)

"I heard somebody mention one time that when you are making a free fall you will start noticing the earth coming up at about 2,000 feet. Above that you don't notice any movement of the earth. So, I was kind of laying on my back and looked over my shoulder, but didn't see the earth coming up. Finally, when I noticed that it was, I pulled the ripcord. The chute popped and I had a good opening and a pretty good jolt. Looking down I noticed a German staff car or command car...an open vehicle with four wheel drive was coming down the road. I thought, 'Oh Shit. They've got me.' But they were evidently talking or something and they went right by as I landed 200 yards off the road in a little woods. I let myself down out of the trees, folded up the parachute and everything

Ed receiving some training in lifeboats, information he'd never have to use. (Sheldon Berlow)

into a little bundle, and put it under some brush. I covered that up with some junk around the woods and started running and running.

"I had a pretty good sense of direction, so I was running southwest going for Spain. I ran about as fast I could run and I must have run most of several hours. I was in pretty good shape in those days. I stopped when I ran across a stream, and I scooped up some water in these little rubber bags. I put a Halizone tablet in to purify it and had my little malted milk tablets.

"The next day I was going pretty fast too, but I stayed away from houses and just crossed the road when I had to. I kept going through the woods, and also dug up some potatoes from a field. I filled my flight jacket up inside (the 352nd pilots had 'escape pockets' sewn inside their A-2 jackets); it was just stuffed up and bulged out with potatoes. I thought I would have a pretty good bunch of them for as long as I needed. There was also some wheat I stripped off in some fields, so I filled up my pockets. I was having a pretty good time there, and I walked and walked and walked.

"After I guess three, four, or five days, I walked into what must have been France. I got into these trenches from World War I, but they weren't straight up and down. They were just hills, some of them maybe 10 feet high. I would go up and down, and think I walked through them for a whole day. There wasn't much vegetation, and there was certainly no water. I was starting to feel kind of bad, even hallucinating. I was thinking of a big bottle of beer, and getting thirstier than Hell. I finally got through the trenches and found a little water. That held me for a while.

"I laid down one night to rest, and the ants and things gave me a heck of a time of it. I picked around to find a better place, and fell asleep. I woke up in the morning as it was just getting light, and there was this German sentry box! A German soldier was standing beside it not 50 feet away from me. I thought, 'Holy Hell! What am going to do now?' So I took out my .45, cocked it, and popped him—shot the bugger. I mean I was going to get caught because there wasn't any place to hide. Then I started running along in the direction I was headed earlier.

"All of a sudden these artillery shells start flying around, and I thought, 'What did I do now?'

Ed Zellner in his crusher cap and B-3 shearling jacket usually worn by bomber crews. (Sheldon Berlow)

I hit the ground, then raised up a little bit on a small rise to see. I had walked through an artillery range. They were shooting over here on one side and there were targets over on the other. I thought that if this is an artillery range there must be somebody back there with binoculars. I got down on my belly and crawled for I don't know how long. Boy, just as low as I could get. I bet I crawled for a couple hours and finally got through to the other side. An SS camp was there and they had this artillery range. It turns out that it was one of the German's biggest Army camps or depots and that's why it had its own artillery range. I don't know how they ever explained this guy getting shot, and they certainly didn't know who did it.

"I soon ran into some woods and there was a fellow cutting trees and he had a little two-wheel cart and a horse. I approached him and gave him my little spiel in French about being an American aviator. (Our escape kits included some simple words in French we could use to identify ourselves, ask for food or water, etc). He was very happy to see me, and took me into town. It was called Suippes, on the Marne River between Chalon and

Zellner is displayed on a horse at the liberation of Suippes, France. (Ed Zellner)

Rheims. I stayed with him for a while. The head of the Underground was like the Mayor of the village and the local wine merchant as well. We had a heck of a big time there for a while, a big celebration, and then settled down.

"A couple of days later I started helping these Underground fellows. They would go out with things called 'cliques.' They were made with two nails twisted around each other in a certain way. Whichever way they fell in the road, one sharp point would stick upward. We filled these little mussette bags full of these cliques and went out on bicycles at night with the head of the Underground. Paul Simone was the fellow I was staying with, and his brother was a wild-eyed 'bird' there too. I think the brother was in the Maquis. They would all kind of work in the same way. He and I would go out with cliques and throw a couple of them over our shoulder along the road so the German trucks would get flat tires.

"Some nights we would also go out cutting down telephone poles or power poles. You know, just get an axe and chop the Hell out of a pole until it falls over. You hoped it wouldn't hit or electrocute you! The French were really sabotaging themselves by disrupting their power and everything else. It took a long time to cut down a telephone pole because it was hard wood, the axes weren't great, and I wasn't in that good a shape for cutting with an axe. Then all these Frenchmen would gather around, (there may be eight or nine of them doing this job) and they'd talk and talk. I couldn't enter into the conversation, and was getting a little bit leery of these fellows. I was getting frightened too. You get caught doing that and you get killed. I didn't have a uniform on either, as they had supplied me with some other clothing.

"We also used to go swimming. There was a 12 year old boy there, their son, and we'd go swimming every day in the afternoon when it was warm. We'd go down to the creek, get a little swimming in, and have a nice time. A couple of times these German trucks would be parked along the road as we walked past. They'd be repairing the engine or something, and I thought I was taking a heck of a chance. A fellow my age should have been doing something else other than just walking around. We had some adventures there, as they'd be cursing at you something like, 'Get the Hell out of here!' After that I said we had better just go around the back way to get to this swimming hole.

"One time I was really scared. The local beans had ripened, string beans, and there was a pretty good-sized garden in the back of the house. We picked a lot of beans and the lady of the house cooked them. There wasn't too much else to eat, maybe a few little things, so I ate an awful lot of those beans. That night I had a gas attack or something like it, and I thought I had appendicitis. I could just see me getting operated on for appendicitis in France someplace! Luckily it went away over the night.

"This local wine merchant would also come to visit me and he would bring a little basket of wine and Calvados. The Frenchmen would sit around talking and I'd sit around drinking. Every night I would get a pretty good buzz on. It wasn't too bad, and it helped kill the time.

"We listened to the Voice of America broadcast on this little radio in the evening. They told how far the Allies had been advancing, and I'd mark it up on a little map. When they came pretty close there was lots of 'binging and banging.' All of these Germans soldiers were retreating through the town and it was really pathetic to see them sometimes. They would have a horse with a rope tied on and maybe ten soldiers would be hanging on just to try to help pull themselves along.

"Then the Americans finally came. There was a little artillery barrage, then these French people came up, put me on a horse and started parading me up and down the street of the town. Remember there were still Germans around too.

"A cavalry troop of the 3rd Army finally came into town. I met a captain there in a jeep with his driver. The captain asked if I wanted to go along. They'd get me back to Paris, but they were going east. I went with them for two days, liberating towns at 30 mph, just going through and looking around. I said, 'Hey! There are Germans back there!' They said 'Ah, we know there are Germans back there. Don't worry about it.'

"About this time the 3rd Army was running out of gasoline, and they were having a hard time keeping their armor going. It was a really tiresome thing traveling in a jeep out in the open. Bouncing around and sleeping outside at night is no fun. Then this gas truck came up with fuel so I just got on the gas truck and said, 'Good-bye. I am going back towards Paris.'

"I made my way back toward Paris and ran into this prison camp for Germans. I helped a doctor with operations there as I was in the animal business back on the farm. I helped him with a

Victory photo with Jack Thornell and Ed Zellner standing, and Earl Abbott and Dick Brookins kneeling. (Ed Zellner)

couple of operations while I waited for a ride and finally hitched one into Paris. There I stole an MP's motorcycle. He just pulled up, parked it, and went into a building. I hopped on, started it up, and away I went. I looked over Paris for a couple days then I parked the motorcycle out at Villacoublay airport and hooked a flight back to London. There I had to go by headquarters where they supplied me with a uniform and various sundries.

"Next I went back up to Bodney to visit the fellows and found out my things I'd left had been sent to the port at Liverpool. I made my way up there to some government warehouse, and found a Britisher in charge. He said he wouldn't release my gear as it was supposed to be sent back to the States. I had a .45 in my shoulder holster, so I pulled it out and said, 'You give me my stuff or by Jesus I'm going to shoot you!' He gave me my stuff. They'd asked me back in London if I wanted to fly or take a boat back home. I said I'd rather take a boat as I flew over to England originally. They put me on an aeroplane—just the opposite, but typical of the Army. I left from Prestwick, Scotland, arrived back in Washington, and that was that.

"An interesting aside to the mission where I went down involved my regular wingman, the late McDonald 'Mac' Godfrey. I was an element leader that day, and later I just assumed he was my wingman. Actually he was tagging along with the 486th Squadron on that mission. Until later I did not know that he was shot down on this same mission. I got back to the States, got married, and about a month later I got this postcard from a Stalag Luft prison camp in Germany. It was from Godfrey, congratulating me on getting married. Years later I saw him at a 352nd FG reunion and he told me the tale. They were coming back from the mission, so he let down a little bit and took his mask off to light a cigarette. His engine stopped, he went down, and was a prisoner for the rest of the war! Strange, and funny, sort of."

Ed's former mount PE-U is undergoing maintenance in Belgium in early 1945.

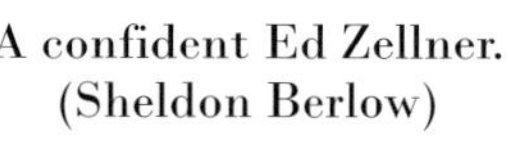

A confident Ed Zellner. (Sheldon Berlow)

American Wartime Interlude
by Wing Commander Edwin H. King, RAF

An American Fighter Group is the equivalent of an RAF Wing. A Squadron is about 30 aircraft and 40 pilots. The 352nd Group first flew Thunderbolts but were re-equipped with the P-51 Mustang in about April 1944. This aircraft was fitted with the Packard-built Merlin engine, basically with the Merlin 61 with a two-stage, two-speed supercharger.

The principal job of a fighter group was long range escort for the B-17 and B-24 heavy bombers. To do this job they needed to carry an enormous amount of petrol. The Mustang, with external tanks, had sufficient fuel to stay in the air, at normal cruising, for about seven hours. It had a radius of action of about 500 miles plus one hour in the target area. It was reasonably fast, about 400 mph, with a climb rate inferior to a Spitfire IX, but quite satisfactory for the job. The diving speed was very good and acceleration into the dive excellent, It could slowly overtake an Me 109. both in the dive and in straight and level flight. Radius of turn using 30 degrees of flap was as good as any German fighter and better than most, but not as good as the Spitfire.

Dapper Ted King strikes a pose by an early Mark Spitfire. (Ted King)

The armament was either four or six .50 calibre machine guns with between three and four hundred rounds per gun. This armament was nothing startling, but was sufficient for the task and superior by far to the .303. The ammunition was armour-piercing except for an odd round or two of tracers near the end to indicate that you've "had it." The guns were mostly trouble-free and gave a good account of themselves when used in anger.

Escort sorties were usually flown between twenty and thirty-five thousand feet with the bombers at about 25,000 feet. The aircraft was fairly comfortable to fly at that height, trim being superb. The oxygen system using the mask was good and better than the RAF issue. Heating was satisfactory and one only needed flying overalls over one's uniform, but the two pairs of sox worn inside the RAF escape boots was advisable. The cockpit was roomy and fairly comfortable. However, one's behind took a beating since it was difficult to sit comfortably with the compressed-air bottle getting in the way. We sat on top of the dinghy and I quickly obtained one of the RAF water cushions which had recent been introduced to put on top of it. This had the advantage of an additional water supply for use in an emergency. The parachute in use was the very comfortable USAAF back-type. There was no provision for a "C" suit.

The radio fitting was an American-built VHF and very satisfactory. Selection of frequencies was by button press arrangement with Button "A" as the Group frequency for airdrome control and local homer. "B" was for Air-Sea Rescue contact; "C" for fighter-bomber communications and "D" was to get a position fix. Everyone flew using "A" channel on missions with one pilot in each squadron designated to listen in on "C." For long range homing one would call Wing Controller on "A," go over to "D" and "pip-in" on the contactor (like an RAF squawk), then back to "A" for the steer. A reciprocal steer could be received at over 200 miles at 20,000 feet. An arrangement such as this was essential when returning from deep inside Germany, perhaps getting low on fuel and not too sure of one's position after many hours above cloud, many times 10/10ths.

Each squadron had its own individual call sign with the Group Leader known by a single word, say "Topsy." The lead section consisted of White & Red flights, the other sections Yellow & Blue. The squadron normally flew as 16-aircraft plus one or two to fill-in if someone aborted, returning to base if not required.

FORMATIONS

These were very similar to ours. The squadrons formed up in flights in line astern, aircraft in finger-four, with No. 2 man on the right and 3 and 4 on the left. This is a tight formation and not difficult to fly. Squadrons formed up with the Group Leader in the centre and the other two squadrons on each side in line abreast. This formation was used until the enemy coast was reached. A tight formation is good for moral, an important point on long, rather lonely missions. It also necessitates little throttle movement and jockeying for position.

As the enemy coast was reached the Group split into battle formations. White and Red flights positioned their aircraft line abreast about 100 yards between aircraft and a comfortable distance between flights. Yellow and Blue Flights opened out in the same way and stepped themselves up in the same way on White and Red, about 1,000 feet down sun. Squadrons within the Group were in line abreast on the Group leaders with one stepped up 1,000 feet up sun and the other 1.000 feet down sun. This formation covers an enormous amount of sky but it did not matter when on a sweep or en route to join the bombers. All turns were the normal cross-over turns.

Over Europe the weather was so often lousy that this was a good formation for taking a group through clouds and it worked well. If the cloud was thought to he thin enough the group climbed through it as squadrons with flights closed up tight in line astern, with the squadron leaders being the only ones on instruments. The Group leader always reported via R/T when he was breaking out on top so the Group could quickly form up again. This procedure obviously required a high standard of instrument flying on the part of the leaders and good formation discipline from the pilots. This was well within the scope of any RAF or RN squadrons.

If the cloud was low on takeoff, say 200 feet, each flight climbed out straight ahead until breaking out above cloud, the leader keeping on his course for two minutes, then making a port 180 turn, the Group forming up on him as he came back.

BOMBER FORMATIONS

A large bomber force of say, 360 bombers, would be split up into ten combat wings of 36 aircraft. The combat wing was the basic formation and bombed from this formation. These combat wings flew in line astern about two to three miles apart so the escort problem was not so much width as length. Since the relative speed of the bombers and fighters varied greatly, the fighters had to escort in relays. A group would fly directly to a rendezvous point along the bomber route and stay with tbe bombers until relieved by another group or as long as possible, then return to base. However, if their fuel and ammo supply was sufficient, the fighters would often go down to attack targets of opportunity en route home.

The long line of bombers was protected not only by the fighter groups assigned for their protection over assigned areas of the route for given periods, but by other fighters en route to or from their rendezvous with other wings of bombers.

The proportion of fighter escort to bomber at any one time seemed to be about one fighter group to three combat wings. The Fortress and Liberators cruised around 160 mph indicated at 25,000 feet and Mustangs usually about 200 to 250 to 270, so a certain amount of weaving was necessary for the fighters to remain with the bombers. Once the rendezvous point had been reached the group leader gave his orders by R/T as to the disposition of the squadrons. If the combat wings to be escorted were at the front of the bomber force, he took his own squadron ahead of the force and swept the area some ten miles in front.

Enemy fighter attacks were often made from head-on as, amongst other reasons, if they missed the front box they could go on to the second or third. By having a squadron ahead it was sometimes possible to break up a formation of enemy aircraft before an attack developed. The second squadron would most likely be split into sections to cover the flanks of the bomber wings and the remaining squadron would serve as top cover for the group, flying some three to four thousand feet above the bombers.

All friendly fighters tried to stay below vapor trail height. Any pilots in the group could report sightings of enemy aircraft (bandits) or unidentified aircraft (bogies) to the leader. If the leader picked them out and there was time, he decided which flight was to attack and gave the necessary orders. If, however, he either could not see them or timing was paramount, the pilot reporting, irrespective of seniority within the flights, could go down on the bounce, provided his leader was made aware so as to give him cover. Things obviously happened so quickly that this was necessary.

Normally, those in the No. 1 and 3 positions in the flight were the initial shooters with Nos. 2 and 4 covering them. It was important that these wingmen not get separated from their element leader. Their job was to protect their leader whilst he was firing. If, however, he overshot, then number 2 would become the shooting man and his leader the cover, i.e., with only two men in a flight as attackers. This was an excellent ruling as the shooter could concentrate on the attack knowing that his back was covered. Good aircraft recognition was of prime concern to all and perfection was the aim.

The tendency whilst escorting was to sweep out from the bombers and try to pick up any big enemy formations as they were maneuvering for an attack position. In the early days they attacked from different directions in small numbers and it was necessary to stay close to the bombers but tactics were constantly changing to match the enemy's innovations. Considerable radio help in locating enemy formations was received from the UK listening posts. A Group leader would be passed information on their activities.

OPERATIONS

I flew with the squadron on fifteen operations, on occasions as Major Preddy's No. 2 or No. 3 and as I gained more experience I became the leader of one of his sections. Through my association and flying with Major Preddy, I came to admire his flying skill and his care and dedication for his squadron. I always felt extremely confidant when in his company and in comparing him with RAF leaders must put him at the top of my list alongside Squadron Leader O'Meara. When attacking an enemy he showed his perfection by only using sufficient ammunition to achieve the kill. This was borne out when he had six kills in one operation. Regretfully I was not flying that day and so missed this epic achievement.

On returning from my first operation we went first to Squadron Headquarters where we were met by the Flight Surgeon who offered us a drink of whiskey (Bourbon or Rye). This I found to be SOP (Standard Operating Procedure). Not having tasted bourbon whiskey before, I opted to try it and clearly remember soon feeling quite "squiffy..." In my defense it must be remembered that I had just completed nearly five hours in the cockpit and it had been some eight hours since I had last eaten. After the drink we were de-briefed by the Intelligence Officer on our own efforts, and on what we could report about the bombing results.

We then stood down and either proceeded to the mess hall for food, followed by a wash and brush up before meeting at the Officers Club to partake of more alcoholic beverages or we returned to Squadron Headquarters to view the previous operations combat films and discuss our tactics, formation flying etc.

OPERATION HIGHLIGHTS

Ramrod 451—July 17, 1944 "This was my first operational sortie in the P-51D. I flew as No. 2 to George Preddy on 'withdrawal support' for B-17 Fortresses which had boomed marshaling yards at Belfort near the Swiss border. There was quite a large response by the enemy. Whenever we moved towards aircraft coming within the vicinity of the bombers they rapidly departed, so the sortie was really uneventful. On the latter part of the return we had to give close support to a Fortress which had a wing and the fuselage damaged and one engine feathered and it was straggling badly. The crew of the bomber fired on us whenever we got too close. However, we stayed with them until they reached the English coast and then we parted company."

Ramrod 456—July 19, 1944 "This was interesting in that it was again a 'withdrawal support' operation in which the bombers were attacking 'Buzz Bomb' factories. We met them at the target, taking over from other friendly fighters. The 17s appeared to have been very accurate on their bomb run with an excellent pattern on the ground of exploding bombs. There was little or no opposition."

Ramrod 458—July 21, 1944 "A most exciting 'Penetration Support' sortie. We were ahead of the bomber stream and so were in a good position to intercept a flight of twelve 109s to which we were directed by the Controller from the UK. We were told that the 109s were forming up line abreast above a lake just to the SW of Munich, preparatory to a head-on attack on the bombers. The enemy were at 35,000 feet and we closed up on them from behind and below climbing up from 30,000 feet, aiming to be unseen as long as possible—I was again flying No. 2 position and managed to get in a short burst of some 362 rounds at the 109 alongside the one Major Preddy had selected as his target, this just as they broke downward. George, closely followed by myself, then chased his 109 to ground level, gradually closing up with it. In the descent we reached speeds well in excess of the permitted 'red line.' I got my first experience of the effect of 'compressibility.' After a hair-raising chase around trees, I managed a short burst at the 109 and saw strikes on the wing, then still following George, I saw strikes from his guns on the fuselage. The pilot then zoomed upward, jettisoned his hood and bailed out. His parachute opened successfully and his aircraft crashed nearby. Regretfully my aircraft was damaged and on return had to go for repair. It failed its air test, and so I flew a variety of 51's until it came back on line shortly before my tour with the Americans came to an end."

Walter Smith, the RAF's Ted King, and Malcolm C Pickering by "Ye Olde Lager Inn"—a play on the words "augering in," referring to a vertical plane crash.

Ramrod 483—August 5, 1944 "We were to escort the bombers to the target area near Magdeburg joining them as they coasted in at Den Helder which meant getting our height over the UK. We could hear the bombers climbing out whilst we were still at briefing and since the weather war poor we had to climb in sections through cloud from 600 to 18,000 feet, joining up as a squadron and then as a group on top. Over the continent the cloud began to break up and there was considerable opposition almost all the way in. I was again Number 2 to Major Preddy and watched him get two Me 109s (confirmed). I only got a squirt at one. After leaving the target area we were relieved and went directly back to the UK without further action except for the flak."

FIELD ORDER 490 — August 7, 1944 "This was a sweep in the Paris area and and I was given the privilege of leading a flight. There was little or no activity. We were detached to look at a pair of unknown aircraft well below which turned out to be friendly RAF Typhoons. I then had the pleasure of leading my flight back to base flying directly over London and successfully finding Bodney."

FIELD ORDER 508 — August 12, 1944 "This was a change from the normal in that it was a planned 'skip-bombing' attack on the marshalling yards at St. Quentin followed by ground attacks on motor transport on nearby roads and shipping on the river Olse. We each carried two 500 lb. bombs under the wings instead of the usual drop tanks. The attack was at ground level. In squadrons line abreast the bombs were dropped on the order of the leader. I claimed to have two German lorries and two barges, a much shorter mission than usual."

RAMROD 522 — August 18, 1944 "This was a most eventful close escort sortie. I was leading a section of four originally, reduced to three when one aborted shortly after leaving the UK coast and after the spares had returned to base. We were flying just north of the bomber force at around 30,000 feet with considerable enemy activity to the south looking like swarms of bees. We were instructed to protect the northern area but allowed ourselves to be jumped by Fw-190s, our first knowledge of them being when they came down on us. We broke to port into the attack. I was still feeding from my drop tanks and executing a maximum rate turn with combat flaps to avoid enemy fire and my aircraft flicked over into a spin. By the time I got rid of my tanks and recovered from the spin I had lost much height and was now at around ten thousand feet and on my own, my number two not having managed to stay with me. When I had climbed back to a sensible height I found myself still alone, well north of the Ruhr. Initially I set course northwards then northwest aiming to cross out to sea in the area of the Waicheren Islands. Since no one seemed to take any interest and as visibility was poor I called for radio assistance from the homing facilities, receiving steers for the UK. I eventually arrived back not having been troubled other than by some fire from Walcheren Island. It transpired that my colleagues were more fortunate, having dropped their tanks and so were in a better position, but were far too occupied looking after themselves than to follow me and assumed that I had been shot down. Much alcohol was consumed that evening."

RAMROD 529 — August 25, 1944 "I cannot remember the route, but we were sweeping ahead or the attacking force in the vicinity of Hredstedt, Maashlm, St. Kirksby, H Basznitz and Anklam. On being relieved, we had fuel and ammo to spare and so carried out a ground attack on Neubrandenburg Airfield. During the strafing attacks, about five of them, considerable damage was done in spite of heavy AA fire. I was flying as Red 3, initially as top cover over the airfield. Red Section was then permitted to take part in the last strafing run before departing for base. I claimed one Me-109 damaged. Considerable flak was still evident, but there was no enemy aircraft intervention. My aircraft only received two hits, one through the spinner and the other behind the cockpit. This was the first mission that the German jet aircraft were reported hut I haven't seen one yet myself."

OFF DUTY

Having a private car I was permitted to use it on duty and to draw petrol from the MT Section. This supplemented the normal ration and allowed for some social journeys. When not rostered for operations or training and since I did not take part in the unit duties, there was spare time so Dick Law and I visited some RAF bomber bases nearby where we had friends. I twice managed to visit my home and on a couple of occasions I was given permission to use the Group's Oxford aircraft. One such occasion was when the Group went on the Russian mission, which, despite many phone calls,

Dick Law and I were not given permission to take part in. So, we went to Culmhead to look up 131 Squadron.

The Officers Club at Bodney was a war time building which the squadron personnel were 'very busy doffing up' and when completed we we had a celebration party. Coaches were sent to near-by towns and hospitals to collect any ladies who wished to come to the party. Sorties took place to find the mwh needed grog and food was provided from the cookhouse. A band was found and the party on Saturday went on all night. As long as there was anyone wanting to dance the band played on. It was arranged for the last coach to leave early that morning, but there were many of the ladies still in camp at breakfast time.

There was no operation on that Sunday, and calls went out over the 'tannoy' stating that 'all women will be off the camp by mid-day.' Nevertheless there were still some in camp when the MPs went around checking up on them. This happened again almost every weekend and having been forewarned, Dick and I took most weekends off if not rostered. We were both given PX cards which allowed us to purchase cigarettes and chocolate and other items from the shop, a much desired favor, especially when goods were taken home for the children.

Much gambling took place in the Club—craps on a special table, bridge and other card games. I joined a bridge game until I found out the amount of the stakes. What I thought to be per hundred turned out to be per point! I then had to drop out as one British pound per point was most often the stake and well beyond my pocket. On the flight line we played volley ball, but however much I tried, I could not get my colleagues interested in cricket.

END OF TOUR WITH THE YANKS

The 30th August 1944 FO 542 saw my last mission with the 487th Squadron and coincided with the return of Lt Colonel John C Meyer and the departure on leave back to the USA for Major George Preddy. Although I had enjoyed my attachment and had learned a lot, I found it hard to come to terms with witnessing these large bombers being shot down and, on so many occasions, being unable to prevent it. That we stopped the slaughter being even greater is without question, but it still hurt.

Ted King in the late 1990's. (Bill Espie)

I came to admire and respect my American colleagues as I did my fellow RAF pilots, some of whom were natural pilots whilst others had to work hard to keep up the high standard, particularly in their formation flying. I was not impressed by one particular pilot whose first action on coasting out from the continent was to light up a cigar thereby letting his attention wander to the peril of all.

And so I left my American colleagues to return to 131 Squadron, a much wiser and more experienced pilot, knowing that the 352nd Group would shortly be moving forward with the advancing armies to continue the fight. Over the following years I received messages of 487th's progress and was devastated to hear that Major George Preddy had returned to the Group from his home leave only to be shot down by "friendly fire," and American fire at that, whilst the armies were busy crossing the Rhine.

By then he had become the highest-scoring 8th Air Force Mustang pilot.

From the Zenith to the Deck
by B/Gen. Francis Griswold

In the Fall of 1944, B/Gen. Francis Griswald, Commanding, had his staff compile a simple, lithographed study of 8th Fighter Command's participation in the tactical phase of the Battle of France through the eyes of some of the top Group commanders and outstanding pilots of the 8th. This treatise was not only a tribute to the courage and aggressiveness of the 8th's fighter pilots, but a combat manual for the fighter pilots who were joining the fray. His introduction to this study follows along with the report of Col. Joe L. Mason, initial commander of the 352nd Fighter Group.

P-51B and D Mustangs of 328th escort majestic B-24's of the 458th BG, 753 BS.

The first priority of our fighter groups is escort of the Fortress and Liberator heavy bombers. It is long range, five…six…seven hour stuff, to Berlin and back, to Poland and back, to Russia, and all of it at high altitude. Yet, on August 12th, at the height of the Battle of France, these same groups flew 45 missions comprising 1,326 sorties on the greatest ground attack on record. Blanketing German supply lines north and northeast of Paris, our pilots bombed, burned and riddled with fifty caliber API (armor piercing incendiary ammo) some 2,616 railroad cars, 359 locomotives, 112 ammunition cars, 464 trucks, 362 oil cars, 9 oil tanks, 9 oil barges, 306 vehicles, 15 bridges, 7 roundhouses, 13 buildings, 4 water towers, 19 aircraft and many other targets. It was a big day, but not unique. On the 13th it was duplicated, and since the first part of March of 1944, the 8th Fighter Command has in this strike mode been burning the enemy's deck.

The privilege of ground attack by fighters, if an operation so difficult can be called a privilege, must be won in the air by defeat of the enemy's air forces. This defeat in a long series of culminating battles in which this Command played a decisive part, has been described as "The Longest Reach" and is already a part of the rapidly unfolding history of our Air Forces. Immediately our fighters pursued the grounded Luftwaffe to its airdromes and there crippled it beyond hope of significant recovery. Not only were our bomber missions freed from more than sporadic attack by this victory, but the enemy's entire transportation system was forced to depend upon, for its defense, flak and machine gun fire from the ground. As severe as this has been and despite the losses we have suffered, this defense has never been enough to deter our fighters in their determined attack.

Enemy locomotives, freight trains, troop trains, truck convoys, barges, oil tanks, ammo dumps, coastal vessels, power houses, bridges, staff cars and communications have been victim of a hitherto unimaginable and unparalleled fighter ground assault across the entire face of Western Europe. To the Command's achievement of 4,009 enemy aircraft destroyed, 283 probably destroyed and 1,339 damaged can now be added the destruction and serious damaging of rail and road transport by the thousands of locomotives, railroad cars and trucks on such a fantastic scale that to put it into figures fails to paint the picture. The real meaning is that from long before D-Day, through it and to the present, the German supply and reinforcement has been destroyed or impeded to a point incompatible with the support of modern armies.

Since the beginning of this war the profit and loss on the proposition of fighter aircraft attacking ground targets has been the subject of professional debate and pilot discussions. Small profit to shoot up two or three trucks or a couple of machine guns for the loss of a valuable pilot and aircraft. Worse still, when two, three or four go down over one well dispersed enemy A/D or, as on the days of our large scale attacks by the whole Command, 25 or more may be MIA. In addition to the loss of these planes and pilots is the unfortunate fact that our best, our outstanding leaders and fighters who had yet to meet their match in any enemy they could see, have gone down before the hidden gunfire or light flak incident to the ground attack. (Men like) Duncan, Beeson, Beckham, Gerald Johnson, Gabreski, Juchheim, Andrew, Hofer, Goodson, Schreiber, Millikan, Carpenter . . .the list goes on. For equal numbers engaged four times as many pilots of this Command are lost on ground attack as in aerial combat. Light flak will ring an A/D or a M/Y. Flak cars open up in the middle of a train. A truck convoy, with sufficient warning, may be a hornet's nest. Every target of special value to the enemy will be heavily defended and may exact its price.

Where then, is the profit? The answer is the successful invasion and the victorious Battle of France. The answer is our flight of many a heavy bomber mission without challenge by enemy fighters, and the presence of our hordes of bombers and fighter-bombers over our troops in Normandy. The roads of France, strewn with enemy wreckage, reply. And an enemy starving for oil, ammo, supplies, re-inforcements could answer with deep feeling. The loss of every single one of our pilots is an individual and personal loss to us but the harsh voice of war says clearly that had the entire 8th Fighter Command been wiped out in the course of its tremendous ground attack, the cost would have been well spent towards the purchase of mankind's victory.

In an all-out war such as this, a successful air operation must pay the most and must cost the least. With fighter ground attack as with other operations, experience has taught many lessons leading to this desirable end. It is hoped that this bitterly gained experience may not be the only possible teacher and that the recording of it in such publications as this, added to the instruction by those who have been through it, will point the way for replacements and new groups in this and other theatres, for pilots still in training, and/or those pilots and leaders of experience who have yet to encounter this special type of fighter duty. Like the "Long Reach," Down to Earth" is a message from the battle at its height, told in their own words by the men who fight it.

B/Gen. Francis Griswold, Commanding

Remarks of COL. JOE L. MASON, C.O. 352ND FIGHTER GROUP

1. Ground attacks on enemy airdromes.

When the mission calls for shooting up a specific airdrome, you should obtain al the information available about the target—flak maps, photos, etc. and look at it on all map scales. You should know what you are going to shoot at before you shoot—if it is possible to know. Flights should be assigned specific targets on the field at specific times. This sort of planned attack works nicely if everything goes according to plan.

First, we work with a few basic fundamentals, namely: we do not feel that losing one of our aircraft just to destroy one E/A on the ground is a good swap. In other words, unless otherwise ordered, we will not attack an airdrome unless we know there is plenty of stuff there to shoot at. We do not aim to pay for each victory with one loss.

After we have decided the target is juicy enough, we go for it, but not knowing how much flak we might encounter we make an effort to split it as much as possible.

This is done as follows: In most cases the German knows we are around but he doesn't know what we are going to do. So we give him lots to see in the way of diversions.

We have one squadron circle above the field at 10,00 to 12,000 feet, just high enough to be poor targets for their light flak. While this is going on the other two squadrons have moved off to hit the airdrome on the deck from two different directions, making only one pass. This fundamental is simple. The Germans have only so many guns; the more moving targets going through their gun range, the less chance he has of hitting any of them with the high squadron making dive bombing feints just prior to the lower squadrons hitting the field, many of these guns are diverted upward and our attacking squadrons can make their pass and get away before those guns can be brought to bear on them. This plan has worked well on a number of occasions with the high squadron also serving as top cover for the attack.

EDITOR'S NOTE: In these attacks, if the high squadron still had their empty wing tanks on, the simulated dive-bombing feints would include dropping these tanks. And, attacking aircraft made every effort to stay under 50 feet as they crossed the field shooting at their selected targets since the German gunners ringing the field would be shooting at each other if they lowered their guns below that, thus providing an "arc of safety" for the attacking planes at that level.

2. Ground attacks on transportation.

On convoys we use the same principle. We know there won't be as much flak as we encounter on an airdrome, so we try to make these attacks from the up sun side and a whole squadron line abreast all shooting at the same time. The second squadron follows the same procedure while the third squadron provides top cover from enemy fighters. If minimum ground fire is encountered, the attacks are repeated until all the vehicles are burning.

On train attacks you always have the factor that a disguised flak car may be in the train of cars and they can be really mean. We lost two pilots to them. It depends entirely on what kind of train you attack as to your procedure. Troop trains are

Joe Mason receives the Distinguished Service Cross.

juicy but will always have one or two flak cars that must be taken care of. Tank cars burn nicely. We started the idea of dropping half-filled wing tanks into the sides of the wooden freight cars and then setting them afire with our API ammo. It works nicely. The tanks split open and spill gas all over the place, making it easy for the next guy in to ignite it, but only works on targets that will burn.

3. Ground strafing

The greatest tendency when shooting ground targets is to waste bullets. When making an airdrome one pass attack, I think it's OK to start shooting way back and shoot all the way through You might waste amm but all these bullets flying around slightly disrupts the ground fire shooting back at you and your fellow pilots.

On trains and convoys when you encounter no return fire, you must make every bullet count. The ammunition is belted with five tracer rounds just fifty rounds from the end of the belt to indicate you are about to run out of ammunition. Our rule of thumb is to not continue shooting at ground targets when you see these tracers. We lost some possible victories over some 109s one day because none of our pilots had any ammo for attacking them. And, good shooting is good shooting regardless of the target and good shooting kills Germans.

4. Bombing with fighter aircraft.

On specific assigned targets I think it's OK. I am partial to dive-bombing as I think it is as accurate as any. Anything less than 1,000 pounds is not much good on bridges. On all the bridges we've bombed we have only been successful in dropping one span.

I'm sorry to have to admit that, but it seems to be the truth. Bombing with a fighter aircraft is 100% personal pilot skill, and it's like playing basketball—the more you practice the more baskets you sink. We have not had the time, ranges or equipment to practice fighter bombing to even approach the degree which could be attained. We are basically an escort outfit and in that we have had sufficient practice to come closer to perfection. The scoreboard shows that. You can mess up a lot of railroad by lying straight and level and dropping one bomb at a time. If done right, one group can break the tracks every ¼ mile for about fifty miles. That should drive the Hun nuts trying to fix it; it's not a permanent injury but I'll bet it makes him mad as Hell. A good fighter-bomber pilot can hit his target from any dive angle but it takes practice.

5. General

When flying on ground missions at medium and low altitudes you are open for attack from both above and below. To get caught by the Hun is plain dumb. We always make one squadron responsible for the top part of the sky. They don't like it because they don't get to shoot much unless we find so many ground targets that we can swap top cover squadron in the middle of a mission when the other two are low on ammunition. But we try to rotate the squadrons' positions so they all get a crack at it.

My boys will tackle any target with pleasure if they know that, if the mission is completed the right way, it will help those boys on the invasion beaches.

A fighter pilot who doesn't want to shoot his guns is no fighter pilot. I continuously have to warn them about non-military targets and the angle they shoot so it will won't kill the French people.

The aerial photos of targets are not much good either because they are mostly too old. But the job of supplying new photos for all enemy A/Ds every week would be impossible. We do the best we can, identifying the fields by nearby landmarks.

We have not run into a truck convoy that was so big that one Group couldn't handle it. The largest was about 26 trucks and we had 22 of them burning when the 56th FG came in to help us finish them off. Incidentally, I saw no German leave any truck from the time I spotted the convoy and it was moving when I first saw it. We must have left a lot of dead Germans there.

In shooting up barges I think they should be targeted according to what higher command says their cargo is and they are supposed to know such things.

I do not think the boys are fully aware of the services they are rendering by these ground attacks. There is no way of assessing or scoring the damage they do.

Shooting up convoys and especially staff cars and dispatch riders is considered good sport.

I think a good poop sheet on the effect of the fighters on that beachhead would be a good thing.

Col. Joe L. Mason, C.O.

Ed Heller, 486th, in April 1944. (Sheldon Berlow)

Glen Moran relaxes in this PR photo. (Sheldon Berlow)

Lt. Colonel J. C. Meyer (Sheldon Berlow)

Mustangs of 328th escort 458th BG, 753 BS Liberators.
"Punchy" Powell's "West by Gawd Virginian" is the lead escort ship.

"Strafing – a gun camera view."

BARGES

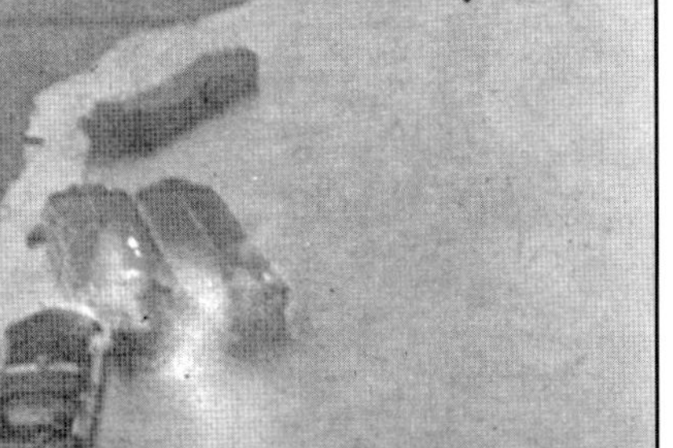

ARMORS

TRAINS

TRUCKS ON D-DAY

"A War Correspondent Gets Initiated"
by Bob "Punchy" Powell

When I completed my two tours as a combat pilot with the 328th Squadron, 352nd Fighter Group in late August, 1944, for some reason or other I was not given orders to transfer back to the States, but got assigned to HQs, 8th Fighter Command, as a Technical Advisor to the Public Relations Office there, possibly because I had newspaper experience. My job was to check publicity releases going back to U.S. newspapers about 8th Air Force pilots and ground personnel. This was a boring assignment after flying combat missions in P-47s and P-51s.

However, one day the Colonel in charge came in to introduce me to a war correspondent from West Virginia, my home state. His name was Howard Chernoff and he was the top honcho for the W.VA. Radio Network. He had come to England to meet and interview West Virginians and record their interviews to broadcast them over his network so that parents, relatives and friends could feel closer to their loved ones overseas. Then, the Colonel "volunteered" me to fly Chernoff to various 8th AF bases to accomplish this mission. Anyway, it turned out to be more fun than pushing papers.

The next morning, Chernoff and I went out to the nearby airfield where I was given a Norseman (a C-78, I think) to fly him around. Well, I had never seen a Norseman before that, but in those days, if you could fly one airplane you could fly them all, so we climbed up into the cockpit. After looking over the instrument panel for a moment or two I opened the window and called to the mechanic, "Hey, Mac, how do you start this damned thing?"

Punchy Powell peers past the backup gunsight post on his P-51. (Punchy Powell)

Howard Chernoff (left) as war correspondent (Howard Chernoff)

328th FS Norseman "Joy Ride"
with crew chief Art Snyder (Art Snyder photo)

After getting his instructions, I cranked up the engine, checked the gauges and called for takeoff instructions.

As we taxied out, Howard, sitting in the right seat, said, "Why didn't you know how to start the plane?"

"To be honest with you, Howard, I've never been in this kind of airplane before," I replied. I noted his shocked look and slightly whitened complexion and decided I would have some fun at his expense.

When we got to the assigned runway, I checked the mags and reported ready for takeoff to the tower. Unnoticed by my passenger, I had rolled the rudder trim tabs full forward so that I could hold the ship on the ground longer. Down the runway I started pretending I was trying to pull the stick back to get into the air. Actually, I was holding the stick forward but rocking my body backwards to make it appear I was trying to get it off the ground. I could feel Chernoff pressing harder against his seat and knew his eyes were getting wider open and his face a little greener as the trees at the end of the runway grew larger and we were still on the ground.

Finally, as we neared the end of the runway I pulled the stick back and we went up like an elevator. We could have done so several hundred feet sooner, of course. I heard Chernoff let out the big breath he had been holding down the runway and I let out a big "Whew!" to make him think that had been a close call.

He didn't say anything but I knew what he was thinking—"What kind of a pilot is this guy they gave me?" And, I knew that his confidence level in my flying ability had just gone to zero. I was having fun.

We climbed up a few thousand feet and leveled off, heading for Duxford, near Cambridge. Having been used to cruising at about 250 mph indicated in my Mustang, it seemed to take forever in this plane flying only slightly over 100 mph. Consequently, I started looking for Cambridge sooner than I should have and pretended to be concerned.

Chernoff asked, "What's the matter?"

I told him I was lost, "Bigger than Hell."

"There are some maps over here in this pocket," he said, and I replied, "Won't help! I never could read those damn things," and continued to search the horizon for Cambridge. After a few minutes more I told him that I was going to have to find me a railroad station to see where we were. Sighting a train going in our direction I told him I would follow the train to a station and read the name of the town on the station and then check the map. So, I peeled off and dived down to about 50 to 100 feet just above the deck flying off to the right of the train. That part of England, East Anglia, is quite flat and about at sea level and since the railroads were built up a few feet above the normal ground level, I could fly almost level with the top of the railroad cars and parallel to their direction.

Now, you have to visualize this a bit to understand what Chernoff was seeing and thinking.

Remember, he is in the right seat and the train was to our left. As we flew along I had my head turned fully to the left as though I was looking at the train and I am waving at the passengers with my left hand. However, my eyes were turned ahead, unseen by Chernoff, so I could maintain level flight and pull up over trees and other obstacles. Chernoff was screaming every time he saw something at our level ahead of us and I just pulled up a little to go over them. I couldn't look at him at that moment but his face must have been totally white or green by then. After a few minutes of this we passed a station and I pulled up and resumed our flight toward Cambridge.

By this time I figured Howard was almost a basket case.

Later, after a few stops at American airfields, we flew into Bodney, my former combat station. When I introduced him to the Squadron Commander he said to the Major, "I really can't understand why they let Lt. Powell fly. I offered him a cigarette and he said, 'No, thanks. I don't smoke, I have a weak heart.' Why would they let him continue to fly with a weak heart?" The Major had heard me make that remark before and he just grinned. He also told the Major that I was the "craziest pilot he had ever met."

Many years later, when Senator Rockefeller was Governor of West Virginia, I got a call from the Governor's aide asking me if I was the "Punchy" Powell who knew Howard Chernoff, and I acknowledged that I was.

"Well," he said, "we are honoring him as a 'Distinguished West Virginian' and we asked him if he wanted to have us invite any special guests. You were the first one he named. Would you be willing to come to Charleston for the ceremony we have planned for him?" I agreed to do so.

At the appointed time I flew into Charleston where I was met by the Governor's aide and taken to a large theatre in the Capitol building and was ushered in and seated on the front row with two other special guests. One of them was Clement Bassett, a 328th Squadron airmen who had also been interviewed by Chernoff at Bodney. Clem was from Welch, WV and we had known each other in high school days. We also attended WVU at the same time. After the war, having obtained a law degree, he had risen to a high position on the Governor's staff.

When the Governor and Chernoff came out on the three foot high stage and Chernoff looked down and saw me, he jumped off the stage and came over and gave me a big hug, remarking, "You S.O.B., I can't believe you are still living." Later, at lunch, the Governor said to me, "Howard tells me you were the wildest fighter pilot he ever met during the war."

And I suppose that could be true. Growing up during the Great Depression and getting to fly WWII's greatest fighter plane free was enough to make this hillbilly a bit overwhelmed and maybe a bit of a crazy fun-loving fool.

A wartime War Correspondent showing typical uniform.

Robert "Punchy" Powell, pilot, author, and editor, displays his P-47 and A-2 at Bodney in the Fall of '43. (Robert Powell)

A Rare Breed

By Marc L. Hamel

"Apparently the German pilot reefed his plane in too tight and it started to snap roll. He recovered but I gained enough ground to get my sights on him. I gave him a burst, and could see hits on his right wing and engine." This description is made doubly interesting when one considers it was penned by a 30-mission bomber pilot.

Henderson achieved a transition from bombers to fighters in May, 1944 before this was widely offered in the 8th Army Air Force (8th AAF). Later in '44, a tour-expired bomber pilot could elect to fly a second tour as a fighter pilot or as a Fighting Scout. The latter ferreted out fighter and weather opposition ahead of the huge bomber formations. However, for Richard the change was driven by a more basic desire—he always wanted to be a fighter pilot.

Richard Henderson poses with the Nilan Jones artwork on "Miss Susie Q." (George Nunemacher)

When asked about this significant transformation, he replied, "I was one of the first ones in my bomber outfit to finish his tour, and there was quite a celebration. The C.O. of our 448th Bomb Group, a fellow named Mason (no relation to Col. Joe Mason of the 352nd FG), indicated that I was to have dinner with him. That evening Mason asked me what my plans were. I told him my plan was always to go over and fly fighters. He said he might be able to arrange that as he had a friend with the 67th Fighter Wing, a General Anderson, I believe. Mason sent word that he had a pilot he wanted to send over, and soon I had a short appointment with the general. Within a week new orders came over and I was sent to Bodney. That was that."

Henderson had gone through flight school at Ontario, Merced and Luke Fields in Class 43-G, showing the skill and inclination of a fighter pilot, and so was sent to single-engine training. Achieving the 3rd highest score in gunnery and being chosen to lead the graduation ceremony formation certainly had him primed to fly fighters. Upon graduation however, he was inexplicably diverted into multi-engine and was thus posted as a co-pilot

Henderson in "Miss Susie Q" PZ-R does a low pass for the boys at Bodney. (Richard Henderson)

to the 713th Bomb Squadron (BS). This 2nd Air Division heavy bomber unit flew B-24's from Station 146 (Seething) from late 1943 through the end of the war. Richard flew across the Atlantic on the "southern" route and arrived in early December of 1943 with the original cadre.

Cater Lee was the bombardier in Henderson's first crew where Richard served as a co-pilot. Cater recalls Richard's initial reluctance at being posted to bombers with, "He was incensed." However, he excelled once he realized the only way out was finishing his tour, becoming a POW, or worse. After flying a few missions, Richard showed such talent that he received his own crew. Cater also allows that Richard was quite the upstanding gentleman during this period, with no "bad habits" such as smoking, drinking or chasing skirts. His one weakness was his love of flying.

After the completion of a hair-raising tour of 30 missions with the 448th in May of 1944, Richard still had the feeling that fighters were where he was meant to be. As detailed above, his request was processed and Richard was sent to Station 141 at Bodney assigned to the 486th Squadron of the 352nd Fighter Group flying the P-51. Cater Lee recalls a humorous incident from this transition time. It seems Richard was careful to instruct his friends rotating back to the States not to tell his wife back home about volunteering for a second tour in fighters. He did not want to get into hot water on the home front.

On arrival at Bodney immediately after D-Day, Richard notes today that his new C.O. Willie O. Jackson did not know quite what to do with him. Jackson wisely assigned Richard to local flights and practice missions for 30 days to get the feel of a fighter again and then had him fly his wing. Richard began flying fighter combat missions immediately after the bulk of the 486th returned from the first three-way Russia Shuttle Mission in the first week of July 1944.

He quickly adapted, and soon became a valued member of "Angus Squadron." With the rest of the 352nd pilots, Henderson moved later to the primitive Y-29 forward airfield in Asch, Belgium before Christmas 1944. Living in tents in the freezing cold and snow of that winter, the 352nd continued to fight the Luftwaffe and heap glory upon itself, peaking with J.C. Meyer's uncanny and justly famous read of the German offense on New Year's Day 1945 (the Legend of Y-29).

December 27th 1944

Just before New Year's, on the 27th of December, Richard proved that his transition from heavy bomber pilot to fighter pilot was completely warranted. Due to the primitive conditions and separation from their intelligence office, no Encounter Reports were filed for this hot mission. Luckily, Henderson recalls the action well today and relates, "We were stationed at Y-29 in Belgium and I went on a mission flying as element leader to Flight Leader Colonel Willie O. Jackson. We were patrolling over the front lines and were

Henderson's gun camera records his first Me-109 victory in December 27th 1944. (Richard Henderson)

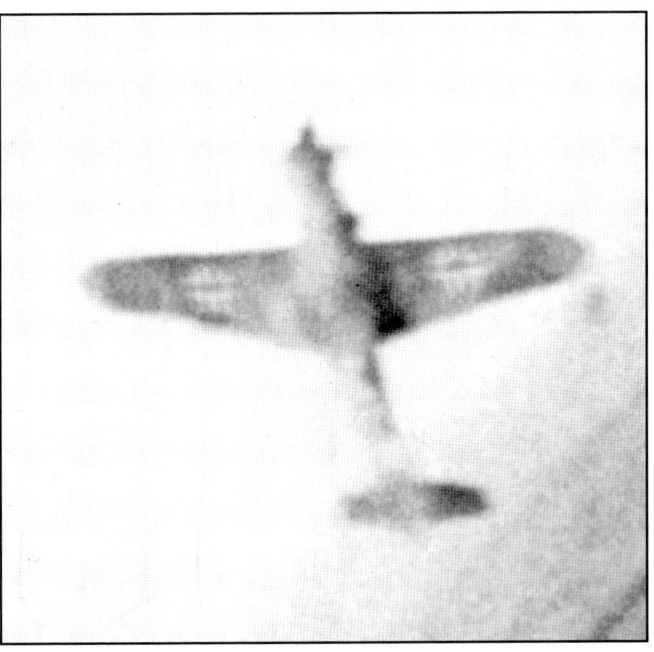

December 9th, 1944 prang of "Miss Susie Q" by David Reichman off the SW corner of the field at Bodney. (Richard Henderson)

in contact with the radar controller for that area. My wingman had some problems and so he was sent back to Y-29. We continued our patrol looking for German aircraft.

"Our radar controller advised us that he was seeing some very low targets going east at our 9 o'clock position, which was on my side of the flight. We were flying at approximately 12,000 feet in a westward direction. I looked down and spotted an aircraft going east right on the deck. I called this in to my Flight Leader and he advised me to take it while he gave top cover. I peeled off making a wide diving turn to my right and identified the aircraft as an Me-109. I intended to line up on his tail, hoping he had not seen me. As I was overtaking him from behind (almost in range) he saw me and broke to his left. I followed him and we both went into a tight turn. I had never tried to out-turn a 109 before and was not sure I could. We were making 360-degree turns and staying about even. I was concentrating as hard as I could to get the maximum out of my plane. After about two complete turns I thought I was gaining a little on him. Apparently the German pilot reefed his plane in too tight and it started to snap roll. He recovered but I gained enough ground to get my sights on him. I gave him a burst, and could see hits on his right wing and engine. He opened his cockpit and bailed out on the right side. We were only about 100 feet off the ground, and his parachute never fully opened. The '109 rolled over and hit the ground. There was snow on the ground, and we were right over a farm with a barn and house. I think he hit right in the barnyard.

"I pulled up to about 3,000 feet and started looking for the rest of the flight, but they were chasing another 109. I spotted another one going east right on the deck and lined up behind him and closed to about 50 yards. I was ready to fire, but hit his prop wash, and so pulled up a little and then gave him a burst. I saw pieces coming off his tail and I was getting hits all over him when he pitched straight down into the trees and exploded.

"I pulled up again and watched Colonel Jackson shoot down his 109. The flight joined back up on the western edge of Bonn. We returned to Y-29 and congratulated each other. Our gun film was sent back to England to be developed and we got confirmation of our kills the next day.

"I have thought about that first 109, and going into a tight turn with him. I had altitude and speed on him, and could have pulled off after he broke to the left, but after I committed to the turn neither of us could break off as the other would be in the position to shoot. Once we were in this turn, it became a matter of which one of us would win.

"Afterwards, Colonel Jackson said something to me about how long it took to shoot down the first Me-109. I agreed, but did point out that the German was not cooperating, and that made it hard."

As Richard noted, the gun camera footage was forwarded to England, and victory credits were established. The 352nd compiled an excellent record that day with the 487th Squadron's ("Meyer's Maulers") James Bateman, Sanford Moats, Russell Ross, and James Wood gaining singles over 109's and 190's; Marion Nutter, and J.C. Meyer scoring doubles, and Bill Halton and Ray Littge posting brilliant "triples."

Additionally Anthony Goebel shared two half credits over 109's, (one with Bill Halton), and J.C. Meyer damaged two other Fw-190's. "Angus Squadron" (the 486th) was not resting on its laurels this day either. In addition to Henderson's two Me-109's, escape-and-evasion expert Bill Reese also scored a double. Pitching in single victories on the 27th were Willie O. Jackson, Earl "Lazy" Lazear, and Frank Greene. Quite a miserable day for the Luftwaffe, and a 22-0 win for the Bluenosers.

Recalling his action in this same air battle, Bill Reese said, "I was flying #4 position with 'Olaf' (Ernest O. Bostrom) as flight leader and Lt. Bill Gerbe as element lead. I was flying Gerbe's wing and saw a pair of Me-109's coming up and behind us. I called in 'bandits' and whipped over and out of formation. Gerbe said, 'Have at them and I'll cover you,' and I did. I got the German 109 wingman in the cockpit and he immediately crashed from about 300 feet above the ground.

"The German leader was a horse of a different color. He was good! He immediately pulled up into the sun (on a nice clear day). The 109 could climb at a sharper angle than a P-51, so I just kept the plane at full bore and just above stall. Lt. Gerbe tried shooting his guns, stalled and spun out, and I didn't see him again until back at Y-29 (he recovered okay, but was so low he didn't have time to get back into the action). After the German flight leader had pulled away from me a little, he did a full rudder turn and came back firing away. Fortunately he was close to a stall and he couldn't pull his nose up enough. I could see his tracers going well under me. The first time we had this head-on confrontation, he passed about 50 feet under me, and I lost him momentarily while trying to get turned around. I found him again climbing back into the sun. The second time he came at me was an exact rerun, except we were passing each other a little closer. Again I lost him under me and then he was climbing back into the sun. I knew I had to do something to break this cycle, and I did. Next time he came charging back at me with his guns blazing, I was still in a steep climb—but upside down. This time I did not lose sight of him for even a second, and when he started to chandelle back toward the sun, I was sitting behind him firing my guns. I was still flying too slow for my sight to be much good, but I managed to get a couple of hits near his canopy. The next thing I knew, the canopy came off and he bailed out."

Richard Henderson finished up his tour with the 352nd while the unit was still at Y-29 in January 1945 (the Bluenosers soon moved again to A-84 at Chievres). After returning to Bodney to gather his gear, Richard was sent home in a C-54 transport via the Azores.

Remaining in the Air Force, Richard was assigned as Supervisor of Flying at Las Vegas Army Air Field postwar before deciding that "the excitement was all gone" in 1946. He left the service and got into the civilian flying business with oil companies. In 1958 he formed his own FBO and Beechcraft dealership in Texas, from which he retired in 1991. Richard only recently folded his wings for health reasons, having attained more than 25,000 total flying hours.

Reflecting on the war and today's youth, Richard allows, "Our generation had the opportunity to test ourselves, to see how good we were. Everyone wonders what he is made of, and the war gave each of us an answer."

From "Gangland" to England

From a personal account by Bill and Betty Reese

Bill Reese grew up during the Great Depression years in a tough neighborhood in Chicago where he developed the aggressiveness he would need later in life as a fighter pilot flying combat against the German Luftwaffe in a bluenosed Mustang he named "Rascal," which he evidently was when playing with his friends as part of their "gang" in the middle of Chicago's "gangland."

His flying career began in 1942 when he was in Junior College in Kansas where his parents had moved. Ironically enough, his instructor was a WWI fighter pilot who demanded the utmost precision by his students, a factor which was most helpful to Bill when he was going through the Army Air Corps cadet program. He had already learned to do precision spins and other aerobatic maneuvers.

Winning his wings in December, 1943, he was sent to Bartow, Florida for RTU and got checked out in the P-51A with the Allison engine. After training for only a few months he found himself enroute to the British Isles, arriving in Scotland on April 17, 1944. There he got his first combat formation training flying the P-51B with its more powerful Merlin engine and four .50 cal. machine guns in the wings. Soon afterwards he was assigned to the 486th Squadron of the 352nd Fighter Group at Bodney, England, a former RAF airfield located in the flat English countryside of East Anglia. He was now at war. His description of D-Day, June 6, 1944, follows.

Bill Reese, with his oft smoked pipe, poses in pal "Olaf" Bostrom's P-51. (Bill Reese)

"We were making a night takeoff from our sod airfield, no lights, just total darkness—in flights of four planes. I was flying in the last flight of only three planes. There was about a 1,000 foot overcast. The second flight ahead of me lined up a little too far to the east instead of northeast and, on takeoff, one of them, Lt. Robert Frascotti, carrying max fuel and ammo, flew into the second story of the almost completed new control tower being built on the rise in the field.

"Suddenly there was a terrific explosion as his plane impacted the tower and his element leader's wingman clipped the volleyball post with his gear, fly-

Reese's second P-51 PZ-Rbar, "Rascal" displaying its unusual bluenosed treatment below the windscreen. (Bill Reese)

ing between the tower and the 328th Squadron pilots Ready Room building. He was momentarily blinded by the explosion but managed to get his bird into the air and landed safely at a bomber base nearby.

"The rest of us had plenty of light for our takeoff—provided by the burning Mustang. My leader aborted after takeoff, calling me to join another flight. Seeing a couple of exhaust stacks glowing ahead of me under the overcast, I joined up just as we went up into the clouds. I griped over the radio about climbing too slow—about 130 mph—but didn't get any response so I staggered along with them until I was on top of the clouds. Only then did I realize I was flying formation with a B-17. I broke off and checked in with our ground controller—a direction finding station (DF) and, although he was surprised that I was a flight of one, he got me over Normandy—and what a sight that was. I don't know how many ships were involved, but it had to be several hundred, and not a single enemy plane to be seen.

"Our secondary mission was to strafe anything moving toward the beaches, and I do mean anything. The French had been warned to stay inside and off the roads. Anyway, the controller sent me 90 miles into France and gave me some erroneous info that took me even deeper. I dropped down looking for targets near a small town and saw an olive drab bus moving toward town and thought, 'Aha, a piece of cake.' I lined up about five miles out and started my strafing run. The bus had stopped and was a sitting duck. I started to squeeze the trigger when the bus just erupted with children coming out the windows and doors, falling all over each other and I didn't fire a shot, just pulled up and came back over them and they were waving and shouting still right beside the bus. I said, "to Hell with this," and called the controller and went home, ending my fifth combat mission.

"The 7 of June was mission No. 7 for me. I was flying as wingman for Major George Preddy, the top Mustang ace of WWII. We lost both planes in our element due to ground fire while strafing, so Major Preddy and I were heading home when I saw an Me-109 drop down out of the overcast. I called the 'bandit' to Preddy's attention and, since he couldn't spot him, he told me to attack and he would fly my wing. I went 'balls to the wall' and was just getting into range when the Jerry spotted me and popped back into the clouds. He came down once more and saw us heading for him and went back in again and that was the last we saw of him. I will never forget Major Preddy giving me a chance at that plane when another victory for him would have further enhanced his fame in the ETO.

"As we headed for the French coast Preddy saw three trucks moving along a road off to our side and immediately setup for a strafing run and I followed with great enthusiasm. He said something on the RT about the time I started firing at the middle truck, which immediately blew up, much to my surprise. I flew through a bunch of debris and rejoined him and that's when I realized the three trucks were ambulances with big red crosses on their topside. I had been too low to notice. However, they were obviously carrying ammo to the German troops on the invasion front, so we didn't feel bad destroying them. It was my first time to fire my guns in combat.

"On June 8 business picked up. We had a relatively short escort mission and then were released for "targets of opportunity." Our flight destroyed three trucks and then I saw a black command car racing down the road, the kind used by high ranking German officers. The low overcast made it difficult to maneuver and I had trouble avoiding some telephone poles and lines, but managed to get in a good shot from the side, literally blowing the car off the road.

"June 10 was really a fun mission except that we lost one of the pilots in our flight when he flew into a tree on a strafing run. I destroyed three more trucks and damaged five others. Since they were carrying either ammo or fuel, the explosions were something to see, and the debris was a mess to fly through. Later, I spotted a tugboat in the canal and three of us set up a strafing pattern until it sank and blocked the canal. There appeared to be no survivors. Again, I was amazed at the damage those six .50 cal. machine guns could do. The side of the tug was just caved in with gaping holes at the waterline. It was all over after two passes.

"I had gotten into a habit of volunteering for aircraft test hops since we flew these without wing tanks and I could give the bird a good wringing out at altitude and then I would hit the deck for some low level navigation practice. This provided a natural high—just enough adrenalin to make the reflexes sharp. What a kick it was to flip the plane on its side to avoid trees and to come roaring across Bodney just a few feet off the ground. On one of these flights I had a bet with one of the sergeants in the squadron for a bottle of Scotch. He bet I couldn't move him off a spot in front of the Squadron Ups building, but he was long gone before I got within a 100 yards of him when he realized I was lower than he was.

"In July, 1944 I was taking another test hop and taking off across the width of the field instead of the length because of the wind direction—actually heading toward our Operations Building. This gave me nearly 2,000 feet of field, plenty for a light fuel load, but just as I got airborne my engine

George Preddy's "Cripes A' Mighty 3rd" armed with bombs for a late August '44 mission.

Reese's ground crew

quit—stone cold dead. Here's what I faced— the Ops Building slightly to my right, a forest to my left and the Group garbage pit (50 ft. deep) straight ahead and then more forest.

"I yanked up the gear, chopped throttle and mixture, hit full flaps and cut the ignition switch just as the plane slammed into the ground at about a 30-degree angle. That sucker plowed along for no more than 50 yards, stopping right side up. I cranked back the canopy and climbed out—no fire—and just about 30 yards short of the garbage pit. Later, when the Squadron C.O. was questioning me about the crash, he asked me how I felt and I said I was fine. He said we can afford to lose planes but not pilots, thanked me for an excellent job, and that was the last I ever heard of it. They salvaged what they could of the plane and junked the rest.

"July was a strange month for me. One of the other pilots asked me to stand down and let him fly instead. He had one mission to fly before he was headed home and that mission looked like it would be a milk run. So, I let him take the mission, with the Flight Commander's permission, in my airplane. The mission was a milk run, but on the way home he had an engine failure over the English Channel and was never found.

"In late June, 1944 our squadron was assigned to participate in the first shuttle flight to Russia where we picked up the bombers over Germany and escorted them on into Russia, landed there, then worked our way back with missions ending in Italy and then France and back to England. I was scheduled to go on this mission but Major Higgins' plane went out and "RHIP," (rank has its privileges), he took my plane and I ended up staying home.

"It turned out not to have been a bad deal for me, however, since there were few left from the squadron at Bodney and I had an opportunity to spend a lot of time in London getting acquainted with the big city. This was during the time when the V-l 'Buzz-Bombs' were coming over. I just couldn't help being fascinated with these things. They sounded like a large outboard boat or motorcycle motor buzzing along, but when they suddenly got quiet as their engine stopped, they came down at about a 45 degree angle and exploded their 1,000 pounds of TNT on impact.

"The British people weren't particularly shook up about these things because you could hear them and see them and they were well-disciplined about getting into their bomb shelters and deep underground in the subways in London. Usually they met the American soldiers coming out to see what was going on.

"It was really fascinating watching these at night, especially when the flak gunners were trying to shoot them down and you could hear the put-put, staccato, sounds of their engines and see the exhaust flare passing through the night sky. Then things would turn quiet and you would hear and sometimes see the big explosion light up the sky.

"London was, of course, totally blacked out during this time. You just cannot imagine trying to get around a city this size at night under complete blackout conditions. You almost developed a sixth sense that someone was in front of you but it didn't seem to be a big problem and there were few accidents. Of course, there weren't many civilian vehicles on the streets and roads. London took an awful beating from the German bombers and V-1's, but the worst was yet to come with the more unpredictable V-2 rockets against which there was no defense.

Reese's first P-51B "Rascal" PZ-Rbar displaying the light blue nose paint fairly common in the 486th. Squadron painters apparently had some latitude in mixing the blue applied to the aircraft. Note also unusual painting of codes and Invasion stripes. (Bill Reese)

DIVE-BOMBING MISSION

"Our first dive-bombing mission occurred during this time period. We had absolutely no training in RTU or OTU in dive-bombing techniques. . Our first mission was to take out a bridge bridge in France. Our first flight's bombs hit a small town about a quarter-mile from the bridge, which gives you some idea of our accuracy. We were the fourth flight and by the time we went in we decided that we should change something. We went in at a higher angle than the other flights and finally did put the bridge out, but it was a fiasco of the highest order.

"On another dive-bombing mission we had figured out a pattern and did much better, but we were no experts by any measure. When it came my turn to drop the bombs, only one bomb released and I went and I couldn't get rid of the other one. So, we joined up in formation and headed for home. About this time, my radio started acting up, so I was just getting intermittent transmissions and I couldn't understand much of what everybody was saying. I was getting lots of advice on how to get rid of it, but nothing worked. One of the big questions to consider was, 'Should I place the bomb switch back in the un-armed position or leave it in the armed position?' I think most of us were under the impression that when you armed the bombs an arming wire was pulled out of the bomb so it was hot from then on, but I thought, 'What the Hell!' and put it back in the un-armed position.

"By the time we got to the field my radio was really out. We made our normal fighter approach in echelon right, coming in on the deck and spreading slightly as we made a climbing left turn, sort of a slanted loop at the top of which we put our gear down and as we completed the circle onto our final we dropped flaps and eased it down onto the field. All four of us could land fairly close to each other since we were landing on a sod field, not on a runway in spread formation. Suddenly the planes slightly in front of me took off in all directions. They didn't quite touch down, but gave it full bore and took off again. I was wondering what all that was about. Then, as I started flaring out to land I noticed that I went over an object that looked like some kind of obstruction in the field, but I went ahead and landed. It was only after I had taxied in that I found out that my bomb had come off when I had pulled up into the loop and had landed on the first part of the field, the obvious reason the others had not landed.

"Strafing ground targets presented other problems. Our six guns, three in each wing, were aligned by our armorers to converge on a target at about 250 to 300 yards as desired by the pilot. This was the optimum distance for maximum effect. Due to trees, or other obstructions, we sometimes had to approach at a rather steep dive angle on these targets but preferred to use a more shallow dive with lower obstructions. Quite often you would find yourself flying through the debris flying up when the target exploded.

"Another pilot problem was called 'target fascination,' concentrating so hard on your target that you would fly right into it. We lost a few inexperienced pilots who flew into their target, were knocked down by down by debris or failed to recover from their dives soon enough.

A FATEFUL DAY

"17 August 1944 was a strange day to say the least. I took off on this mission as a Second Lieutenant and was shot down as a First Lieutenant on what was expected to be a milk run—striking targets of opportunity inside France. We found a marshalling yard near a small town with three or four freight cars on a siding, so we set up a pattern and started shooting them up.

"What we didn't know was that for some reason Jerry had some heavy machine guns and quad-fifties set up in the area and, after my first pass, I got clobbered—smoke coming out of the engine and I was losing power. I zoomed up to about 500′ and tried to roll back the canopy, but couldn't do so as it must have been hit and was jammed. I wasn't very enthusiastic about bailing out anyway, so I looked for a place to belly-in.

"The only spot I saw was a tiny green meadow. Everything else had trees or some other obstructions. About then, the engine quit so I started my descent and turned toward the field. It, too, seemed to be completely surrounded by trees. I could see the field as long as I was in a turn, but once I started in on my final approach, all I could see was smoke and flames and I couldn't side-slip because I was too low—couldn't even see out the side of the canopy because of the smoke, so I just held a gentle glide until I felt the prop tickling the ground and then held it there until the plane settled in.

"The initial touchdown was smooth as glass, then came two sudden, severe jolts and dead silence, with me and the plane in a cloud of dust, My green field was mostly dirt and, when the dust settled, I tried the canopy again and, lo and behold, it cranked back just as it was supposed to do. Then I realized I had not come through unscathed. My Mae West life preserver and my lap were covered with blood—mine. After taking off my helmet and goggles it felt like my nose was all over my face—cheek to cheek. A flap of skin was hanging down over one eye and that was when I found I had a fair-sized wound on my forehead.

"As I climbed out of the plane I noticed my metal gun sight mount was broken, proving once and for all how hard my head is! The other three planes in my flight were buzzing around overhead for two reasons, one, to give me a chance to put some distance between me and the plane with no interference from ground troops, and destroy my plane as that was SOP. It had been established that on several occasions Jerry had repaired fighters and then used them in sneak attacks on our bombers.

"Anyway, I started walking away. Strangely enough, although I was in somewhat of a daze, I felt no pain at all and when I realized I still had my parachute and Mae West on I ditched these in some bushes and started off toward a wooded area. About this time my guys had to leave since they were getting low on fuel and, as soon as they were gone, two German trucks came out. They had been hiding under a small bridge and were loaded with soldiers. To give you an idea of my mental state, I just stood there and watched as they drove up to the plane. I guess they didn't see me or assumed I was a farmer. I finally turned around and walked into the wooded area.

"When I approached the other side, I realized I wasn't being very cautious. I heard someone chopping wood, so I crept up to the clearing that was in front of me and saw this farmer chopping away. I watched him for about 30 minutes and didn't see anybody else, so I decided to approach him to possibly get some help. I really needed a drink and some cleaning up. I had a hankerchief tied around my forehead, was dressed in my A-2 jacket, olive drab wool shirt and pants and combat boots, and for the last time in this war, I was unarmed—no gun, no knife.

"The farmer spotted me as I stepped out of the woods, but he didn't run, just kept on chopping. As I got closer, he got a good look at my face and came toward me saying something in French, and all I could think to say was, 'I'm an American.'

He understood the American part, but could not speak one word of English — which made us even so far as communicating. It's surprising how fast you improvise with sign language. I got the idea over to him that I needed something to drink. He shook his head and showed me his empty wine bottle and then brightened up and got me to follow him to a small clump of trees in the middle of the field. Eureka, a spring! He held the empty wine bottle under the water until it filled, but as I started to grab it he held me off and held the bottle up to the sky, shook his head, poured the water out and filled it again. This time I wanted to look also, and you could see some little wiggly things swimming around inside. Finally, after four fillings, we couldn't see anything and I thought, 'What the Hell. I've had all kinds of shots for just about everything,' and had my first and last drink of water in France—and no ill effects.

"I got out my 'escape and evasion' kit which included a few hundred French francs, silk maps of Europe, compasses including a small one we referred to as an 'asshole compass' since that's where you hid it if you thought you were about to be captured. I knew I was in France, but just exactly where escaped me. The farmer, after much rapid-fire French, finally oriented himself and pointed to a name, Epagne, so I decided I had better head northwest and hope for the best. I shook hands with him and left the field and went through a wooded area before coming out on a paved road that seemed to go more or less in the direction I wanted to go.

"I followed it for awhile before coming to a bridge with a small village on the opposite side. Knowing how exposed I would be crossing it, with no possibility of jumping off to the side to hide as you could do on the road. I carefully studied it for about 30 minutes without seeing a single soul, then stepped out just in time to meet a mother and daughter who came over the hill behind me. Just as I got across the bridge I met a little girl on a bicycle. She took one look at me, turned around and rode like a bat out of hell away from me.

"I thought at that time that I was really in trouble, that she would alert every German in the vicinity. Little did I know I had just had my first meeting with the French Maquis (French underground force).

FOUND BY THE PARTISANS

"The only thing the Germans feared more than the Maquis were American planes. I walked about a mile past the village and saw a man coming riding a bicycle. He didn't look too threatening, so I stayed on the road. As he came opposite me, he slowed and said, "American?" I said, "Wee, I'm American,"

"He got off his bike and led me off the road into some bushes, all the while yakkety-yakking in French, but he did get the idea over to me that I was to stay hidden there and he would be back. Sure enough, in a little while he came back with another man and a spare bike which I got on, and the three of us took off down the road and away from the village. We came to another small village which had a tavern, and they took me into the kitchen in the back of the tavern, had me lie down on the floor, fixed up a pillow and got some blankets. I was beginning to feel like it had been a full day and, believe it or not, they got a doctor to attend my wounds. Several in this group could speak some English, including the doctor, who put sulfa powder on my wounds and bandaged my forehead and said my nose had been broken lengthwise but thought it would heal okay.

"It was dark by now (I had been shot down about 2:30 p.m.), but the day wasn't over by a long shot. I was lying on the floor opposite a door going into the barroom. We could hear a lot of commotion in there, and one of the Maquis—and to this day I can't believe he did this—opened the door a crack so I could see into the barroom and it was full of German soldiers in uniform drinking beer. I'm sure the smile on my face would have been described as sickly.

"About an hour later one of the Maquis rushed in and with the usual burst of French and things started happening. It seems the Germans had originally thought I had perished in the plane, but then found my parachute and Mae West. They found my tracks leading to that wooded area and had lined up about 100 soldiers abreast all armed with machine guns and, in the dark, marched through the wooded area gunning every dense thicket they found and were now searching all the villages in the area.

"The Maquis took me to a barn near the tavern, got me up in the hayloft and hid me behind several bales of hay, telling me the Germans usually searched haylofts by using pitchforks to check for hiding places and for me to just stay still. They felt there was too much hay in front of me for the Jerries to find me. Well, I thought I might as well make do, so I made a makeshift bed out of the hay and went to sleep. The Germans did search the tavern but never did come to the barn or if they did I didn't wake up until the Maquis woke me up the next morning.

"We then took off on our bikes for the little village of Epagne where I was told I would be staying with the Mayor and his family, and what a family this turned out to be.

"Papa Doiziet, Mama Doiziet and sons Andre, 15, and Henri, 17. I lived with this family for almost four weeks and was treated like a favorite son. Every day I witnessed feats of bravery and love that I will remember and cherish to the end of my days.

"Papa Doiziet was not only the mayor, he was also the head of the Maquis for that entire section of France. Andre, the youngest son, could speak a little English, but no one else in the family could say anything other than the word "American."

"When I first arrived, I sensed that Papa and Henri seemed sort of unhappy and, at first, I was afraid it was because of me. I couldn't have been more wrong. Fortunately, Andre was able to tell me the problem. It seems that the same night I was at the tavern, Papa and Henri were on their way to do a little sabotage on a German train parked near a small town waiting until after midnight to move on. On their way, they saw a German truck moving slowly along the road with its blackout lights on (just small slits with the rest of the headlights blacked out), so they thought they would use one of their four sticky bombs (these were like big hand grenades covered with a soft sticky substance so they would adhere to whatever you threw them at and explode a few seconds later).

"Henri threw one at the truck and it stuck to the windshield. Scratch one Jerry truck and crew. However, they had not noticed the two trucks following with no lights on, and these two trucks each had ten soldiers in back and two in front so they had to throw their remaining three bombs and use their Sten guns to mop up. The old man and young boy killed 26 Germans that night! I couldn't understand why that would make them unhappy until Andre explained that they had received specific instructions by radio from England to use these bombs only for sabotage against trains.

"Mama Doiziet greeted me with a big meal served with outstanding red table wine. Two things I really remember about Mama—her cooking and her love for everybody around her and especially for Papa. She knew that each morning she said goodbye to Papa might be the last time she would see him, so her goodbye kiss in the morning was a sexual experience by itself—two old folks kissing up a storm. And, of course, when Papa came home, the reunion was even more enthusiastic. One thing was obvious to all, and that was the deep physical and emotional love these two had for each other.

"I also remember some wonderful meals that Mama whipped up on one of her two stoves, one of which used coal, the other wood. Poached eggs for breakfast with homemade bread, squab, roast beef or fish for dinner and, of course, table wine at every meal. I did not drink another drop of water during my entire stay with this outstanding French family. I certainly remember the first fish dinner, a six or seven pound fish of unknown lineage. The meals were cooked to perfection and the French fries melted in your mouth. I gained about 10 pounds during my ordeal although all was not fun and games.

"In the evening if something special happened that day, Papa would bring out a bottle of Schnapps and fill little shot glasses for everybody. This Schnapps was something elsel A drop on the furniture would go right through varnish or paint—dip your finger in it and pass it over a candle and it would flare like lighter fluid. This was not a sipping drink—one gulp and you finished it off and then you could keep track of it as it wended its way through your stomach.

"About the only visitor we had during the day was the local gendarme. The first time I saw him he was in the house before I was aware of it and, of course, I thought I was in big trouble, but he laughed and gave me a hug instead. About then Papa came in and the gendarme handed him a small bag he had brought to Papa. Inside the bag

Becoming A Fighter Pilot in 1943
by Ray Mitchell

So how did I get to be a fighter pilot? Well, as World War Two got underway in 1942, I turned 21 and had just finished a year of college and was eligible for the draft. Like most young men in those years I thought it was my duty to enlist and I wanted to enlist in a program that would have rigorous training and provide some knowledge that hopefully could also be used after military service.

With that in mind, I took a battery of test being given for the Aviation Cadet Candidate program and was accepted I was sworn in as an Aviation Cadet in June 1942, but because of the backlog of training spaces available, it wasn't until January 3, 1943 that I received orders to report to Kelly Field, Texas for training.

After a rugged physical training course and a crash course in essential navigation, Morse code, aircraft engines, etc., only two thirds of us passed the course and made it to Primary flight teaming. I was assigned to Sparton School of Aeronautics at Tulsa, OK, I had never been in an airplane, so when the instructor learned this he made sure that I was properly belted in this open cockpit PT-19 Fairchild trainer and flew to about 3500 feet, immediately spinning the plane down and pulled out at about 1500 feet. He looked back at me through the mirror to see what my reaction was. I just smiled at him. Guess that satisfied his question. After about 8 hours over the next few days the instructor got out of the plane one day and told me to take her up solo, the ultimate thrill for a student pilot.

After completing primary I was sent to Independence, Kansas for basic flight school where I flew the North American BT–13 which was called the "Vultee Vibrator." By this time I was doing acrobatics and learning the military style of flying, After Basic, I was sent to sent to Eagle Pass Texas for Advanced Training where I flew the North American AT-6. It had retractable wheels and a 650-horsepower engine, making it a good aircraft for aerobatics and gunnery practice.

Before being commissioned and getting my wings it was necessary to fly the Curtis Hawk P-40 eight hours. These were old P-40's from General Chenault's China era war and still had Tiger teeth markings on the nacelle. They were in bad shape and as luck would have it, on my first flight in one, the electrical system went out just as I left the ground and retracted the wheels. This meant that I had no radio, and could not lower the flaps or wheels electrically. I had to use a little hand hydraulic pump that took a long time to get prepared for a landing and without radio instruction. I finally mastered it and landed without a major problem. That incident could have finished my flying career before it got well started.

Ray Mitchell with his "Carol" PE-Mbar and crew in Belgium, early 1945. (Ray Mitchell)

were four more sticky bombs, so now I knew how the goodies were delivered.

"I should digress a moment to explain something. The Germans had every intention of staying in France forever. They realized that to do this, at least two groups of people would have to be treated reasonably well—the farmers, so they would keep on farming and supplying food for the populace, and the police, so they could help to maintain order. These two groups, in most cases, formed the backbone of the Maquis resistance. With luck, daring, courage and raw guts, they were able to successfully operate right under the Germans' noses.

"An example comes to mind. Papa was coming home from a meeting of his district commanders. He was carrying papers identifying him as a Maquis leader, and he was carrying a .45 pistol when he was stopped by a group of German SS soldiers who were retreating on foot to the east. They were lost and the day was cloudy and rainy. They asked Papa just one question: 'Which way is East?' Without hesitation, he pointed west. They thanked him and went on their way toward the west. If they hadn't been in such a hurry, if just one soldier had a compass, if...anyway, this is the kind of guts these partisans had.

"One evening about two weeks after arriving at Epagne we were about to have some Schnapps when there was a knock on the door and a couple of the local Maquis came in with another 'evader,' a young Canadian. He had been a tail gunner on a British bomber. His plane was on a bombing run at night when it was hit and exploded immediately, and the young Canadian found himself sitting in the air with no airplane around him. Fortunately, he was wearing his parachute and made a successful landing and once again, the Maquis got to him before the Germans. He was the only one to survive out of his 10-member crew.

"His introduction to the Doiziet family was to join us just as we were having our Schnapps. He said he was thirsty, but something was lost in the translation and, since they were out of shot glasses, they handed him a water glass full of Schnapps, which he started to chug-a-lug, thinking it was water since Schnapps is clear like gin. After his third gulp, he realized something was amiss, turned a rather pretty shade somewhere between purple and red, and ran outside where he solved the problem. He survived and we really did enjoy each other's company.

"Having someone to talk to did make the time pass faster. A couple days later, Papa came home with champagne for all and a very dry cigar for me. Patton had broken out at Normandy and could be in our area in just a few days. The cigars and champagne had been buried at the start of hostilities and now Papa felt the time had come to start enjoying them. So we had champagne with every evening meal, and I had one very dry cigar every evening.

"Something else happened about this time that really lifted our spirits. We heard some aircraft one afternoon and ran outside to find some 9th AF P-47s working over a German truck convoy just a few blocks away. We all ran to a nearby hill, sat down and really enjoyed the action. The P-47's had about ten trucks trapped on the road, having destroyed the lead and tail-end trucks and now were setting up a pattern to take care of the rest. I was really amazed at the damage eight .50 caliber machine guns could do. These were large trucks which would literally be moved sideways off the road when the 50's hit them, completely turned them to junk. The destruction was complete, and I know the Thunderbolt pilots got a kick out of the people up on the hill jumping up and down and cheering them on. When they left, they flew over us and rocked their wings.

DEFENDING THE BRIDGES

"Things started happening pretty fast after that. The Maquis were ordered to try to hold the two cement bridges in our area, one at Epagne and the other at the little village upriver, the one I crossed the day I was shot down. The Germans were pulling back and trying to slow the Americans down by destroying bridges, etc. Papa decided that the time had come for the Maquis to come out of hiding. They would have to man the bridges and fight for them. Two things had to be done first—get all the women and children in as safe a place as possible and capture a German female spy who had been living in the area since 1939 and working as a prostitute.

"Just recently they had realized what her real job was, and now they wanted to hold her for the Americans to interrogate. Andre was able to make me understand most of this, and then came the shocker. They wanted *me* to be one of the guards for her and, if she tried to escape, I was to shoot her. As a matter of fact, I was the only guard that would have a gun, which I was then given. It was a really nice Llama .45 automatic with several clips of ammo. Andre said that it was Papa's own weapon.

"I had two teenage boys to accompany me, although neither could speak English. They picked the woman up that night, and Andre said we would have to spend the night in the woods. He decided to take the place of one of the other boys since he could talk to me. Also, the next day we would walk down into the forest and stay where the women and children were stashed.

"Well, that was some night. We walked about four or five miles in the direction of the hideout, then settled down for the night with a rope around my wrist and the woman's ankle. Her hands were bound, too. She was about 30 years old and could speak some English and was quite confident that this was all a mistake. Thank God she didn't try to get away.

"Early the next morning we started out for the hideout, which turned out to be the darndest setup. It belonged out in the wild west of the U.S.A—a stockade with a 10-foot log wall around what looked for all the world like a fort. It was really large with stables and barns on the ground floor and living quarters above, and the place was alive with about 60 or 70 wives and assorted children. Some of the women were armed, and all were very curious about us.

I was an instant hero right on the spot—the first American to be seen—so now they could believe the Americans were really coming. I had breakfast, lunch and dinner all in one meal. I noticed some of the older women really enjoyed watching me eat. They were all smiles and yakety yakking in French. Andre said they were the cooks.

"About this time our spy said something to Andre, and he said she wanted to wash up, etc., so we took her down to the river. As she finished, our little gendarme came strolling up, said something to Andre and smiled at me (this is the gendarme who gave Papa the sticky bombs), took one look at our spy, let out a scream, and piled into her with both fists going. I couldn't believe my eyes—what the hell was this? I finally decided this was my prisoner, grabbed him by the collar (they were on the ground by now), lifted him up and off her and said, "No!" He immediately calmed down and I let go. He brushed himself off, smiled, saluted and left. Our spy was a mess. He had pulled both pierced earrings off, she had a bloody nose and mouth (by the way, we had untied her when we arrived at the hideout), so we took her back to the river and let her wash off again.

About this time, another Maquis came running up to us from out of the woods, acting pretty excited. He jabbered to Andre and pointed to me. Andre came over and said he had some bad news. There were Germans in the area and Papa was afraid that if they captured the spy, she now knew who many of the Maquis were and all would be killed by the Germans.

"I was told to shoot her immediately as this would be an American soldier shooting a spy in wartime. A day later I would have killed her and not thought a thing about it, but I hadn't really been exposed to the real horrors of the war yet, and I pretended I didn't understand what Andre was saying. While we were talking, the Maquis went up to her, shot her five times, dumped her body in the river, said something to Andre and took off.

"Andre said we had to get back to the bridge and help hold it, so we took off, too. On the way, I asked why the gendarme had acted the way he had. I was told that the spy had turned his brother in to the SS as being a possible Maquis. They had tortured him until he had died. When his relatives buried him, you could no longer tell that he had been a man, but since no other Maquis had been arrested, it meant he had not talked.

"The Germans couldn't make a move without the Maquis knowing about it, so we knew that a force of about 100 German soldiers was retreating toward our villages. This was the group that caused Papa to order the death of the spy. He was also sure that the Germans had orders to destroy the two bridges, and his intelligence indicated they would probably hit the bridge north of Epagne first, so most of the men and guns were dispatched to the other bridge.

"They set up along the bank of the river, about 30 men, and, of course, all the women and children from the villages were already at the stockade, except for four or five old folks who couldn't travel and one 11-year-old girl who wouldn't leave her grandfather. There was no way for the Maquis to defend the village and the bridge, so they had to position themselves at the bridge. The rest of us positioned ourselves around the Epagne bridge, and a motley crew we were. I couldn't believe some of the weapons the teenagers had—pitchforks, a shotgun with a broken stock held together with wire, some rusty antique rifles with only three or four cartridges. I had the .45 caliber pistol and several hand grenades, Andre had a Sten gun and the two Maquis with us each had good rifles and ammo. The Canadian had a rifle of sorts and a handful of ammo. I set up in some bushes next to the road approaching the bridge since I was the only one with grenades and we had been told that the Germans might have a light tank traveling with them, and we felt our only chance against that thing would be the hand grenades. Lucky me!!

"We settled down and waited. About 1400 that afternoon, we heard heavy firing from the other bridge; however, neither friend nor foe—not a single soul came down our road. About an hour later, a Maquis came running down from the other bridge and said we were all to come up and help at the other bridge. So off we went. Suddenly, just before we got there, all shooting stopped. We found that several Maquis had been wounded. Thank goodness, Papa and Henri came through okay, and we were told that the Germans had pulled back and were heading upriver in a hurry to get out of there.

"We cautiously approached the village and our worst fears were realized. The old folks had all been shot and the little girl tied to a bed and repeatedly raped, blood all over. She was in deep shock—conscious, but her mind a complete blank. We cleaned her up and the Maquis took her to the doctor that fixed me up, but the last I heard, they did not expect her to live.

"I realize now that this whole experience had a profound effect on me. From this point on, every time I turned my machine guns loose, I would be smiling, really enjoying it, no matter who or what was on the receiving end, as long as I thought it was German.

"The next day we saw our first Americans. A jeep came rolling into Epagne with two American soldiers, and I was immediately brought out as the whole town turned out. There were no more Germans to the west of town. The two soldiers were engineers scouting for Patton's tanks. They had become lost and were very surprised to find we had two bridges intact—a boon for Patton's tanks. At the moment, the tanks had outrun their gas supplies so were waiting for fuel about a day away. The soldiers said they would take me to the tanks and they in turn would get me back to an airport so I could fly back to London.

"It was really a happy and a sad farewell with much hugging and kissing as I really did feel like part of the Doiziet family, and I felt that they considered me their American son. My Canadian friend also went along with us, and that evening we arrived at some type of field headquarters the tankers had set up. It was here that I met the chaplain who got word to Betty that I was alive and back with friendly forces.

"I must digress again for a moment. This was one depressed tank outfit which had had more than their fair share of bad happenings. The day before I arrived, a young boy of about 10 or 12 years had come into their camp and said he knew where a German Tiger tank was broken down and could be captured. So they sent five of their old Sherman tanks to get the Tiger. The kid led them to a wooded area with a small clearing inside and, sure enough, there was a Tiger tank sitting there with hatches open and nobody visible. All five Sherman tanks rumbled into the clearing and the trap was closed with Tiger tanks in the woods on three sides. All five Shermans were destroyed with their crews, a total wipeout and only one Tiger lost. All the German crews got away.

"Then, the night they arrived there they had another bad deal. Their cook was on guard duty, heard something, called halt and shot at the same time, killing one of his own people. I was really glad to get out of there the next day to the city of Lyons, which had just been liberated that week. There I was turned over to an Army captain who was to look after me until a flight could be arranged. This was to happen ASAP since the intelligence folks really wanted to chit-chat with me in London. The Army captain was a really first class type and suggested a party that was really needed, a celebration by the whole town, and wanted us to join in—and we did. Two days later, I finally made it back to London, rocky but whole."

London, September, 1944—"I really don't remember exact dates here. I believe I must have arrived in London sometime during the second week of September, 1944. I spent about 10 days in the city. I really did get to know London again during those days. First off, I had my debriefing. This was the first time the debriefing team learned of the two good bridges in the area of Epagne; it was also the first time it could be confirmed that the area I was shot down in was under American control."

A BIG DECISION

"Since the debriefing was out of the way I was on my own until they decided what to do with me. I had decided that I hadn't accomplished much to date, so even though I had orders in hand telling me what boat and berth I was assigned to for my return to the U.S., I still felt that I had to stay. The fact that the area where I was shot down and the Maquis with whom I had been involved were now safe behind the American forces plus the fact that my squadron C.O. and the Group C.O. really went out of their way to get me back allowed me to be one of the first evades allowed to return to his unit for combat duty.

I sure had some conflicting feelings about this. I sure enough wanted to see Betty again, but I now had an entirely different viewpoint about this war. By turning down my chance to come home I felt almost fatalistic about ever coming home, but I had to do it, and I did.

"I returned to the 486th Squadron in September as a very different person than I had been in August. I had a different viewpoint on my mission over there as I had discovered a lot about myself, including my reaction to fear.

"Many people carry that little rascal with them and it jumps out when facing hazards. Some freeze up and cannot act; some just muddle through, sometimes successfully, sometimes not. But for me, fear became an ally which pumped up copious quantities of adrenaline and made me react faster than I thought possible, think clearer and see even better. I no longer worried about how I would react under stress—a big factor in building your confidence..."

EDITOR'S NOTE: Bill Reese continued to fly with the 352 until WWII ended and his military awards included the Silver Star, the DFC and the Air Medal with six Oak Leaf Clusters and the Purple Heart. He was officially credited with 4 enemy aircraft destroyed. The citation for his Silver Star reads as follows:

WILLIAM H. REESE, 0-801163m First Lieutenant, Army Air Forces, United States Army. For extraordinary achievement while participating in fighter sweeps over Continental Europe, 27 Dec. 1944 and 14 Jan. 1945. Observing two (2) Me-109s coming up from astern of his Feflight, Lt. Reese turned into them and with the first burst from his guns destroyed one (1) of the enemy. Then, executing a brilliant maneuver, he gained a vantage position and after scoring numerous hits on the second plane, observed the pilot bail out. On 14 Jan 1945, Lt. Reese was proceeding with his Flight when he sighted an Fw-190 far below. He immediately dived on the enemy, setting the aircraft on fire and forcing the pilot to abandon the plane. Climbing to rejoin his Flight, he encountered approximately fifteen (15) Fw-190's and made a bold attack. Despite the overwhelming odds, he split-up the enemy formation and destroyed one (1) enemy aircraft before withdrawing. The zealous, cool and determined manner with which Lt. Reese sought out and destroyed the enemy on these two dates exemplifies his disregard for personal danger, Entered military service from Missouri.

By command of General Doolittle

"And There I Was..."

by Walter Starck
As Told to Marc L. Hamel

Those words are familiar to many of us since they relate to so many situations in our everyday lives. "And there I was, out on a hot humid beach in the bright sun without my sunglasses." Or, "There I was, leaving the office after a hard day of deciphering patterns on schematics, only to find it raining torrents and not having an umbrella." Or, to a group of fighter pilots lounging around in the ready room and listening to one of their lads telling his tale of horror: "And there I was at 30,000 feet turning in a tight Lufbery trying to get inside of the enemy aircraft I was engaging so that I could make my shots most effective. As I was closing rapidly on him, my engine coughed and then failed. I soon realized that I, the hunter, was about to become the hunted."

Those opening words would get the attention of those fighter pilots whose interest became acute to find out how the speaker got out of that unhealthy situation. So it is in our daily lives that we find ourselves in situations wherein we could easily have said, "And there I was..."

Having been a fighter pilot and flown in combat in the European Theatre during WW II, I can readily relate to those attention-getting words, not just once, but many times. How easily you can tell your story by beginning with those words, "And there I was...!" They give you a natural opening."

Wally Starck with fresh medals in front of 487th Operations.

With those words we are introduced to Walter E. "Wally" Starck, fighter pilot and ace from the top scoring squadron in the ETO—the 487th Fighter Squadron. Wally's dashing good looks and fearsome tenacity in aerial combat certainly fit what many would consider to be the "fighter jockey" stereotype, but he bucks the norm with his genuine modesty and quiet humility. It was not without some effort therefore that Starck was finally persuaded to share with us his recollections.

Wally recalls an interest in airplanes from any early age, even building "planes" from scraps of wood and metal. He allows, "From as early as three years old, I wanted to be a fighter pilot. As years went by, that desire continued to grow until it almost became an obsession. In 1930, in Jetmore, Kansas, I saw my first airplane. In the summer, on weekends, a 'barnstormer' would fly his single engine biplane to a pasture outside of the town, giving people the thrill of watching him put his plane through acrobatics and varied maneuvers. Then he would take people for a ride for a small fee. Naturally, I did not have money to buy a ride but I sure enjoyed watching. What a thrill, just seeing that 'bird' fly.

"Throughout my years as a teenager and a young man, I had a desire to become a pilot. Not just any kind of pilot, but a fighter pilot in the US Army Air Corps. At that time, entry into the US Army Air Corps Flying Cadet program required a minimum of at least two years of a college education. Even though I was earning money as an usher in a theatre and even became an Assistant Manager, the earnings were insufficient to attend college and provide support for the family (*his father had passed away in 1933*). The Army did allow admission to the program providing the applicant could successfully pass a written examination, an equivalent to a two year college education. To this end I worked. A retired college professor provided instruction in nine major subjects. Learning sessions were held three days per week, four hours per session. This effort was expended for a period of about one and a half years. At about that time, the University of Wisconsin Extension Division established an accelerated night school program designed to assist young men like me to learn sufficient materials needed for this exam. It was while we were in class that we heard that, with the declaration of war on Japan, the US Army reduced requirements for entry into their flying program. In essence, you needed but to successfully complete a lengthy general purpose exam of about five hours and satisfactorily pass a rigid physical exam. When accepted, you enlisted in the US Army Air Corps and were immediately transferred into the Aviation Cadet program. The night school class was discontinued. Applications were made. Testings were completed and on 28 January 1942 I enlisted in the US Army Air Corps, was assigned to the Aviation Cadet program, and was on my way to getting those coveted silver wings of an Air Corps pilot."

Starck's career began at Kelley Field, San Antonio, Texas, progressing on to Primary Flying School at Vernon, Texas. PT-19's were the aircraft of choice, and of his first flight he enthuses,

Henry Stewart confers with Bill Halton at the "Aces Club."

"Flying. What a treat. The instructor flew the aircraft off the airfield and into the sky. This was my first flight. What a thrill. Here we sat, looking out on a small wing that you know is supporting the airplane and letting it fly. Amazing. And, to look around the countryside from up above was a thrill in itself. Soon you were given the controls and told to fly straight and level, make turns, climb and dive and do things to prove that you were in control of this machine. Too soon, the flight was over and we returned to the airfield, landed, and taxied to the ramp. There are no words to describe the elation I felt following that first flight. I was on cloud nine. Wow!!!

Second in line is Wally's "Starck Mad/Even Stevens" HO-X in Spring 1944.

"After about ten hours of dual instruction and upon returning to the field, the instructor got out of the aircraft and said, 'Go take it up yourself, Starck. It is all yours.' The moment had arrived. I was to solo. I did all the things I was taught, lined the aircraft into the wind, pushed the throttle forward, and in a few moments the craft was airborne.

"As I turned away from the airfield, I felt so elated that I actually yelled out loud. Kind of theatrical, I am sure, but it felt good. I was soloing. Flying alone, in complete control. Upon return to the flight line, I thanked the instructor for the confidence he had in me, to which he replied, 'Starck, you were ready to solo and you knew it.' I did.

"Needless to say that during my entire stay at Primary School, I let it be known that I wanted to be a fighter pilot. My flight instructor, my academic instructors, my fellow cadets, all got the word that I wanted to be a fighter pilot. Then the orders were posted. Wow! Starck, Walter E. was assigned to Goodfellow Field, San Angelo, Texas. That was a single engine school, a prelude to going on to single engine fighter aircraft down the road. Things were looking up!" At Goodfellow, Wally flew BT-13's, and he then graduated to Advanced Pilot Training at Moore Field, Mission, Texas. "The aircraft there was the AT-6, a single engine aircraft, designed to train pilots in the skills required for fighter pilots. Again my prayers were answered." November 10th, 1942 was graduation day. "Now I was an Officer with gold bars on my shoulders and silver wings on my chest. My assignment was to a Fighter Group at Westover Field, Massachusetts."

Little did he realize his good fortune at the time, but he was assigned to the 34th Fighter Squadron (soon to be renamed the 487th FS in May of '43). Wally continues. "After a short visit at the old homestead, I proceeded on to my new assignment. The Commanding Officer was Captain John C. Meyer (just returned from Alaska). The Operations Officer was 1st Lt. I.B. 'Jack' Donalson, an Ace with five victories to his credit, having flown P-40's in the Pacific Theatre before being reassigned to our squadron and group. Our aircraft was the P-47 Thunderbolt, with a massive R-2800 radial engine. It carried eight 50 caliber machine guns that fired 600 rounds per minute. Although a heavy machine, it was quite maneuverable." Soon, capable Starck was assigned Assistant Operations Officer and shortly after, promoted to 1st Lieutenant. "That sure helped the ego of this young fighter pilot," he recalls.

"During our time in the northeast, several happenings occurred that raised the eyebrows of local citizenry. The George Washington Bridge was a beautiful thing to see from the air. It spanned the Hudson River between New York and New Jersey. As we flew above the river, 'buzzing' as it were, one often was tempted to fly under that bridge—not only once, but many times. Of course, I never did *(wink, wink – author)*. But, according to the newspapers, some fighter aircraft did. Unfortunately, people who saw the feat were unable to get the aircraft identification numbers or letters. I wonder how high the adrenaline flowed through that pilot's veins? Mighty high, I am sure."

After intense Operational Training, the 487th headed for England with the rest of the 352nd in June of 1943. "The date we arrived at our base was 6 July, 1943. It was raining, a phenomenon of England we would get used to." Their base was, and still is, a grassed field in East Anglia southwest of Norwich. Named after a small parish near Watton, Bodney would soon gain fame as the 352nd FG took the fight to the Luftwaffe over Europe starting in September.

Living accommodations? "The Officers Club was also our mess hall and a good feed was laid out for us. We were initially billeted in a nearby manor, Clermont Hall, until our officers' huts were built. Eight officers were then assigned to each Quonset-type Nissen hut, our quarters. Single cots with G.I. olive drab blankets. No lockers for clothing but horizontal rods were fastened off the wall on which to hang our clothes. One small pot-bellied stove was located in the middle of the hut. This would burn coke and provide sufficient heat for creature comfort. Within a few days following our arrival we were able to organize our new home into something livable. We would be there for some time to come. I don't want forget one very important building. This was not a Nissen Hut. It was built of wood and looked just like an outdoor privy. That's what it was. It was most heavily used just after a mission briefing and before pilots prepared to take off. We called that operation 'the nervous pee.'

"*And there I was*....on a fighter base in England. Ready and eager to get into combat—but we had no airplanes. Soon our aircraft began arriving however—the mighty P-47 Thunderbolt. What a thrill, watching them arrive, engine exhausts roaring, flying low over the airfield, then peeling off overhead in a tight turn, wheels dropping down, flaps lowered, and in several cases making a three-point landing. We were pleased to get our contingent of sixteen aircraft (plus a couple of spares), and we could now become operational and get into combat. Much had to happen before that day arrived. The aircraft maintenance personnel had to go over every aspect of the aircraft, including painting squadron colors and the individual identification letters. Since there was a senior officer with a last name beginning with "S," I was reduced to select another letter of the alphabet. My personal identification letters were HO-X and I was most proud of that. Its serial number was 42-8684. In addition, I personally taped my initials "WES" on the bottom of the oval engine cowling so my crew could identify my aircraft as I taxied off the airfield toward the revetment. Naturally, I had my name painted on the left side of the cockpit just below the canopy opening. My crew chief, assistant crew chief and armorer's names were placed on the right side below the windscreen.

We learned tactical flight formations by squadrons as well as the entire group of 48 aircraft. We knitted our flight operations to a finely honed machine so that the flight leaders, element leaders and wingmen knew their specific role in every maneuver. Not easy, but rewarding to both the leaders and the individual pilots. We were proud of our performances, individually as well as collectively. We were ready. Then the day came. 15 September 1943." The 352nd went operational over enemy-held Europe.

"The primary mission of the fighter groups was escorting bombers as they lumbered along their route to varied assigned targets. In 1943, the P-47 did not have pressurized external auxiliary fuel tanks. It did have a fuel tank that was fastened to the underside of the fuselage, making the P-47 look like it was pregnant. It provided sufficient fuel to get us to about 20,000 feet at which time the auxiliary fuel tank was jettisoned. That occurred about the time we reached the coast of France. From that point, we flew the remainder of the mission on internal fuel. We would pick up our assigned bomber group just after crossing the French coast and escort it as far as our fuel would allow, generally to the German border or a few miles beyond. Except for occasional bursts of flak along the route, nothing really exciting happened. But upon arriving at the border or our turnabout point, we would see German fighter aircraft well ahead of our location. As soon as we began our turn, the German fighters would begin their attack on the bombers, hitting them from head on positions as well as from the sides and rear. There was

Keith Stevens, Wally Starck, Bob McKinney, and Bill McCurtain in an excellent photo with HO-X "Starck Mad" in July of 1944. (Sheldon Berlow)

nothing we could do to defend the bombers. To do anything would have depleted our fuel supply to a point where we would not be able to return to England. It hurt us to watch the relentless attack on the bomber force as it continued on to the target. The attacks would continue after bomb release and along the return route to England until other allied fighters could pick them up and escort them back to the English Channel. After encountering strong headwinds, my wingman once just made the runway where he rolled only a few feet before his engine quit from fuel starvation. That was close. Incidents like that made believers out of those who wanted to stay with the bombers for just a few minutes more than that calculated return point.

"Another mission assigned to our fighter group was called a 'fighter sweep.' On these missions, the group, consisting of three squadrons of 16 aircraft each, would fly into France at an altitude of about 30,000 feet and would spread out and search for enemy aircraft. These missions often resulted in shooting down one or more German aircraft. It was on a fighter sweep on 10 February of '44 that I encountered my first fight with an enemy aircraft. We were headed home, near Arnheim, when I saw a fight going on between an Me-109 and a P-47.

"My wartime Encounter Report for the day reads, 'I was leading Crown Prince Blue flight at 30,000 feet when we sighted a lone E/A attacking a lone P-47. I went down to attack and engaged the E/A at about 25,000 feet. The E/A attempted to dogfight but I was able to turn inside of him so he broke down. I followed him down and closed rapidly. I held my fire until I was about 150 yards from him and then let him have it. He was taking terrific evasive action but as I closed on him I saw hits all around his fuselage and wing roots. The E/A started pouring smoke and clouds of white vapor. He suddenly stopped his evasive action and as I pulled away at 14,000 he went down in a steep dive taking no evasive action whatsoever. I lost sight of him as I pulled up to rejoin my flight circling above, and we climbed up to 25,000 feet and came home. On the way home we made a pass at another Me-109 which had attacked a P-47 but we lost sight of him and did not fire our guns. On return to base many pieces of canopy glass were found in the cooling fins and oil cooler of my engine. Claim: 1 Me-109 Destroyed.'

"I read through my wartime mission reports recently. Boy, but they are simple and direct. None of them however leave any room for questions about anxiety, sweat, or tears. Just words. I will use them as a framework, and embellish them a bit with thoughts as I recall them.

"Regarding the 10 February mission, I recall I took my flight to that location and noted a P-47 going down in flames and the Me-109 getting on the tail of a second P-47. We were at 30,000 feet. I drove into the fray and found myself in a Lufbery with the Me-109, it directly across from me in a tight circle. I recall thinking, 'Starck, you fool. You saw this guy just shoot down a P-47 and then get on the tail of another P-47. Now you are in one-on-one combat with that same pilot. You better prove you are a better fighter pilot than he is.' We went round and round with me holding the circle ever so tight, reducing power on the engine, dropping a few degrees of flaps, increasing power, pulling back on the stick to make the turn tighter, until finally I was able to gain a little advantage. Soon I closed on him, and shot a few rounds that scattered on his wings and tail, but I needed to gain more lead to make my rounds effective. A few more rotations and the Me-109 wobbled momentarily in a high speed stall. That stall enabled me to get the favorable lead and I opened fire. All eight .50 caliber guns fired, throwing rounds into the engine, cockpit, wings and fuselage of the Me-109. The aircraft straightened out, diving to the ground with black smoke trailing out behind. I discontinued the attack at 14,000 feet. The pilot did not get out and the aircraft crashed into the ground. Slowly I took my hand from the joy stick and flew the craft with my left hand. My right arm seemed frozen in that position I held so long trying to gain the lead needed. I climbed up to where my flight was circling and we headed for our home base. In my first encounter, I met the enemy and the victory was mine. Upon arrival at Bodney, and taxiing to my parking spot, I noted my ground crew was grinning and giving me the 'OK' sign. They could tell something had happened because they saw the gun ports on HO-X were black with powder residue. Naturally, I had to tell them all about that encounter. By the next day, my 'bird' had a German Cross painted on the fuselage just below the canopy opening…my first victory."

On May 27th, 1944, near Strasbourg, Wally encountered his next opponent. He bases his memories on his Encounter Report, saying, "I was leading Yellow Flight in the lead squadron escorting the lead box of bombers at 25,000 feet. Suddenly the sky up ahead and above was filled with what appeared as a swarm of mosquitos. A large mass, all heading toward us. It looked to be a large 'glob' of airplanes. They were Fw-190's and Me-109's....100 plus, all headed for the bomber force we were escorting. Within moments, we and the bombers were engulfed in a wild melee of whirling, churning aircraft.

"In moments, it was all over, except that I and my wingman found ourselves alone flying in the middle of a Lufbery of approximately 12 Me-109's, appearing to be flying in three flights of four aircraft each. I was reefed in hard, actually making a smaller circle inside of the 109's. To do this meant alternately partially lowering flaps and airspeed and flying close to a stall. Whenever I would try to get my sights on an E/A, I would see other 109's ready to come in on my wingman and myself and would have to break off the attack. After several rotations of this type, one of the 109's in front of me appeared to stall. In a flash, all aircraft disappeared, leaving only me and my wingman in the lonely skies behind a 109 which broke away and headed east. I tore out after him, staying a little below so that he would not see me and thus it gave me a chance to close on his tail and let him have it. It seemed we flew at wide open throttle for about five minutes. When the 109 pilot did see me, he turned sharply to his left. As I tried to cut him off, he turned the opposite way and we were back where we started. Then the E/A made a sharp diving turn to the left and then the right. I swung my ship around and as he turned to the right, I managed to close to about 150 yards from him, measured off about 20 degrees deflection, and opened fire. Strikes showed up all over the cockpit in one brilliant mass. The E/A caught fire, did a half roll, and spun slowly towards earth where it crashed and burst into flames. The pilot did not get out.. I went down to take a picture of the wreckage and then headed back upstairs and went home, all alone." Scratch one Me-109.

Wally next got a "probable" on June 20th northwest of Magdeburg. His typically brief Encounter Report allows, "While escorting bombers out I saw a flight of three E/A coming in behind and to the side of the box of B-17's. I took my flight and swung around behind and below the A/C. They saw me and pulled up into a steep chandelle to the left. We pulled up too and turned inside the last two E/A. The first did a couple of snap rolls and hit for the deck. The second hit for the deck too. I took off after him and closed rapidly. At 300 yards I opened fire and hit him in scatters on the wings and on the side of the fuselage. At this time he saw me and broke off. I went after him but was shooting above him. At about 25 to 35 yards I stopped firing to swing out and come in again, but as I let loose the trigger my right outboard gun ran away, and I, thinking someone else was shooting at me, broke off combat. In so doing I lost my E/A. My wingman last saw him going straight down from 8,000 feet, and did not see him pull out when he disappeared from sight. Judging from my hits soon before breaking off and what my wingman saw, I claim a probably destroyed Fw-190."

Adolf Bielock supported Starck with, "I was flying Red 2, wingman to Captain Starck. I saw him fire at the E/A and saw it going straight down at 8,000 feet as set forth in Captain Starck's statement above."

On the 23rd of June, 487th pilots Starck, Goodwin, Drisko, Bowers, Berkshire, Moats, Bielok, and Fowler did some strafing, with the combined re-

Wally Starck posing on his P-47 HO-X. (Wally Starck)

port reading, "I was leading Yellow section when we entered area Number 2 where we were to strafe anything moving on roads or railroads. At Melun (France) and in the near vicinity were two trucks which the flight strafed. One truck was destroyed and the other damaged. At the marshalling yards were two freight trains with engines which were strafed. The engines were engulfed in steam after our attack. Nothing happened to the freight cars as we fired at them so I assume they were empty. The trains were heading in a southeasterly direction. After strafing trains the Op Commander called everybody out and we climbed to altitude and went home. Claim: 2 Locos destroyed, 1 truck destroyed, 1 truck damaged."

"On July 1st I was leading the Group into France on a strafing mission, after our bomber-area support in Pas de Calais had been cancelled. We went directly to our area when the controller called us about something in the air 30 miles ahead. When we arrived we found 20-plus Me-109's and a group of P-47's. I spotted two '109's and immediately attacked. They turned into us but to no avail, so they immediately broke for the deck. I tagged onto one and closed rapidly to 100 yards and opened fire. The 109 caught fire after hits on fuselage and engine and the pilot bailed out at 12,000 feet, almost hitting my wingman. I then saw another 109 on the cloud below and went after him. He saw us coming and ducked into the cloud. I played "cat and mouse" with him until he broke out into the open and then dove down onto his tail. I was then at 450 mph and could not slow down to get sights on him. I did manage to get sights on him at about 100 yards astern but my overtaking speed was so fast that I almost rammed him before pulling over him. As it was I managed to get a few strikes on his fuselage behind the cockpit and on the wingroot. Lt. Heyer, my wingman, said he was smoking as he went into the clouds and disappeared from sight. We stuck around waiting for the 109 to come out, but he did not. We then pulled up. We were at 4,000 feet and headed back to altitude to get into the scrap going on at 30,000 feet with other parts of the group. At 28,000 feet I noticed fuel was getting low so I called the entire group out and we headed home." Make that one Me-109 Destroyed and another Me-109 Damaged.

A "Ramrod" penetration on July 21st ended up with another score for Starck east of Munich. "I was leading Yellow Flight and flying on the port side of White Flight. We saw some bogies high to 9 o'clock and went to check them. They were 10-plus Me 109's. Major Preddy, leading White Flight, came in behind them from below and broke up the formation. I picked out one of the E/A that broke down and away and went after him. The E/A apparently did not see me coming and as he pulled up I was able to close to about 230 yards. I opened fire and strikes appeared on the cockpit, right side of the engine and fuselage and on the right wing root. The E/A began to burn and smoke badly and coolant and gasoline poured from it. I ceased fire and pulled away and watched it as it went down in a crazy uncontrolled spiral. I then pulled back up and with my flight continued escort. Lt. Heyer, my wingman, witnessed this encounter." Another Me-109 destroyed. Sheldon Heyer confirmed this with, "I was flying Yellow 2 on Captain Starck's wing and witnessed the above encounter. I saw Captain Starck get hits on the wing roots and canopy of the E/A, it burst into flame and went down spinning. I think the pilot was killed."

Wally now shares some non-combat information with, "Early in training, a fighter pilot is injected with the fighting heart and is told, 'Practice makes perfect.' During our stay in England pilots were always practicing maneuvers that would increase their effectiveness. Having just returned from a 30 day Rest and Recuperation leave back home in Milwaukee, I needed some flying time in the P-51 to regain my skill. It had been three months since my last flight in a P-51 and I was mighty rusty. Shortly after returning to the base in England, I made several flights to regain confidence in the '51 and hone my skills. It was on one of these flights that I saw an RAF spotter aircraft, similar to the US Army's L-5, flying at a low altitude. With a 'Tally-Ho' attitude, I split-essed from some altitude and tore after my quarry. As I came upon the light aircraft, I noticed it had British markings, and he was ready for me. You can imagine the 'battle' that ensued. A 400+ mph fighter chasing around after a little 70 mph A/C. It was fun. The British pilot had great skill and managed to break away from my thrusts time and again. Of course, I was not to be defeated. I did all sort of things to get on his tail—chopped throttle, reduced airspeed, alternately dropped and then raised flaps to make sharper turns. It all worked great. I was amazed at how skillful I was in dogfighting such a small light airplane. Then 'it' happened. I noticed that the control column was loose, as though all the cables had come off. No resistance of any kind. I glanced at the airspeed indicator and saw I was flying at 62 mph. The P-51 does *not* fly at 62 mph. INSTANT SWEAT. I immediately pushed the power on to its fullest and milked the flaps up slowly, all the while making sure I maintained straight and level flight. I think I even stopped breathing so as to not shake the airplane. Slowly, the airspeed came back to 100, 150, 200, 240 and I began to breathe again. I realized I had reached a point where I could have stalled the Mustang and it would have crashed. Oh Boy…and there I was. How stupid can a grown man get? And you know what? I never told anybody about that stupid stunt until now.

Capt. Walter Starck

"On November 21st, a little while after that 'stunt,' I damaged an Fw-190 30 miles Southwest of Leipzig. My EC read, "I was leading White section. We were stooging around under the haze layer around the target area and found nothing of interest. We found a hole and went up through it. Above the haze we saw three small formations of contrails which shortly after merged into one large formation above the vicinity of the target. They were headed about 240 degrees. As we were about 8,000 feet below them we lit out ahead of them, gaining altitude. As we reached about 31,000 feet the formation of about 35-50 Fw 190's passed underneath us. Apparently they were on their way to hit the tail of the bombers. As they passed under us they made about a 120 degree turn to the right and headed back into Germany. I brought the squadron around and attacked from the rear. Unfortunately, I turned a little too late and we had to use full throttle to catch the formation. I finally tagged onto the tail of a 190 and bore down on him. He was flying perfectly straight and carrying a belly tank. When I got about 300 yards from him I opened fire. Only once did I observe strikes on his fuselage and wing roots. Then I was flying in his dense contrail and prop wash. I flew on firing all the time, until I noticed the contrail was very thick. I pulled up very sharply and zoomed right over the 190. Had I stayed behind him a moment more I would have rammed him. I could not see what damage was done as the formation of E/A's went into the heavy haze and disappeared. Claim: 1 Fw 190 Damaged."

Wally next tells us about the beginning of his last wartime mission. "On November 27, 1944, our Group departed Bodney on a mission to Northeast Germany, in the vicinity of Leipzig. It was a fighter sweep, a search for targets of opportunity. It was a beautiful day, a fighter pilot's dream. Clear skies. A white cloudy undercast up to about 10,000 to 15000 feet measuring about 3/10th sky cover. The background would show enemy aircraft that might be below you and it could be a safety haven should you need a place to hide in if chased by an E/A."

Captain William T. Halton turned in the following Encounter report for Wally upon Halton's return from the mission: "I am the Squadron Commanding Officer of the 487th Fighter Squadron. I make this encounter report for Captain Walter E. Starck who was M.I.A. the day this encounter took place. Shortly after I broke off combat I heard Captain Starck (who was leading Red Flight) call on the R/T and say that he was hit and was heading for friendly territory, but that he thought he was going to have to leave the ship. He wished all the boys luck and said he had gotten two 109's. Nearly everyone in the squadron likewise heard Captain Starck say this. Knowing Captain Starck, who was the Squadron Operations Officer, it is my opinion that he did get the 2 victories he claimed. William T. Halton. Claim: 2 Me-109's destroyed."

An attached supporting statement was made by Al Rigby, who wrote, "I was behind Captain Starck when he attacked the first 109. I saw him get strikes and both wheels of the E/A came down at 10,000 feet. The E/A flipped over and went down through the overcast at 6000 feet in a wide uncontrolled oscillating spin. I did not see his second encounter. 2nd Lieutenant Alden P. Rigby."

First Lieutenant Henry M. Stewart added, "I was also near Captain Starck when he attacked his first 109. I saw the wheels of the E/A come down and saw it flip over and start spinning."

Captain Seymour Joseph supported this with, "I am the Squadron Intelligence Officer and have been for the past 2 years. I have known Captain Starck, the Squadron Operations Officer, for this time. I know him to be scrupulously fair and honest and in my opinion he would not have claimed two E/A destroyed had he not obtained the victories beyond any question of a doubt. Moreover, the first claim is supported by the statements of Lts Rigby and Stewart. Unfortunately no one saw his second, but knowing him, his word is sufficient."

As we can see, there was some confusion about what actually happened to Starck, as well as his victories. We are fortunate to have Wally's first person account of that mission. "At about 1145 hours, the MEW Controller requested we proceed to an area where he thought something might happen. We were at 30,000 feet in a tactical formation. As we turned to the heading specified by the controller, we saw gaggles of enemy aircraft coming from different directions, apparently bent on gathering in a mass formation like they would use to attack a bomber formation. The Group Commander opened fire on them and they split into their original flights, turned and headed for home. I took my flight and headed for one of those flights, about twelve to fifteen Me 109's. They were decending rapidly so I applied full power to catch them.. They still held their tight formation. I called my flight to follow me and told my Group Leader my intentions. The enemy aircraft were about 5000 feet below me when we started after them. With a full throttle setting, it was not long before I caught up with the gaggle. I adjusted my K-14 gun sights to best fire upon the 109 and as I was about to pull the trigger, the E/A canopy flew off and the pilot bailed out. The E/A wobbled a bit and then flew haphazardly without the pilot to control it. Then things started to happen to it. A wing came off, then the tail separated and it fell towards the earth. All this took only a few moments and I went back to the task at hand. I set sights on the next 109, adjusted my deflection and opened fire. I only fired for a few seconds and I noted bullet strikes on the engine and along the side of the aircraft. It immediately fell off and started to spin (apparently this is the attack Rigby witnessed). Soon pieces were coming off the A/C as it spiraled down. I never saw the pilot leave the aircraft nor was there a parachute.

"The enemy formation was just beginning a turn to the right when I shot the next 109. The aircraft near the tail of the formation was my next target. My closure rate was great and I was very close to his tail when I opened fire. It was only a short burst because I saw most of the strikes were on the tail of the E/A. At that point the main tail section (or at least a large part of the tail section) came off, and as I made a skidding turn to stay with the E/A, that tail flew back and hit my aircraft. My aircraft shuddered and I immediately broke off contact. I noted the engine temperature rising and the oil pressure starting to drop. Moments after I pulled away, engine coolant was spraying on my windshield and along the side of my aircraft. That signalled to me that the underside of the nose had received damage from that E/A tail section and that I was in for all kinds of trouble. I throttled back while turning to a heading that would take me home to England. I called the Group Leader and told him of my troubles and advised him I had destroyed 2 E/A and was heading home. Why I did not tell him there were 3 E/A, I do not know. When I had time to think, I recalled all three but that was of little concern to me at the time.

"I was alone. No other aircraft were around. I kept checking for E/A but saw none. I am sure I was trailing smoke or at least engine coolant and nothing is a better target than a crippled bird. Soon live flames were flitting past my right side of the canopy. My temperature guage was stuck on the red line. I thought about what to do and decided I had best leave this machine and use the parachute. After disconnecting the seat harness, the seatpack dinghy, and the earphones, I blew the canopy. I then rolled the trim tab to a full down position, stood on the seat in a crouched position, pushed the control stick far to the right to put the aircraft onto it's back, let go of the stick and away I went. I guess the aircraft flew up and away from me. I then let my body free fall for what seemed like 4 or 5 minutes, not wanting to provide a target for somebody who might like to shoot people coming down in parachutes. I delayed pulling the rip cord until I estimated I was about 5000 feet above the ground. When I did, I felt a mighty jolt which was then followed by another jolt. Then all was quiet.

"I opened my eyes and noted I was near a large farmhouse. Glancing down, I saw my feet were only 2 feet above the ground. Looking up, I saw that my chute canopy was caught on the top of this large tree. That caused me some concern. Was I 5000 feet high when I pulled the D-ring? Or was it only 600 feet? Shortly after, my jaw started to ache. I soon realized that I had been knocked unconscious by the harness buckle, which raised up abruptly as the chute opened, and smacked me on the jaw. I never heard the birds singing nor the wind whistling though the port at the top of the canopy, something I heard so much about. Now, when I had had the chance, it hadn't happened to me. And, so ended my days as a combat pilot.

"I soon realized there were quite a few German civilians standing around. We were told we must try and escape, but any thought of that evaporated when I noticed some of these people had dogs and shotguns. They took me to a farmhouse and stripped me down, searching for a gun. A young German girl there was attempting to be an interpreter with her limited English, but it was very difficult to communicate. They next hauled me to a fire station where I was locked in a small room, and then on to a Luftwaffe airfield around 3 or 4 p.m. where an attempt was made to interrogate me. The interrogating German officer accused me of 'flying bombers and killing innocent women and children.' I was finally taken to a train station in Osnabruck around midnight along with four captured American bomber crewmen. At least one of them thought I was a German 'plant' or spy as I was considerably cleaner than they were after they'd evaded in a forest for several days. Unshaven and dirty describes them.

"I later arrived at the 'Durchgangslager der Luftwaffe' (Dulag Luft) at Oberursel which was 13 kilometers northwest of Frankfurt." Many captured airmen and other POW types transitioned through the interrogation center – 29,000 or so during 1944 alone. The "Master Interrogator" Corporal Hanns Scharff worked the tough-guy routine on Wally for 19 days, but received only "name-rank-serial number" in return. Scharff had file folders on most of the 8th AAF pilots, and would regularly stun incoming prisoners by sharing with them parts of what he already knew about background and unit. This array of "gen" would often cause the prisoners to "drop their guard" and tell other bits of information, assuming that all was already know by the Germans. As we see, it didn't work on Wally. Finally tiring of the little he was gleaning from Starck, Scharff released him to the Stalag Luft POW camp system.

Wally was sent to the officer's camp Stalag Luft I at Barth, Germany on the Baltic Sea coast. And there he was. Fellow Bluenoser and 486th CO

487th pilots relax by their Nissen's on the southeast edge of the field.

Luther Richmond had earlier joined the Stalag Luft I ranks in April of '44 *(See Richmond's story of "Stalag Luft I" later in the book for details.)*

Wally requested permission to meet Richmond in another part of the camp, and they were able to spend a short time catching up on events. Wally endured the hardships of the camp, and dropped from his fit and fighting weight of 145 pounds down to 110 due to the lack of adequate nutrition. In one of his few allowed letters home to his mother, he asked that she contact his CO at Bodney and tell him to chalk up that last victory he'd failed to mention when downed. Starck was "sprung" from the camp two days before Mother's Day 1945 by the Russians, and was transported to Camp Lucky Strike in France. To his amazement, he met his brother (and B-17 gunner) Carl there, who had been imprisoned in a camp in Poland.

Wally decided to remain in the post-war Air Force, and after marrying his sweetheart Millie in late 1945, was sent on occupation duty to Schweinfurt, Germany. He also participated in the Berlin Airlift, and retired as a Colonel in 1965 after a stint with Strategic Air Command.

This article is dedicated to my tireless and faithful crew: Keith F. "Steve" Stevens, Crew Chief; Robert L. "Bob" McKinney, Assistant Crew Chief; and William R. "Bill" McCurtain, Armorer. When Major General William E Kepner referred to the 352nd Fighter Group as "Second To None," I can't help but feel he was also referring to my crew.

WALTER E. STARCK
Colonel, USAF (Retired)

Sad Mission of July 21

Sanford Moats (Berlow)

July 21st was an extremely painful day for the 328th Squadron of the 352nd Fighter Group. The mission was an escort to Regensburg and the Group rendezvoused with four combat wings of the 3rd Air Division right on schedule.

The 486th FS ("A" Group) escorted the front of the bomber formation and the 328th ("B" Group) escorted the rear formations. Just as they were assuming their escort positions they spotted a flight of ten Me-109s and Major Preddy led the squadron after them. Two of the bandits were downed in the initial attack by Capt. Walter Starck and Lt. Malcolm Pickering and a third was shared a few minutes later by Major Preddy and Lt. Sanford Moats. Pickering was brief and to the point in describing the victory.

"I was flying as White 3 when we attacked about ten 109s. I picked out one on the right of the formation. He was in a slight climb at about 28,000 feet so I held my fire. He then started a dive and I opened fire at about 300 yards. I noticed strikes immediately in the engine and cockpit. He caught fire and trailed a huge cloud of white smoke. Then he spun out of control and crashed. The pilot did not bail out."

While this air battle was taking place the remainder of the 352nd Group continued its escort to the target area and observed excellent bombing by the B-17s. The B-17s escorted by the 328th did not hit their primary target but did hit targets of opportunity near Nurenburg with good results. Afterwards, the squadron broke escort and headed toward a date with fate.

David T Zimms and Francis Horne

Things started going wrong shortly after Lt Francis Horne radioed "let's go home." Most of the journey was through a thick overcast and heavy headwinds later estimated to have been in excess of 70 mph further hindered their flight west across the Continent. Fuel consumption increased sharply as the Mustangs struggled against the elements and the whole squadron began experiencing low fuel readings.

Their troubles took a turn for the worse as the 328th made its way over the North Sea with fuel gauges now down to the critical point and land was nowhere in sight. It became obvious that the unusually high winds aloft had carriend them way off course, causing them to fly almost parallel to the English coastline still invisible to them.

As their fuel supplies gave out, three of the Squadron's newer pilots, Lts. Richard Casper, Robert Lampman and James Lanter disappeared into the North Sea, never to be found.

Fortunately for the Squadron, flying with them was an exchange pilot from the Royal Navy's Fleet Air Arm who had been assigned to the 352nd as a liason officer. He was Lt. Richard B. Law, who had experienced flying in such weather during the Battle of Britain and had actually been rescued from the cold North Sea when shot down. Dick called Lt. Horne and offered his help in making a course correction from about about 270 degrees to about 250 degrees, more into the wind. According to the other pilots, this adjustment was instrumental in getting them back on track.

Even so, the 328th's trials and tribulations were not over. The engines of Lt. Dave Zimms and Elmo Dubay's Mustangs were the next to sputter and shut down. Luck was with them, however, as they bailed out over ships patrolling the North Sea and were quickly rescued by the crew of a U.S. Navy minesweeper.

Glenn Clark by "Mom and Dad II" (Berlow)

As the other planes managed to make landfall, two more 328th pilots found themselves in trouble and had to crash land along the shoreline, one of them landing just short of a mine field on the shore. Lts Charles Rogers and Glenn Clark both got their planes down safely without injury.

Lt. Elmer Smith became the "miracle man" of the mission, however, by keeping his plane airborne long after his tanks should have been dry. In his struggle for survival he found that constantly priming his engine pulled the last dregs of gasoline from his tanks and kept his plane aloft for twenty minutes longer than the flight manuals said it would, getting him back safely, too.

Of the fifteen Mustangs which flew the mission, only eight returned to Bodney. Three pilots were lost, five planes destroyed and two others badly damaged. Ironically, the weather and not the enemy had dealt the 328th Squadron its worst defeat of the war. The 352nd stood down on the next day and flew only three uneventful missions during the next six days.

Note: Dick Law later returned to the Royal Navy's Fleet Air Arm and retired as a Captain. After his retirement he was employed by the British aircraft manufacturer, Hawker-Siddley.

Lt. "Frenchy" Dubay and crew (Berlow)

British Fleet Air Arm pilot Dick Law and Major Earl Abbott discussing tactics while sitting on Abbott's "Flossie II" (Berlow)

Lt. Elmer Smith at wing root of his PE-Ebar. (Elmer Smith)

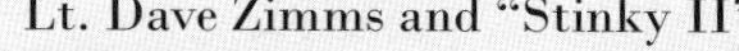

Lt. Dave Zimms and "Stinky II"

Ray Littge, The Little Known Ace

By Tom Ivie

Although Capt. Ray Littge became the third highest scorer in the 352nd Fighter Group with 23.5 victories (10.5 air and 13 ground). his exploits have remained relatively unknown in the years since World War II. This may be due to the fact that he was a replacement pilot viho niissed out on the major air battles over Europe during the spring and early summer of 1944 when the more well known aces were gaining the free world's attention. At the lime of Littge's first missions. beginning in July 1944, the opportunities for frequent aerial combat had lessened somewhat. But from a wingman's point of view he watched the handiwork of the 352nd's best, Maj. George Preddy, and learned his lessons well.

Like George Preddy, Ray Littge developed an interest in aviation at a young age, but his road to becoming a pilot was a "little longer and rockier." His brother Ralph tells of Ray's younger days:

Ray Littge and Cy Doleac (displaying his 487th "Little bastard" insignia) share some humor in the 487th area.

"We were eight kids born to impoverished parents. Our father died when Ray was only four and we were raised by our mother and older sister, Altheda. By the time Ray was sixteen, the family was able to buy a bike for Ray and his younger brother, Vernon, to share. Shortly afterwards, Ray set off on a trip of 110 miles to visit his sister in St. Louis. While there he got to see a large passenger plane operating out of Lainbert Field and immediately became obsessed with a fanatic propensity for airplanes and flying and made a solemn pledge to himself to someday become a pilot.

"He read every aviation magazine he could get his suds on and, with his scanty allowance, bought and built many a model airplane. At seventeen he landed a job as a farm hand during the summer harvest and earned enough to buy a worn out 1931 Chevy. With some money begged his mother unrelentingly for permission to take flying lessons. When she finally gave in he started taking lessons at a small field about forty miles from his hometown of Altenburg. Missouri, earning his certificate in short order. After finishing high school in 1942, he enlisted in the Army Air Force and the rest is history."

He won his wings and commission in December, 1943 and got his orders to head for England in May, 1944 assigning him to the 487th Squadron of the 352nd Fighter Group at Bodney, England. He flew his first mission on July 1, 1944 and this trip over occupied Europe gave him a good preview of what the air war was like.

During the mission his squadron encountered a gaggle of 109s and claimed one destroyed and three damaged during the engagement. Littge's next few missions were uneventful, but on his seventh mission on 18 July he got to watch Major George Preddy at work. The mission was a Ramrod (escort) mission to the Kiel area and the 352nd encountered swarms of enemy fighters. When the engagement had ended the Bluenosers had scored 21-0-11 against two losses. Littge recorded it in his diary: "Group got 21 enemy aircraft—Lu-88s and Me-109s. My flight leader got two Me 109s. Lots of excitement. Saw at least seventy enemy aircraft and Major Preddy got four. Elhson got three. Fowler got two—all so fast I didn't even fire my guns—I was his wingman."

During the next several missions the 352nd engaged the enemy in dogfights and claimed numerous victories but Littge was not scheduled to fly on those days. His luck finally changed on August 25, 1944 during an escort mission to Germany, when he claimed his first two victories. His diary records this achievement:

"Bomber escort to Northern Germany. Flak, but no onemy fighters. We hit the deck and strafed an airfield at Neubrandenburg in Germany. Got two Me-210s on the first pass and damaged a 109 on the second. The bombers bombed field between passes." He added, "Their airfields are pretty strongly defended by flak, and it's not very healthy to stay long in one spot. When we hit an airfield we get down as low as possible—right on the ground, below the trees, etc. going about 350 mph. After hitting it we usually keep right on going and seldom make another pass on a field until the next mission." These victories occurred on his eighteenth mission. He raised his total to four destroyed on his twenty-third mission on September 11th by destroying two Ju-88s in another strafing attack.

During the next two-and-a-half months Littge flew a mixture of escort missions and sweeps. Included in the late-September missions were the air cover sorties over the parachute drop zones in Belgium. It was not until his forty-sixth mission, which was flown on November 27, 1944, that Littge was able to test his skills in aerial combat. In that engagement he engaged Me-109s in two separate dogfights over Hamelin, Germany and destroyed two of the enemy aircraft.

Ray commented on these victories and on becoming an ace in his diary and in a letter to his fiance, Helen Fischer. "We ran into approximately 400 Me 109s and Fw 190s. I was flying with Major Halton. I got two and the second gave me quite a tussle," In his letter of December 15, 1944, Littge says. "Helen, can you believe it, I became an ace on November 27! I got my fifth and sixth planes that day—they were Me 109s. The going has been quite rough here lately; but that is the thing we live for—I mean we live for the days when the Luftwaffe shows up."

By the time Ray Littge got the opportunity to engage the enemy in aerial combat again, the 352nd Group had been moved to a forward aerodrome near Asch, Belgium in order to help support the ground troops which were engaged in the Battle of the Bulge. This action occurred on December 26, over Olheim, Germany when his flight engaged a gaggle of twelve Me 109s. Littge's account of his seventh kill was very brief. "...I was flying White 4 when we bounced 2-plus Me 109s at about 10,000 feet. They were in a turn to the left, and we bounced them about in the center of their formation. This left two 109s behind me. causing me to break to the right. I then bounced a 109 that was shooting at a P-51. This e/a broke for the deck immediately and I started shooting at 800 yards down to 150 yards, seeing many strikes and setting the left wing on fire. The fire stopped after a little while, and he climbed to 6,000 feet and bailed out."

One day later Littge raised his total to ten by destroying three Fw-190s in one swift dogfight. In this engagement, Littge's flight bounced eight Fw 190s on the deck, and the e/a immediately went into a loose Lufbery circle. Littge and his wingman, Lt. Russell Ross, fell in behind the last 190, and Littge downed it with a long burst. Seconds later he downed a second 190 in the same manner. While this was going on, all but one of the remaining 190s were either shot down or dispersed. Spotting the one remaining e/a trying to escape, Littge turned and closed on it, describing his final victory of the day as follows: "I got strikes on him several different times. He straightened out, jettisoned his canopy and started pulling up. Then an unidentified P-51 came down from above and got several strikes as the pilot of the e/a bailed out."

The 21-year-old ace's next victories, numbers 11 and 12, occurred on January 1, 1945 during the 487th Squadron's historic defense of Y-29 airfield. Both enemy planes were Fw-190s and Littge downed the first one shortly after taking off. He caught this enemy aircraft with a long burst which struck its wing roots and the cockpit and sent it crashing to the ground. The second took a little longer. Littge hit this 190 with several bursts as he chased it at a very low altitude and witnessed a trail of black smoke emitting from it. The enemy pilot then began violent evasive actions that threw off Littge's aim, and he ran out of ammo without further strikes. Nevertheless, he kept trailing the e/a, which continued nearly to Paris before its pilot pulled up and bailed out. For his part in the heroic defense of Y-29, Littge was awarded the Silver Star for gallantry in action. Even though Littge certainly qualified as a battle-hardened veteran, he seemed to indicate in a letter to Helen that he was surprised by his own success. In a letter dated January 9, 1945, he described the New Year's Day battle and his two victories, and ended the description with this passage, "You know that makes 12 victories now! Gee, when I

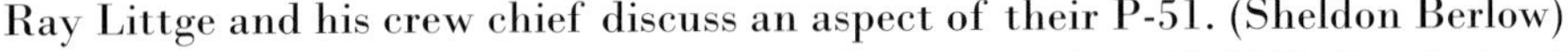

Ray Littge and his crew chief discuss an aspect of their P-51. (Sheldon Berlow)

first came over here I thought to myself, 'Golly, I hope the Luftwaffe isn't wiped out before I get a chance to fly against them.' No kidding, I thought that, Helen!"

On January 24, 1945, the 352nd Group, still operating under the control of the Ninth Air Force, was patrolling in the Bonn-Koln-Dortmund area of Germany. In the vicinity of Wahn aerodrome, two Me-109s were spotted stooging around the airfield. White Flight went after them. The first was shot down by Major Bill Halton. A few minutes later, Littge caught the other one in the landing pattern and shot it out of the sky for victory number thirteen. The enemy pilot tried to bail out of his stricken plane at 200 feet but his chute didn't open.

In February 1945, the 352nd moved to a new base, Y-84, near Chieves, Belgium and returned to Eighth Air Force control. Littge's final aerial victories were scored while flying from Y-84. The first of these was shared with Lt. Jim Woods when they downed an Me–109 over Berlin. His last aerial victory came one week later and it proved to be a memorable one. Littge had long been yearning for an opportunity to battle an Me 262. He had encountered them before but never had the chance to engage one of the speedy jets. On March 25, 1945, his dreamed-of scenario presented itself. He described that encounter as follows:

"I was leading Yellow Flight when jets were called in at the target area of Hitzacker. I flew over where the smoke bombs were and immediately spotted an Me 262 there. He was below me about 5,000 feet going in the other direction I kicked my tanks off and rolled over after him. My initial speed brought me within 1,000 yards of him and I stayed there for a while. He gradually pulled away and after about a fifteen minute chase, he was out of sight. When I pulled up I was directly over Rechlin aerodrome near Muritz Lake. As I was circling it, an Me–262 came over the field, presumably the same one I had chased before, and peeled off. As he was lowering his wheels I made a 100-degree pass at it, seeing no strikes. He leveled off then and I got behind him, fired several long bursts, seeing quite a few strikes, several of which set his right jet on fire. His evasive action consisted of gentle left and right turns. He jettisoned his canopy, pulled up to 2,000 feet and bailed out. His chute did not open."

With the Rhine bridgehead secured and the Allied armies advancing steadily into Germany, the need for keeping the 352nd on the Continent no longer existed and the Bluenosers returned to their home at Bodney. From there they flew the group's final missions of the war. Two of the most spectacular of those missions, in terms of victories, were the missions of April 16 and 17, 1945, when the Bluenosers claimed 39-0-27 and 66-0-24 strafing victories. Littge, now a captain, played an important role in both of these missions. In both, his was a leadership role. The attacks he directed accounted for approximately one-third of the total claims. From the standpoint of individual scoring, Littge excelled on these missions, too, by personally claiming nine enemy aircraft destroyed and five others as damaged. On the mission of April 16, Littge claimed 3-0-5 and followed this up on April 17 with claims of six destroyed in a most courageous assault against Plattling aerodrome. His combat report read in part: "I was leading Red Flight consisting of Lieutenants Reed and Reynolds, F/O Frum and myself. After leaving the bombers we came upon Plattling aerodrome with seventy-five-plus single-engine and twin-engine aircraft parked in dispersals and around the field. Red Flight made several passes at flak positions first, effectively silencing them. During these attacks my oil tank was hit and I lost most of my oil, one of my guns was shot out and two electrical lines and the manifold pressure lines were hit. *[Not mentioned in the report. Littge also lost one and- a-half feet of his left wingtip to flak.]* We then went in to strafe. I made seven passes. My first two passes were at an Me-262 on the northeast corner of the field. It blew up after my second pass. I then attacked and set fire to an Me 109 on the north side. On each of my next three passes I set afire Me 109s in revetments on the south side of the field and on my last pass I blew up another Me 262 at the northeast corner of the field." With these final six destroyed, Littge raised his total to 23.5 and earned himself the Distinguished Service Cross for the extraordinary heroism he displayed on that date.

Even though his shooting war was now over, Littge had another unusual encounter with

Little did Ray Littge know that his plane "Miss Helen" would still be flying as this book is written. Robs Lamplough owns the plane currently in England.

an Me–262. Through the efforts of his brother, Ralph, who served in military intelligence, Littge was selected to interrogate a German Me–262 pilot, Hans Fay, about the plane he had surrendered to the Allies. As it turned out, this was Littge's ticket back to the States. The USAAF decided to send Fay and the Me–262 and Littge back to Wright-Patterson AFB to evaluate the plane. Unfortunately, due to a snafu in his orders when he arrived in the States, Ray was sent off instead to Perrin AFB, Texas, and missed out on the 262 evaluation program.

Later that year, on November 25, 1945, Littge married Helen Fischer. They had two sons. The first was named George Preddy Littge, after the man Littge so greatly respected, and the second was named Ray II. Sadly, Capt. Raymond H. Littge was killed on May 20, 1949 while leading a flight of F-84s from Hamilton Field, California, to Coulee Dam, Washington. During the flight, Littge's F-84 fell out of the formation and plummeted to earth. Apparently his oxygen system had malfunctioned and the unconscious Littge fell to his death.

A sad footnote to this was supplied by Littge's brother, Ralph, who wrote in an article prepared for the *Perry County (Missouri) Republic* in 1987, "It is indeed ironic to add that the Captain's son and namesake, Captain Raymond H. Littge II, who served with distinction in the Vietnam conflict, was killed in 1979 at age 30 while engaged in air combat maneuvers in an F-4D Phantom jet over Nellis Air Force Base training range in Nevada."

The "Lost Shepherd"

By Bob Powell

Early in the 1990's, as 352nd Historian, I was contacted by a Major Thomas W. Clarke, (USAF, Ret.) a man obviously obsessed with trying learn the identity of a pilot whom he claimed had saved his life and the lives of an unknown number of B-17 air crews single-handedly when they were attacked by a large force of German fighters on September 12, 1944. It was his intention to obtain the Medal of Honor for this pilot if he could be identified. This is the story of what we learned from him about what happened on that mission.

General "Tooey" Spaatz, with General Jimmy Doolittle's approval, had decided that it was time to decimate the Luftwaffe's fighter force and the September 12th mission was the first of several missions planned to accomplish that objective. It was designed to bring up the German fighters in the defense of multiple targets so that the 8th Air Force fighters could attack and destroy them. To open this action he decided to put up a small bomber force and every fighter he could put in the sky.

The bomber force consisted of the 351st Bomb Group of the 94th Bomb Wing stationed at Polebrook in East Anglia, and when the 351st pilots assembled for their briefing they looked at the route map in disbelief, describing it as a "Cook's Tour of Europe." They were amazed at the number of targets they would alert. These would include Wilheimshaven and Emden; then Hamburg and Bremen; then Hanover; then Magdeburg and Berlin. Then they were to turn southeast down the Polish-Czech border to Plauen/Lauta. That target was a key chemical plant in the Lauta area. This looked like a "navigator's nightmare," and Major Thomas Clarke, who was to be the lead navigator, wrote me the following when he was trying to learn the identity of the courageous fighter pilot who saved many of them.

"But for the 'lost fighter pilot,' our Group, spearheading the mission, would have been cut to pieces. We spotted some 50 to 80 German fighters at 12 o'clock high moving into position to attack us head-on. And that's when this lone fighter pilot went into action, sacrificing his life to save us. I often remember his courage and particularly on Memorial Day and Veterans Day.

The mild expression on Joe Broadwater's face belies the ferocity with which he fought his last mission. (Sheldon Berlow)

The Battle Story

"Our P-51 escort group came up the First Division line and arrived overhead. 'Last Blue' was their call signal and they were there to defend the three lead groups. Our tail gunner suddenly reported, 'All Hell's breaking loose back there.' Someone in the rear boxes had called on the Command Channel to report they were under attack by some 100-plus fighters and a desperate call came through to our fighter escort, "We could use some help back here."

Our escort fighter commander requested permission to go help them and they were released to do so. The P-51 commander reported, "I'm leaving one of ours above you to fly with you," and they turned and dashed toward the rear of the bomber formations.

Major Clarke continued his account of what was happening:

"Since we were not yet in our target area I had a chance to stand up and look back over the tail from my radar position. All Hell was breaking loose back there—both fighters and bombers were going down. Now we were alone out front of the spearhead with only the one 'Lone Shepherd' guarding us and we still had a long way to go.

"Suddenly specks appeared in the sky at 12 o'clock high and the lone fighter pilot above us called, 'I see them, too, and I'm going up for a better look.' Then we heard him call out to his fighter group, 'Ballast Blue—over 70 enemy fighters ahead—Fw-190s and Me-109s, too.' His Group leader replied, 'We're turning and coming back.'

"Later, the top gunner reported, 'Our lone P-51 flew straight over the Jerry fighters and then rolled over and dived right into the back of their formation and exploded two of them but his forward speed took him through the middle of their formation and they blew him away.'"

But now, with their attacking formation destroyed, they had to attack at a different angle and were limited to only one pass from a less desirable angle, undoubtedly saving the lead aircraft from being shot down. All alone, he had attacked some 70 German fighters to disrupt the German attack until his squadron mates could get back to get in the fight. Six of the 16 bombers in the lead group were lost on the spot and another crashed in France. The other groups lost 15 more. However, the fighters took a huge toll of the German attackers, knocking down 60 on the day's mission. Of these, 13 were scored by the 352nd FG with a loss of only one—the 'Lone Shepherd' Major Clarke still hopes to identify. Major Clarke said they continued on their mission and hit their target.

Clarke's research included the September actions of the three possible fighter groups which could possibly have been involved and it strongly pointed to the fact that their lone shepherd was from the 352nd Fighter Group, particularly since one of the B-17 crew members said their escorts were "bluenosed Mustangs," and pointed out that the escorting fighter group's call sign for that mission was "Blue Ballast."

So, who was the "Lone Shepherd?" Extensive research strongly indicates that the lone pilot was 2nd Lt. Joseph A. Broadwater, a new replacement pilot of the 328th Squadron of the 352nd FG. He was the only pilot lost by the Group on September 12th. And, since the 352nd only lost one pilot on that mission while the other fighter groups lost four, it certainly appears to be true. The Group's mission summary report indicates that he was last seen over Altruppen, Germany at approximately 1100 hours, about the same time the 352nd was involved in the first of three air battles that day.

Lt. Joseph Broadwater was from Sistersville, West Virginia and was a sophomore at West Virginia University when he entered the USAAR as an aviation cadet in 1943. He received his wings and commission on February 8, 1944 just six days after his 20th birthday.

Although official confirmation of the identity of the "Lone Shepherd" may never happen, Major Clarke said he will go to his grave convinced that Lieutenant Broadwater was the pilot whose brave attack on a superior force of attacking Luftwaffe fighters saved his life and that of many others on that day. Major Clarke's efforts to positively identify the 'Lone Shepherd' was to recommend him for the Medal of Honor for his courageous effort which undoubtedly saved some of the B-17 crews from capture or death.

Jack Thornell's famous PE-Tbar "Patty Ann II" in which Joe Broadwater was KIA

"That Kind of Day"

by Sam Sox, Jr.

Six August 1944, 10:00 am, White Flight is east of Berlin and heading toward Bradenburg to escort B-17s.

"Do I feel rough. Stomach rolling, churning. Great party! Too much to drink. That $1200 pot I won will help the group out in the Bond drive. Bed at 4:00, up at 4:30. Should have let Meyer take the mission as he had offered, really shook up Mason when I had trouble staying on the briefing platform. Let's see, 2500 rpm. Manifold pressure 44 inches. Oil temp in the green, indicating 300 knots, running great. No better ground crew than Lunn and McVay. Should be rendezvousing with bombers soon. Good wingmen, Doleac and Heyer. Stay in close. Mmmh, rolling again. Heck of a spot to get sick! Six hours from Bodney and 26,000 feet. Heads up! Big bomber boxes up ahead. Huns, thirty of 'em coming in line astern on that third box of '17s. Oh, no. Not here. Now? Got to clear my mask, must attack. Damn!"

What proceeded, following Major George E. Preddy's throwing up in his oxygen mask, was later described on a radio interview with Bill Shadel of CBS news in London.

"Our job was the usual one, escorting the bombers. We were meeting them north of Hamburg, escorting them on to their targets east of Berlin and part of the way back. About fifteen minutes after picking up the bombers, we saw thirty Bf- 109s come in for a stem attack on the third box of B-17s. Our flight of three was the only protection around at the moment, although I was leading the group escort. We had altitude on the Jerries, so we went down after them. They saw us coming and although they showed no intention of leaving the bombers alone, they did pull out to the left in evasive action.

"We came up on their tail and each picked a Jerry. From about 300 yards I gave mine a burst and set him on fire. I saw him go down in flames (score one).

"They were trying to hold their formation, but were sending singles and doubles out to slip up on us from behind. At about 26,000 feet, one that slipped out and was trying to get on our tail had to be dealt with by one of my wingmen, Flight Officer Cyril Doleac of New Orleans. Lt. Shelton Heyer of Pearl City, Illinois, and I kept after the main formation. I got in another burst and the pilot bailed out (score two). Lt. Heyer, to my right and below me, set another one on fire. Then Heyer had to leave me and take care of those strays from the enemy formation. By this time, another single that had slipped out was down below me with the idea of coming up under me. I throttled down and slipped behind his tail just as he started up. Right after I gave him a burst, he went into a spin and then fell apart (score three). Luckily, just after that, four more of our fighters came in. It was now five of us against about 20 that were left in the formation after they sent out their decoys and counted their losses. The Jerries still intended holding to formation. I got behind my fourth target and shot him down in flames (score four). Then they gave up on the bombers and went down fast in steep turns. I got one on the outside of a turn and closed in on him for a shot that sent him down (score five).

"As they started breaking their formation for the first time our fighters scattered out for the chase. As a Bf-109 pulled up to get on my tail I had to turn into him—that was the first time I left off chasing the formation—and by this time our other fighters were off on their own. I got a short burst at the Jerry but missed. It was just the two of us then, and after I missed he had the edge on me. I went into a steep climb into the sun and as he followed me I could see him firing everything he had. He couldn't stay with me in that climb and dropped off. Then it was my turn again. I got behind him, and he must have known that his game was up. Before I could get a shot I saw him fumbling with his hood above his head, trying to get out. When I hit him, the plane quivered a bit and started down, but he was out by this time, so I saw him float down without bothering him (score six). By this time I was almost to Berlin, so I set my course for home."

Thus culminated the youthful desire George E. Preddy, Jr. had expressed upon the completion of his first airplane ride taken at the age of 18 in 1938 when, in his diary, he wrote "Took my first airplane ride today and it was the greatest experience of my life. I must become an aviator." Did he ever, as we shall see! Shortly after his first flight, he took his initial lessons, soloed and, along with his instructor, Bill Teague, purchased half interest in a WACO 10 in which they would spend the next two summers barnstorming the state of North Carolina. Unknown to him, this flair to "show off" would follow him and get him into trouble throughout his career.

George, often described as a competitive intellect, doubled up on his high school courses and graduated at the age of 16. He enrolled at Guilford College outside Greensboro, North Carolina, and stayed two years. In the fall of 1939, his desire to enter the military came to a peak as he tried three times unsuccessfully to enter the Naval Air Cadet program. Having failed their entrance requirements, he successfully completed the Army Air Corps entrance exam a summer later, only to be told that he would have to wait as all the classes were full. The United States was still neutral and the Japanese had not yet bombed Pearl Harbor. As the war in Europe was beginning to escalate, he was advised to enlist in the National Guard. He was within a matter of hours from shipping out with his Coastal Artillery unit to Puerto Rico, when he was advised that he was to report to Darr Aero Tech at Albany, Georgia, for primary training.

Like many others before and after him, he received his commission and wings at Craig Field, Alabama, on 12 December 1941 and was assigned to the 49th Pursuit Group which was already on station in Australia.

Upon arrival at Darwin, he was assigned to the 9th Squadron, which would later produce such greats as the Ace of Aces, Richard L. Bong and, in addition, Gerald R. Johnson, John D. Landers, and Andrew J. Reynolds, the first ace in the SWPA. Preddy was assigned P-40E number 85 which he named Tarheel in honor of his home state North Carolina. At some point later, his barnstorming background would prompt the addition of a fire-breathing dragon to the aircraft's right side.

He had been at Darwin only three weeks when he got his first opportunity to test his newly acquired skills on the enemy, only to watch the Zero in his sights escape destruction due to his failure to turn on his gun switch. A week later he would face the enemy once more. He reported, "Our flight of seven intercepted seven bombers and three Zeroes. I didn't get any, but our flight got five of the bombers and two Zeroes. I think I was given credit for helping destroy two of the bombers which would give me 1/3 of a kill. We lost one pilot" A few days later, Preddy faced the Japanese in two separate fights but couldn't get a decent target. His fourth encounter was almost his last. He got jumped from above as he was attacking a flight of bombers only to be bounced by Zeroes which drove him away. His P-40 was hit three times.

On 12 July, fate would end his career with the 49th Fighter Group and possibly the opportunity to have been a part of Bong's legacy that followed. On a training mission in which Preddy and another pilot enacted the role of enemy bombers, a fellow pilot, Lt. Sauber, while preparing to make a pass at him, misjudged his initial breaking point and crashed into "Tarheel," striking it behind the cockpit. George bailed out, only to severely injure his calf, hip, and shoulder upon landing. Lt. Sauber did not survive. George was to spend the next twelve weeks in a military hospital in Melbourne recuperating. During this time he reviewed his career and recorded 13 rules he would strive to live by. They were:

Messerschmitt Me-210 heavy destroyer fighter.

1. *No smoking at any time or under any circumstances.*
2. *Drink intelligently and sparingly.*
3. *Eat sensibly.*
4. *Exercise regularly and diligently.*
5. *Learn all possible about flying or any other job at hand.*
6. *Always be willing to go out of way to learn something new.*
7. *Always try to give the other man a boost.*
8. *Fight hardest when down and never give up.*
9. *Don't make excuses but make up with deeds of action.*
10. *Learn from experience.*
11. *Listen to others and profit by criticism.*
12. *Live a clean life.*
13. *Trust in God and never lose faith in Him.*

After examining his record and reflecting on the reputation he established with his fellow comrades, it is evident he did take these goals seriously and pretty much lived by them.

Healed of his wounds, he arrived in New York and at Mitchell Field on 3 January 1942, where he was to join a training outfit, the 320th Fighter Squadron, only to run into several buddies from his old unit at Darwin. They informed him that a new unit, the 352nd Fighter Group, was forming and would be going into combat in the P-47 Thunderbolt after several weeks of training. Wanting to get back into combat, Preddy requested a transfer and with the assistance of his old 9th Fighter Squadron buddy, I. B. Jack Donalson, the unit's commander, Lt. John C. Meyer okayed the transfer. He recalled his first impression of meeting Preddy during an interview at the Pentagon just prior to his death in 1975, "I remember the first time I met George Preddy, I was quite disappointed. He had come to me due to the recommendations of my Squadron Operations Officer Jack Donalson, who had known him in the Pacific. Jack talked of him in such glowing terms that I not only expected to meet a bull fighter, but I expected him to look like one. I was initially disappointed."

"George was small (5 ft 7 in, weighed 125 lbs) and had rather grotesque ears. He was somewhat Latin in appearance and his drawl was so slow that it at first appeared to be an affliction. I doubted that he even had the strength to successfully punch his way out of a wet paper bag. I called Donalson and asked, 'Are you sure this is the right guy?' Jack reassured me that he had correctly described the man and that time would prove my selection as being a wise one. He had a keen sense of duty and patriotism and was the most unselfish man I have ever known. His entire thinking was based on the best interests of the nation, of the group, and of his comrades. His loyalty to commanders and subordinates alike was unimpeachable.

"It became apparent that he was highly skilled in his profession, but more than that he exuded the dedication of purpose that was more noticeably different and superior to others than his pure professional skill. As a matter of fact, he and I spent a great deal of time together on individual practice with each other on aerial combat maneuvers and tactics, mostly in gunnery, using gun cameras and would dogfight interminably. We'd fly extra missions, more than were scheduled, including Saturdays and Sundays. We would spend hours in the evenings going over our gun camera films. I would say that it is fair to say from my personal knowledge that he and I spent more time practicing aerial combat than any other individuals. I think that between the two of us, we could shoot better than most anybody over there with the exception of possibly Sailor Milan, a South African with the Royal Air Force (Adolf G. Milan, RAF, 35 victories). While this was a development of a professional skill, his motivation and dedication to this was superior to my own and to a large extent my own was provoked by him; he needed a partner. He stimulated me to do a lot of it that I wouldn't have otherwise done, for which I will be eternally grateful" Meyer would eventually score 24 aerial victories in Europe, ranking him fourth among ETO aces.

Low-level aerobatics was strictly forbidden, but Preddy's barnstorming background caught up with him on 17 April following a gunnery mission. He just couldn't resist and got caught by his group commander for "demonstrating the excellence of low-level maneuverability of the P-47" and was summarily chewed out.

Preddy, along with the 34th Fighter Squadron now redesignated the 487th, completed its training, and sailed for England aboard the Queen Elizabeth on 1 July.

The unit arrived at Bodney, England, on 8 July. After waiting for almost two months, the group received its first aircraft. Two P-47 Thunderbolts were the first to arrive, but one was damaged in its initial landing. The pilot got too far forward on the main landing gear and tried to use the prop as a ditch digger. The second no sooner arrived than Preddy had gotten into the cockpit and was about to play barnstormer yet again. Bodney was a typical British airfield: a large grass field crowned in the middle. Preddy unleashed the rumbling Thunderbolt and let it accelerate down the runway pulling off and immediately began a slow, climbing barrel roll. What airmanship! What control! How skillful! Everything went great, except, in his haste to get back into the air again, he had failed to retract his landing gear! It would take months before his buddies would let him forget it! By now I he had acquired the nickname of "Ratsy" because of his big ears.

The squadron flew its first mission on 14 September. By 25 November, Preddy had flown 20 missions. The next day, Capt. Meyer would score the unit's first victory. Though Meyer and Preddy never outwardly competed, George did not let this event go unnoticed. In his diary, he noted his suppressed disappointment, "I know that my day is coming and I am going to do everything possible to be ready when I do meet the Luftwaffe. Starting right now, I am going to get into physical and flying condition." This was to be his last entry in a diary that he had kept since leaving for Australia, almost two years before.

His day did come! On 1 December, he completed the following encounter report: "I was leading Crown Prince Red Flight in a section of twelve ships. We came in over the bomber formation at 30,000 feet and went into a left orbit. I saw one Bf- 109 behind the rear box of bombers about 3000 feet below me. I started a quarter stern attack and when about 1000 yards from the enemy aircraft, it started a steep spiral dive to the left. I followed, closing to 400 yards. As I closed from 400 to 200 yards, I fired and saw strikes on the wing roots and cockpit. The airplane began to smoke and fell out of control at about 7000 feet. I fired another burst closing to about 100 yards. After I broke off the attack, the enemy aircraft disintegrated. The Bf-109 was carrying a belly tank which he did not drop. Claim: One Bf-109 destroyed."

His next two victories would also be especially memorable, except for very different reasons. On 22 December, he would score his next victory and, in the process, be recommended for the Distinguished Service Cross. However, in the combat he would lose his wing man. The mission was to hit the marshaling yards at both Osnabruck and Munster. He reported: "I was leading Crown Prince Blue Flight. As we made rendezvous with the bombers, Yellow Flight (led by Meyer) bounced a Bf- 109 and my flight gave them top cover. Shortly after that, I noticed three Bf-109s coming in to the rear of the B-17s. I bounced two of them and they immediately went into a dive straight down and I went into compressibility following them. I pulled out at 8000 feet and sighted one of the enemy aircraft just above a cloud layer. I gave him a short burst and he went into the clouds. No damage was noted.

"My wingman, Lt. Richard L Grow, and I began climbing back up with everything to the firewall. When we reached 15,000 feet, I noticed another Bf-109 above us positioning for an attack. He made an attack on the two of us and we turned in to him. We battled him for almost fifteen minutes, getting short deflection shots but we were

Lt. Grow and his P-47D Topsy II

unable to gain an advantage. He finally broke off the engagement and disappeared in the clouds below us.

"We resumed climbing and picked up Lt. John Bennett, our Blue 3. We sighted the bomber formation about 25 miles west of and above us. We continued climbing toward the bombers and leveled off at 26,000 feet on the down sun side still quite a few miles out. I saw a B-24 straggling to the left and below the formation. He was being attacked by six to eight Me-210s with about ten Bf-109s giving close top cover. I made an attack on the Me-210, but they saw me coming and immediately dispersed. (Lt. Bennett stayed up as top cover for Preddy but was chased into the clouds by 109s.) I began closing on one of them and fired from out of range with 80 degrees deflection. I saw no damage before he ducked into the clouds.

"I pulled back up and attacked another Me-210 which was attacking the B-24. I started firing at 60 degrees deflection from 400 yards. I came down in stern continuing to fire and closing to 200 yards. I noticed many strikes on the center section, fuselage and engines. The enemy aircraft began to disintegrate with large pieces flying off and he went down into the clouds in flames. I broke back up and Lt. Grow called that a Bf-109 was on my tail. I threw the stick in the left corner and saw the enemy aircraft behind me out of range. I continued down skidding and slipping. Grow then called that an enemy aircraft was on his tail but I was unable to locate him. I told Grow to hit the deck, then I went into a cloud and set course for home on instruments. I stayed in the clouds for about fifteen minutes and broke out over the Dutch coast at 3000 feet. I could not contact Lt. Grow and did not see him again. Claim: One Me-210"

A few weeks later, Preddy would be awarded the Silver Star for his gallantry in defending the B-24 from sure destruction.

On 29 January 1944, Preddy's assistant crew chief Cpl. J.J. "Red" McVay was permanently assigned to him. Wanting to do everything possible to make a good impression, Red took exceptional care to see that the P-47 was in top condition. After sweating the runup and takeoff, he settled down to await Preddy's return. The mission was withdrawal support for the bombers returning from Pelm. Preddy later reported, "I was leading Crown Prince Yellow flight and we were escorting two boxes of bombers. The group leader called for everybody out and I started to join him when my number-two man, Lt. Bill Whisner, called that the bombers were being attacked. I turned back coming in behind the bombers and saw an Fw-190 below and behind them. Whisner had started a bounce on another enemy aircraft so I went down on this 190. He went into a steep dive and I closed to about 400 yards and started firing. I was closing rapidly and saw a few hits and a little smoke before I broke off. I lost the enemy aircraft momentarily but picked him up again on my left at about 4000 feet. I started after him and he made a steep turn to the left. I turned in to him and started firing at 300 yards and 60 degrees deflection. He straightened out and started down at about 45 degrees. I got a good long burst at 300 yards and saw hits all over the ship. The engine was evidently knocked out as I closed very rapidly after that. The last I saw of him, he was at 1500 feet going down at an increasing angle to the left.

"I made a steep climbing turn to the left and saw Whisner. He joined me and we climbed back to 10,000 feet. It was past time to go home so I picked a heading as we didn't have enough fuel to do anymore fighting. We went below the clouds and came out on the deck crossing the French coast somewhere north of Calais. A concentrated barrage of flak opened up. I began kicking the ship around but felt hits. She began smoking but did not lose power so I climbed to 5000 feet and gave a Mayday. Shortly afterward, my engine cut out. I bailed out at 2000 feet. A P-47 pilot spotted me and I was picked out of the drink by a Walrus (RAF rescue aircraft)."

The life expectancy of anyone in the English Channel during the month of January is about 20 minutes. Fortunately, Preddy was able to get into his dingy. The hour he waited must have felt like an eternity.

Back at Bodney, McVay waited and waited. Aircraft were returning all around his hard stand, but not one marked HO-Y. He began to question himself as to what he could have done incorrectly to have caused the delay in his return. Finally, as dusk was setting in, someone approached him and asked for whom he was waiting. He said, "Lt. Preddy." It was then that he was informed that George had ditched in the channel but had been rescued. Several days later, Preddy was to get another '47 which would become the first of many aircraft he would name "Cripes A' Mighty," his favorite expletive.

On 11 March, the 352nd flew a Rhubarb (attacking of ground targets) to Amiens, France. The unit shot up machine gun emplacements, flak towers and trucks. He later found out that this mission was a test of enemy firepower in the area of Pais de Calais and provide intelligence prior to D-Day. The enemy fire power was terrific (Hitler believed this area to be the site of the forthcoming invasion).

Preddy would fly 31 missions in his P-47 coded HO-P without scoring an official aerial victory. He did, however, claim two aerial victories on a mission carried out on 10 April. They were never confirmed. He scored a ground victory on an He-111 the next day.

During the second week of April, the 352nd FG would convert to the new North American P-51B Mustang. On his first mission, Ratsy would score his fifth victory, a Bucker Jungmeister 133 trainer, expending only 87 rounds, and it was on the ground! You can be sure that at sometime later he had to think about what a fine ship this would have been to have continued his barnstorming after the war!

Preddy took immediately to the nimbleness and agility of the P-51 as he would double his score in the next five weeks, scoring a double kill on 13 May and getting 2-1/2 more on the 30th. His total score now stood at twelve. Cripes A' Mighty (Preddy named his first Mustang simply Cripes A'Mighty, no number 2) would now display 12 crosses (he used full crosses on shared victories of which there were two).

Though the troops had not yet been informed of the exact date of the upcoming invasion of Europe, they were becoming more and more aware of its nearness by the increase in activity and security around the base. George Preddy, now promoted to the rank of major, assumed command of the 487th Squadron in Lt. Col. Meyer's absence who was taking a 30-day leave back in the States.

"A day or two before D-Day," Red McVay recalls, "we were ordered to paint alternate black and white stripes around the wings and fuselage of our aircraft. For me it meant adding them to the major's new P-5 I D, "Cripes A'Mighty 3rd." We assumed something big was up.

"Nearly every enlisted man who didn't actually work on the planes was placed on guard duty while the painting was going on and left there until after D-Day. They were placed strategically around the field so that no one either got on or left the field. The painting of the stripes was to be kept a secret. Most of the civilian employees were Irish and since Ireland had not joined the Allies in declaring war on Germany, they were not permitted to go home or tell their families why.

"Unfortunately, the guard assignments were done in such a hurry, that no one knew exactly where they all were, which would become a problem later in the day and in the days to come."

The briefing for Major Preddy and the rest of the 352nd took place around midnight. Following the briefing, he and the rest of the squadron were to take off at 0250. The aircraft made their way to the takeoff area in total darkness, turning either to the left or the right to line up for their takeoff run. By the time the second flight reached the final staging point, the first flight of four from the 486th FS had destroyed the small set of temporary runway lights. Blind takeoffs were now the order with compasses being set on 05 degrees. Unfortunately, the second flight set their compasses on 50 degrees which would carry the fourth man through the volleyball court uprights. The flight had expanded into a flight of five, picking up Lt. Robert C Frascotti who had failed to take proper position due to the confusion. When the ground crews heard the roar of the accelerating aircraft, they began to applaud and cheer. This was the beginning of the end for which they had all been waiting.

Suddenly all hell broke loose. The jubilation was shattered by a tremendous explosion. A huge fireball erupted from the area of the newly constructed control tower. Frascotti's '51 had crashed into the tower. The fear of a night takeoff was no longer a problem for the rest of the group for it was now as bright as daylight.

The 487th flew three patrols on 6 June. Preddy flew two of them spending a total of 9-1/2 hours in his new '51. As the Luftwaffe was not to make an appearance that day, targets were limited to motor vehicles and armor.

Ground crews were informed that aircraft were to be in the air as much as possible that day

487th ace George Preddy (right) with Lieutenant Berkshire returning from flight line.

which meant new drop tanks and spark plugs were to be acquired and prepared for the fighters' return. The crews began to go to chow in shifts. When Red returned from the mess hall, he noticed ambulances in the 487th area. He watched them go past on the perimeter road heading to the hospital. One had traveled only about a hundred yards, when it came to a screeching halt. A GI burst open the door dropping his trousers at the same time and then assumed a most unflattering position. It turned out that many of the men had a very powerful type of food poisoning causing the afflicted to void at both ends. Men were dropping in their tracks in excruciating pain. The hospitals became so full that buckets were passed out to those who could not be admitted. The security around the base became very threatened as guards and Officers of the Day were immobilized where they were positioned. It was rumored that the incident may have been caused by the civilians that were restricted to the base, but this was not the case. The unit flight surgeon discovered that it was caused by a spoiled batch of bread pudding.

John C. Meyer's initial assessment of George E. Preddy was later verified in a Letter of Commendation Preddy prepared for his unit on 9 June following this near disastrous episode. He wrote, "The first few hectic days of the initial phase of the invasion have passed. In preparation for that momentous event and in the accomplishment of the mission of this squadron, in the great offensive thus far, it has been necessary that each of you, pilot, ground officer and enlisted man alike, to work long hours with little or no sleep and extend yourself to the limit of physical endurance to perform your assigned duty. Without exception all of you have carried on with enthusiasm, cheerfully and uncomplainingly doing your part in the accelerated operations of the squadron attendant upon D-Day. In the critical times immediately ahead you will be called upon to keep up the increased tempo of your work to ensure the ultimate success of our forces. It is a pleasure and a privilege to have you serve under my command. In the words of the Supreme Commander, General Eisenhower: I have full confidence in your courage, devotion to duty and skill in battle. We will accept nothing less than full victory. Good luck. And let us all beseech the blessing of Almighty God upon this great and noble undertaking."

Preddy flew eight more missions after D-Day before he would be able to try out "Cripes A'Mighty 3rd" on the Luftwaffe. He scored his first victory in his new bird on 12 June, Uffz. Helmut Rosenbaum, 5/JG 53, (WIA) with additional single victories following on the 20th, 21st, and 29th.

On 18 July Preddy and "Cripes A'Mighty 3rd" really got their act together. It appeared that his ground crew, comprised of Sgt Lew Lunn, Chief; McVay, Assistant Chief, and Sgt M.G. Kuhaneck, Armorer, had fine tuned the '51 to its ultimate. Preddy later reported, "I was leading the squadron on a sweep south of the bombers and heading north to intercept the bombers. Yellow Leader called in bandits at four o'clock low. I made a right turn up sun of the enemy formation which consisted of a mess of Ju- 88s with many Bf- I 09s as top cover. I took my flight of three—Lts. Vickery and Greer and myself—to attack the Ju-88s, about 50 in number, while the rest of the squadron dealt with the top cover. As we approached the formation, I saw a single Bf-109 ahead of me and attacked it from quarter stern. I opened fire at 400 yards and drove up his tail. The enemy aircraft was covered with hits and went down burning and failing apart.

"I continued in to attack the 88 formation and opened fire on them, knocking many pieces off and setting the plane on fire. Lt. Greer then called a break to the right as a Bf- 109 was pulling up on my tail from below. After we broke, the 109 stalled out and went back down. During this maneuver, Lt. Greer became separated. Lt. Vickery and I then made a 360 and launched another attack on the main formation and I damaged one with a few hits and drove up the rear of another getting hits all over this one. I believe the pilot and crew were killed as the enemy aircraft began smoking badly and went down out of control with parts of the ship falling off. I broke off the attack and pulled out to the side before the third attack on the formation. I came in astern again. In this attack I plastered one Ju-88 causing both engines to burn, and the enemy aircraft disintegrated. I got a few hits and damaged a second Ju-88. I was out of ammunition—or so I thought, but later

SSgt. Al Geisting and TSgt. Jesse Hubbard pose with the crew side of their HO-V "Josephine." (Al Giesting)

learned that my guns had a stoppage on one side and I had been hit in the engine by the rear gunner in the Ju-88s. My ship was covered with oil sprayed from the enemy aircraft which had been shot down, so I set course for home with Lt. Vickery. Claimed: Three Ju-88s destroyed, one Bf-109 destroyed, two Ju-88s damaged" (should be noted that Luftwaffe records indicate that Preddy's victories were with Me 410's, not Ju-88s which were operating in the area of intercept). The destroyed claims would later be adjusted to a total of three and three damaged.

Preddy's expertise in the P-51 D-5-NA would be noticed not only by his unit but also by North American Aviation technical representatives who were in England. The biggest complaint about the new bubble canopied P-51 was its loss of some of its lateral stability due to the new full-vision canopy. Although the new canopy eliminated the blind spots that were present in the earlier P-5 I B, it was not as stable a gun platform. Preddy's multi-victory successes made it evident that he had learned how to compensate for this instability and was having great success with the new model. Technical representatives from North American appeared at Bodney shortly after the 18th and debriefed him thoroughly about his impressions of the D model Mustang. It now appeared that he could not miss. On each of the next four missions, 20, 21, 29 July and 5 August, he scored victories, bringing his total now to 17½ aerial and five ground. Of these dates, the victory on the 21st is particularly interesting for it illustrates the knowledge and trust pilots acquire of each other when they have trained and flown together as Col. John Meyer and Preddy often had done. Meyer was leading the group that day with Preddy acting as Squadron leader. Northeast of Munich, Ratsy spotted a single contrail and, upon closer examination, discovered eight additional Bf- 109s slightly below. He took the squadron to intercept them while Meyer lead the rest of the group to rendezvous with the bombers. Preddy and his flight were now at 27,000 feet and the 109s were 3000 feet higher. The blue-nosed 51s climbed toward them from four o'clock and came in dead astern. Preddy recalled that their closing speed was very great considering a climbing attack. He opened fire on one of the enemy aircraft at the left of the formation from 300 yards and 30 degrees deflection but noticed no strikes. As Preddy passed him, he pulled over to the right and fired at another from 100 yards, but was thrown into a spin by the prop wash of the enemy aircraft. He recovered at 15,000 feet and saw a 51 (Meyer) being pursued at close range by a Bf-109.

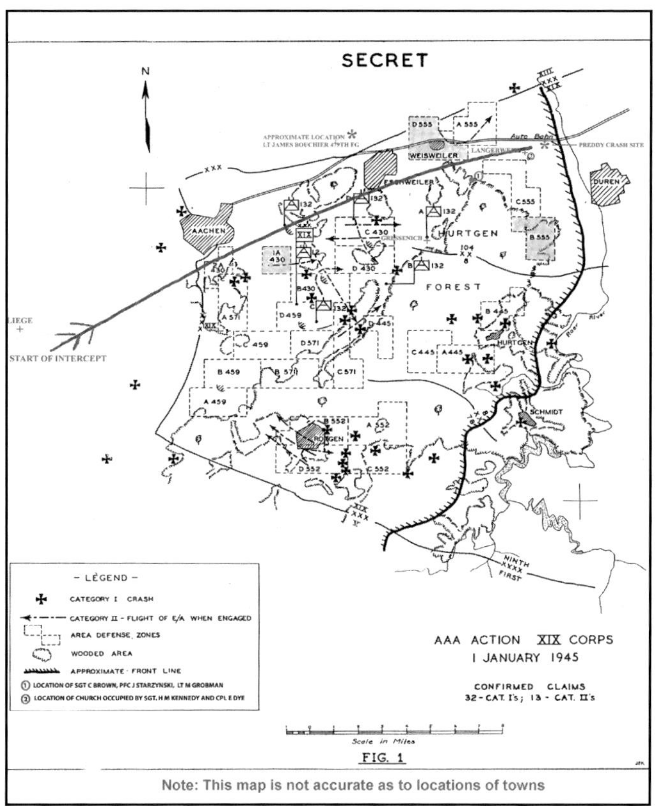

FIG. 1

Note: This map is not accurate as to locations of towns

As Meyer and his flight approached the bombers, they encountered a large flight of 109s determined to split up the formations. A dogfight developed and, suddenly, Meyer had a 109 on his tail which he could not shake. He put his 51 into a tight Lufbery and the 109 followed. Neither gained an advantage. It was a stalemate with the enemy firing short bursts at him. However, Meyer knew that the thought of his running out of fuel put him at a great disadvantage. He realized that he could not hold this maneuver as long as the German, but then he noticed that another 51 had slipped up on the 109 undetected. He could tell from its blue nose, that it was one of his own. In what seemed like hours, he finally made radio contact. He could tell from the slow drawl that it was Preddy. George was still out of range and was shadowing the 109 from six o'clock low. Meyer decided, based on his confidence in his friend—gleaned from those many hours of training together. while at Hamilton Field to pull back on his throttle to give the enemy an opportunity to close his range. More importantly, this would give Ratsy an opportunity to close in on the enemy. The 109 was still unaware that Preddy was on his tail. It worked. As soon as Col. Meyer cut his throttle a little, Preddy was able to gain an advantage on the Bf- 109 and with one short burst scored hits causing him to break for the ground. Preddy fired again from dead astern and knocked his engine out. Meyer credited this action as saving his life and was forever grateful to his friend.

On 6 August George E. Preddy, Jr. would be in the right place at the right time and with the right skills for a record-setting feat. As described at the beginning of the article, he destroyed six Bf-109s in less than five minutes, bringing his score now to 28-1/2. For this achievement, he was recommended for the Medal of Honor. On 7 August he performed a re-enactment of his victorious return to Bodney for an 8th Air Force photo team. He did not let the occasion for "legalized" barnstorming go by without fully taking advantage of it. The troops really enjoyed watching him wring out the colorful 5 1. Shortly thereafter, on the 11th, he was awarded the Distinguished Service Cross, the nation's second-highest award. He would not fly Cripes A'Mighty 3rd again, although she would continue scoring victories with other pilots at her controls. Initially she was renamed "The Margarets" by Capt. Malcolm Stewart ultimately serving under the name of Sexchunate (section eight). In addition, she is was flown by or assigned to Col. John C. Meyer who scored four ground victories; Lt. Marion J. Nutter, three aerial victories; Lt. F.C. Reading, Jr., and finally by Lt. Walter Padden, (two unconfirmed ground victories) who was killed in her while on a strafing mission on 16 April 1945. Cripes A' Mighty 3rd had more aerial victories scored in her than any other P-51 serving in WWII: 18½. It is also possible that she had one of the longest combat careers in the ETO, serving continuously for more than 10½ months.

After spending 30 days in the States on a well-deserved leave, Preddy returned to the 352nd early in October. On the 28th of the month, he was made commanding officer to the lowest-scoring squadron in the group, the 328th. Major Preddy was a winner. He believed that in order to be an effective fighter pilot, you had to know your aircraft thoroughly. In order to operate effectively as a unit, you had to have proven leadership and his service record denoted his qualifications.

Preddy was a man of few words and was not one for idle chitchat. It was noted that when

A 455th Bomb Group B-24 Liberator.

he spoke, he usually had something of importance to say. The conditions within the 328th Fighter Squadron were not unlike those that faced the British ace Douglas Bader when he assumed command of a Canadian Hurricane squadron. Morale was low, with little esprit de corps being present. Shortly after arriving, Bader strapped on a Hurricane and really wrung it out for all the squadron. They knew from the display what was expected from them and that Bader could deliver. Art Snyder, Preddy's 328th crew chief, recalls that shortly after Preddy assumed command he too strapped on his new Cripes A'Mighty and put on a really spectacular show for the squadron. Leading by example, his efforts to improve the performance of the unit began to pay off.

On 2 November, the 328th, with its confidence now restored by its new leader, destroyed 24 Bf-109s. This set a new ETO record of enemy aircraft destroyed in a single mission. Preddy got another 109. The 352nd Fighter Group scored a total of 38, also a record.

Preddy led his new unit on seven more missions during the month of November in his new '51 named Cripes A'Mighty, which was crew chiefed by the 328th's barber, Art "Snoots" Snyder. Snoots left his unique signature" on all the aircraft he crewed. These included red and white stripes on radio antenna, trim tabs, and red scalloped gun ports with textured aircraft names, i. e. red, yellow, and blue poka-dots, white pin stripe around blue nose of 51. White sidewall tires were added on at least one aircraft and, finally, red and white barber poles on the right sides of his aircraft with the prices of his haircuts duly noted! Advertising, the American way!

The mission on the 21 st was the unit's most notable. The 352nd scored 22-1/2 victories with the 328th getting 8-1/2. Lt. Bill Whistler of the 487th tied Preddy's record of six victories in a single mission while Preddy got an Fw-190 making him the highest scoring active ace in the ETO with 25-1/2 aerial and five ground victories.

The next eight missions were mostly uneventful with the enemy not challenging the Blue Nose Bastards of Bodney. The winter of '44 settled in rapidly in the ETO, making bomber and support missions difficult. Ronstadt capitalized on the poor flying conditions by striking deep into the Allies' lines in the area of Liege, Belgium, on 16 December. The 9th Air Force, already operating from the continent for months in providing close ground support for Allied armor and infantry, found itself much in demand and greatly overworked. The 9th sent an urgent request to the 8th Fighter Command requesting two additional fighter units to come to its aid. On the 23rd of December, Preddy led his 328th Squadron along with the 487th and 486th to a small remote 9th Air Force field located at Asch, Belgium, denoted as Y-29. The field was so close to the German lines that aircraft in the landing pattern were occasionally fired upon by enemy antiaircraft units.

The 352nd was not accustomed to the tough living conditions it now faced. Living in tents was a far cry to the palatial town houses the pilots occupied at Bodney. Most of the troops thought they would freeze to death the first night. They placed all their blankets on top of themselves instead of insulating their bodies from the cold below. American ingenuity came to the rescue when it was necessary to find a way to heat their tents, but not without high risks. An oxygen bottle was scrounged from a wrecked B-24. A petcock was attached to copper tubing extracted from the same wreck. The bottle was filled with high octane aviation fuel which was then suspended from a nearby tree. It is a miracle that the little stoves did not explode when the raw fuel was introduced to the tiny burners! But the troops stayed warm.

The next day was spent getting the unit settled down and assembled. The ground crews who were transported in C-47s became lost and arrived a day late. The first mission from Y-29 was a milk run, no action. Christmas found flyable ceilings and two missions were scheduled that day. Preddy was to lead his unit on the second one, a support mission into Germany with the bombers from the 8th. Preddy led in Cripes A'Mighty with Lt. J. Gordon Cartee, a substitute that day, flying his new "K" model named Steph-N-Jane in the number-two slot. The rest of the flight of four was comprised of Lt. Ray Mitchell in "Carol"and Lt. Duke Lambright in "Geraldine II."

Cartee recalls, "After tooling around for a while, due to no action, we were vectored to an area close to Koblenz, Germany, where enemy aircraft had been encountered. Preddy receiving the call said, "They've started without us, let's go join them and immediately turned into that direction. Just as Mitchell was about to peel off, he looked up and spotted two 109s coming down on him and Lambright" He called to Preddy for assistance, but there was so much chatter on the radio that Preddy never heard him. Mitchell believes to this day that, had Preddy heard his cry for help, he would never have placed himself into the series of events that were to follow.

Cartee continues, "Preddy spotted two 109s and got into a Lufbery with the first one. Neither were gaining much advantage when all of a sudden another 109 cut in front of him. He eased up on his controls just enough, gave it a short burst, blazed it and then resumed his pursuit of the first one. The 109 lost his concentration seeing his buddy flamed and Preddy nailed him. (See "Letter from Fritz Koal.") Preddy's score now totaled 27.5 aerial and five ground victories. Moments later, Preddy and Cartee were vectored to an area southeast of Liege where it was reported that enemy aircraft were strafing Allied ground troops.

As they neared Liege, they were joined by a white-nosed Mustang from the 479th FS, Lt. James Bouchier, who had become detached from his squadron. From the initial intercept point, approximately 3 to 4 miles SE of Liege, Preddy, from a height of about 1,500 feet, began to accelerate, having picked out a long-nosed Fw-190 in the distance heading northeast. He radioed a "tally-ho" to Control and was immediately cleared to make the intercept. There was also some talk between Control and Preddy about intense flak in the area of intercept an its being halted so the attack could be made. Unknown to Preddy, Cartee and Bouchier, their line of flight was taking them over the quad .50 cal. AA of 'A' Battery of the 430th AA (attached to the 258th FABN XIX at that time) positioned on the west side of a large clump of trees about two miles south of Aachen, Germany. As they neared the AA gun positions, Preddy was hit first by ground fire, followed by Cartee and Bouchier. Cartee saw Cripes A'Mighty begin to lose coolant, the canopy come off and Preddy's plane begin a chandelle maneuver to his left. Cartee also noticed that a tracer that had entered his own cockpit and was smoldering on the floor of his plane. Without getting it out of the way, it could start a fire at his feet, so he kicked it around while still trailing Preddy. Lt. Bouchier's Mustang also received hits and began smoking. He, too, broke left, climbing to about 1,000' where he realized he would have to bail out of his severely damaged plane. He released his canopy, rolled the '51 over and dropped out and landed safely in the

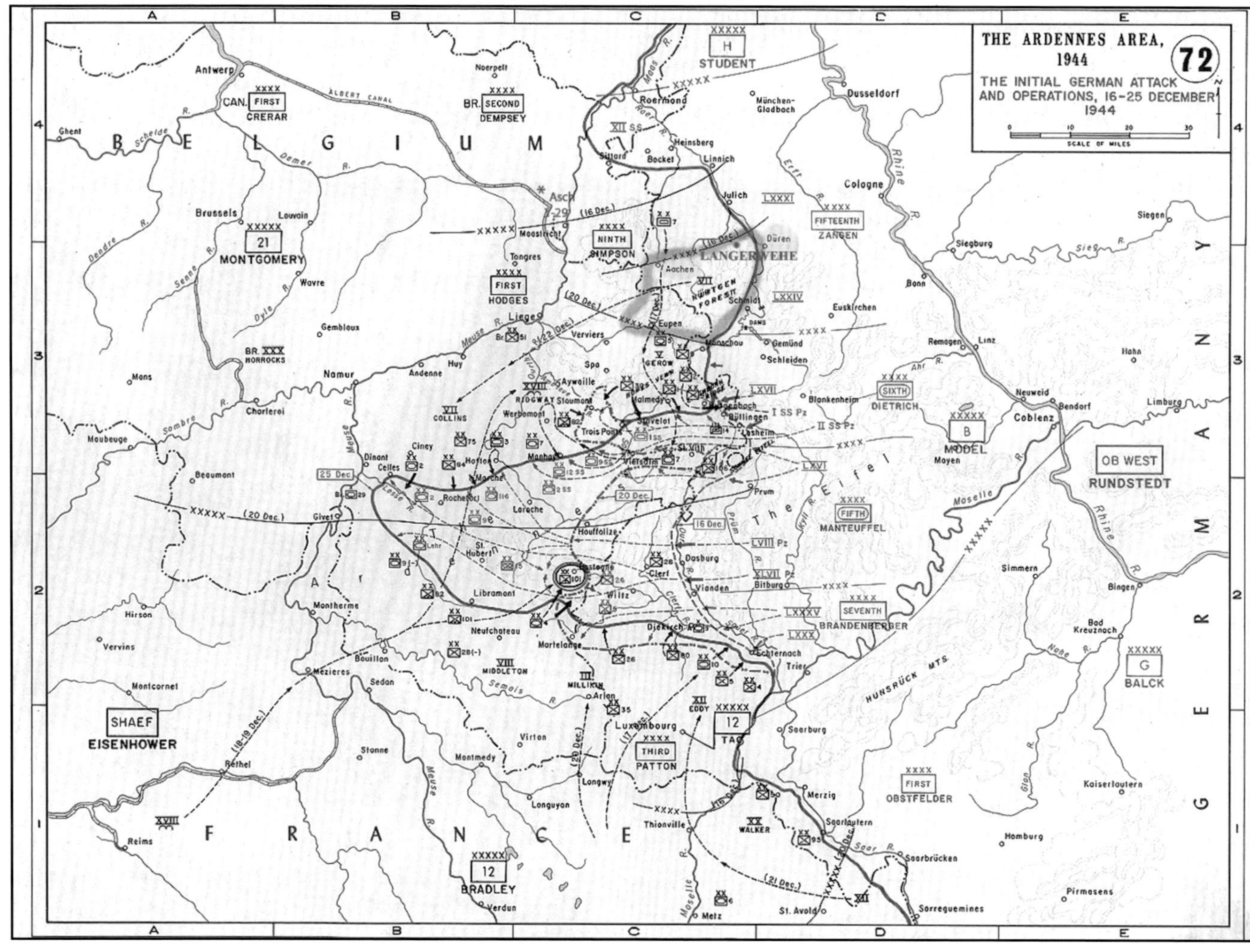

British sector several miles north of where he had been hit.

Further up from Preddy and Cartee's line of flight, now a couple miles south of Weisweiler, Pfc. Charles Brown, Pfc John Starzynbski and Lt. Murray Grobman (258th FABN XIX Corps) were standing at the NE edge of a very large wooded area about 2.5 miles SW of the large church located in the little town of Langerwehe. They were startled by the sound of a sudden burst of quad .50-caliber guns mounted on a half-track behind them to their left. The burst lasted three or four seconds. When they looked to their left, just coming into their view was Preddy's Mustang, now upside down and approximately 200 to 300 feet above the ground and in a 20 to 30 degree nose down attitude.

Up in the steeple of the church in Langerwehe, were Sgt. Harold M. Kennedy and his buddy, Cpl. Elmer L. Dye, both of the 104th Inf. Division. While the Battle of the Bulge raged just a few miles away, it was relatively static in their sector where the Division had dug in on the chance the Germans might veer in their direction. Division headquarters had been setup in a large steel foundry just north of Langerwehe. Dye and Kennedy had spent quite a few hours killing time by posting themselves in the church tower with binoculars to watch the considerable air activity along the front.

Cartee recalled there were three Mustangs passing over a wooded area with a large church in their flight path. The woods NW of the church were occupied by elements of the 555th AAA (AW) BN (see footnote) which was located on the northern edge of the German penetration. Their weapons were 40mm anti-aircraft guns and quad .50 cal. machine guns. They were assigned to protect U.S. troops from low-flying German aircraft. The ground was frozen, covered with snow and the sky was filled with snow and heavy clouds, making it easy for the German armor to move about. The troops had lined up for a hot Christmas dinner consisting of turkey, mashed potatoes, cranberries and pumpkin pie. T/3 Leo J. Thoennes, of "B" Battery recalls that he had just taken his mess kit of food and walked to the nearby gun section #4. Suddenly, before he could eat his dinner, what was thought to have been a P-47, actually an Fw 190 D-9, and a P-51, came over with their guns firing. The NCO in charge of the battery ordered his guns to return fire. Cartee remembers they started receiving heavy ground fire.

Kennedy recalled that as the Mustang passed over the church, firing from the 555th batteries became heavy and continuous. Lt. Mitchell, some distance away, recollects seeing multiple tracer rounds that gave every appearance of being "a whole field of golf balls," so intense was the antiaircraft barrage.

From their vantage point looking NE, Pfc. Charles Brown, Pfc John Starzynbbski and Lt. Murray Grobman saw Preddy fall from his Mustang at about 200 feet, his parachute not deployed. Brown recalls that within minutes of the crash, two Me-109s flew over the same path as Preddy without any of the AA guns being fired.

Both Kennedy and Cpl Dye went to the crash site of the Mustang and noted that the largest portion remaining of the Mustang was the engine. Kennedy recalled seeing a piece of the fuselage on which swastikas had been painted.

Lt. Cartee returned safely to the field at Y-29 and made an uneventful landing.

Back at Y-29, Art Snyder, Preddy's crew chief, waited patiently for his ship to return. Aircraft were landing and taxied by his hard stand. He noticed that as different pilots went by, they would give an accounting of how many victories they had gotten with their fingers. Shortly, a pilot came by him and when their eyes met, he gave him a thumbs down. Art knew then that his friend and commanding officer had been killed.

Lt. Mitchell, having successfully disposed of his attacker, landed and, after parking Geraldine, was heading for his tent contemplating the loss of his leader when a fellow pilot asked him if he would join him for Christmas dinner in the mess tent later in the day? Mitchell replied, "Christmas dinner?" In his effort to survive, he had completely forgotten what day it was. It had been that kind of a day.

To top off their Christmas meal, a keg of beer had been delivered to the group via "air mail." In tribute to their fallen comrade, it's spigot was opened but 'nary a cup was drawn.

George E. Preddy, Jr. has been labeled by many as the fighter pilot's pilot. The late Gen. John C. Meyer said of him that he was the best fighter pilot to have ever looked through a gunsight. Other historians have speculated that had he lived he could have become the highest-scoring ace in the European Theater of Operation. One thing is certain, he did indeed "become an aviator!"

Notes:

There were 4 Batteries in the 555 AAA, A, B, C, and D. Each battery had 8 gun sections of a mixture of 40MM and quad 50 cal. weapons. A total of approximately 32 guns. It is believed guns from B and D batteries initially did the firing. Before it was over, all guns from the 4 Batteries were firing.

After the war, the Victory Credit Board reviewed all victories claimed in all the various theaters of operation. All ground victories were disallowed. Preddy's official score was adjusted to 26.83 victories. (Shared in half a victory with one pilot and a third in another.) He is ranked as the third highest scoring ace in the ETO and the highest-scoring P-51 Mustang ace in the Second World War.

Credits: Joe W. Noah, author of Preddy's biography, *Wings God Gave My Soul* and *George Preddy, Top Mustang Ace*, artist, Darrell Crosby, J.J. "Red" McVay, Art "Snoots" Snyder, Raymond Mitchell, Ralph "Ham" W Hamilton, J. Gordon Cartee (deceased), John C. Meyer, General (deceased), Ignacio G. Marinello, Tom Ivie, Robert H. Powell, I.B. "Jack" Donalson (deceased), C.E. Griffiths, C.L. Fuhrman, Jim Bleidner, Ron Putz, (deceased), Jeff Grosse, Harold M. Kennedy, Thomas J. Thoennes, Charles "Chuck" Brown, Stephen Finnigan, Joe Hutton, John R Beaman, Jr and to all the fallen eagles who made the supreme sacrifice for our freedom.

Flushing the Luftwaffe

by Sgt Maurice Barrington, *YANK* Field Correspondent
Originally published in *YANK*, the Army weekly
Edited and revised by Bob Powell

On November 2, 1944, an epic air battle took place over Germany—more allied fighters opposed more German fighters in a limited space of sky than ever before in the European theatre. In light of subsequent enemy activity, the significance of this ferocious encounter has become increasingly important since it was this mission that disclosed the increased production of German fighters and proved once again that the German pilot is nobody to fool around with.

ENGLAND—The mission of November 2 was not a trap. That's official. The Intelligence Office (IO) explains that no trap was planned. Nothing of the sort. It just happened that the bombers went straight to their target.

As for the air battle that occurred, we had always been ready for that, fully aware that the Luftwaffe fighters would come up in force sooner or later. Since it was SOP (Standard Operating Procedure) that our big bombers do not go directly toward their objective, but zig and zag, the chart indicated that on this mission, with a lack of caution, the bombers were going straight to their targets.

Some 600 B-17s and B-24s droned into Germany toward a point just north of the Ruhr, then drew a long bead on Merseburg, home of the Leuna synthetic oil plant.

At Nazi air bases scattered in the area, a new Luftwaffe was poised. It had hardly been seen for weeks, but it was there. Our intelligence officers were aware of where they were and how many there were. They knew they were never as dead as we had imagined a few months ago. On D-Day, the world, reading communiques out of the teletype machines, wondered where they were. The enemy had made a hard decision, pulling its aircraft back into Germany to better defend the homeland. The German leadership had realized that the hope of the Reich lay in concentrating their aircraft production on its fighter force, a Luftwaffe that could do for Germany what the Spitfires and Hurricanes once did for England. The Allied strategic bombers had to be stopped. They set out to quadruple their production of fighter planes.

Our reaction was a more intense bombing campaign, which included a major attack on the aircraft factories at Regensburg, the England-Africa shuttle mission of August 17, 1943, the great attack on a components plant at Anklam, and the Folke-Wulf plant at Marienburg, but the weather greatly aided the Germans. From November through January, American bomber crews sloshed through English mud to their planes for many missions that had to scrubbed.

In late February the break came, a streak of CAVU (ceiling and visibility unlimited) weather. The Eighth Air Force muscled up and went into action with the Fifteenth Air Force in Italy and, the enemy sent up everything it had—flak, fighters, rockets.

Two hundred ten bombers, 138 from the Eighth, went down between February 20 and 26—a loss of more than 2000 men plus 38 fighters, while the Nazis lost more than 600 planes in the air and a large part of the German aircraft industry lay in ruins.

That is why, on D-Day, the German air force wasn't there. They had lost control of the air over Europe and had to disperse their fighters to many small airfields to avoid having them destroyed on the ground. Those five days in February had placed the Luftwaffe in a bad spot. Our bombers were now at work on the ultimate target, the German oil industry, the synthetic gasoline and lubricants made from coal. The GAF could not join the battle in France and still engage the heavy bombers, requiring them to pull their planes back into Germany. The Huns were running out of time and they had to do something and it had to be good.

At 1221 on November 2, a pilot in a Mustang flying over Belgium caught a radio conversation, faint and distant, from deep in Germany.

"Hello, Topsy, this is DITTO BLUE LEADER. Fifty plus Bandits are headed for the BIG FRIENDS from the east."

"I am now approaching the target," came the reply.

"Bandits are also approaching the target," the P-51 pilot reported.

Switching to C-Channel, the message was relayed back to the control room in England, where officers with headsets sat in an amphitheater, looking down on a great map of Europe. Below, at the map table, a sergeant with a Croupier's stick moved a Swastika block into position near Merseburg.

There was no trap, but there was a plan, which designated that no bomber was to be without fighter protection at any time. The FORTS and LIBS were loaded with high explosive bombs, their targets one of the vital organs of Hitler's war monster. They were lining their sights on the LEUNA oil plant.

A grim SNAFU developed soon after 1115 hours, however, when two groups of bombers, impacted by a 60-knot wind from the west, flew wide of their briefed course, only later to reach Merseburg on a line from the north. For half an hour, these groups flew unescorted. All over northwest Europe, German fighters had positioned themselves south of the bomber course to take advantage of the sun. Then, the German fighter force, estimated to be 400-600 planes, struck the bombers as they were leaving their targets, knocking down 21 of our bombers.

For some unknown reason, the enemy stayed at a distance until after the bomb run of the 3rd Division bombers. Minutes later, as the smaller wave of bombers pulled over the target, the enemy fighters came in. At 1221, Blue Leader, Major George Preddy, was calling his C.O., warning him of the attack. In just seconds, one of the war's great air battles was joined—more fighters opposing more fighters than the sky had ever known. One moment just contrails in the distance, the next 1000 combat planes diving and shooting in a fantastic battle.

German radar stations, estimating the tonnage on our bomber fleets, knew this was an unprecedented engagement. Our pilots could not have known it. They knew they were in a fight, which is all a fighter pilot has time or vision to know. He drops his tanks, checks his gun switches, pushes everything forward—throttle, prop control, mixture—and the prop howls as the air speed climbs. Things happen fast. They happened fast to 23-year-old Captain Don Bryan, whose gun-camera film later comfirmed his claim of five enemy aircraft destroyed and two damaged.

"I was leading Yellow Flight at about 28,000 feet. Just before reaching the target, a lot of contrails were called in coming from the east."

This is how his combat report read, "I heard Preddy say, 'Watch those contrails to the east.' He was leading a roving patrol to give us warning. He alerted us on A-Channel that there were lots of planes coming up."

"Those bastards tried. They went right by me headed for the bombers. I cut loose my tanks and took my flight right into the middle of them. They forgot the bombers and went right on down. I closed to about 100 years on Me-109 but got only a few strikes as I overshot him and dived under him. As I did so, I saw that his coolant system had been shot out, so I claimed this one as damaged."

Another pilot who had ridden his plane down to the deck on the tail of an Fw-190 got this groundside view of the carnage:

"At 2000 feet, I spotted at least 25 fires on the ground, burning planes. Once I looked up and saw a bomber tumbling out of the clouds, a solid mass of planes. All around it were fighters falling, some streaming smoke, others just spinning out of control or flopping like falling leaves."

In the middle of this fury, the bombers made their turn, holding their formation steady and compact, their gunners tracking the enemy whenever he came close. Few came close. The Mustangs were too many and too quick.

Bryan reported, "My element leader called an E/A moving in on me and as I rolled, I saw him positioning himself on me. I did a snap roll turn and lost him, and the rest of my flight as well. For the next 10 minutes, I was in at least five separate combats. I attacked the last E/A in a string of them and saw a lot of hits on him. He went down in flames. I claim this E/A destroyed.

"In about 20 minutes it was over. The enemy had seen 38 of his planes shot down by our P-51s. The 352nd Fighter Group, whose C.O., Col. Joe L. Mason, was the 'Topsy' mentioned in that first warning message, set a new group record of 38 enemy aircraft shot down in one day, 24 of those by the 328th Squadron."

Miles to the north, enemy interceptors had fought and attacked the wandering heavies who had been blown off course. On their own, Liberator gunners fought the good fight, knocking down the bulk of the 30 aircraft credited to the bomber crews while losing 12 of their own.

In the escorted bomber force, however, only one bomber was shot down by an enemy plane. We could hardly ask for a greater victory, whipping the enemy better than 10 to 1 in fighter planes. But our high brass at headquarters did not jump for joy. They were relieved, of course, knowing what could happen.

Our air superiority was a deceptive thing. We're not 10 times better. We may not even be two times better. Mustang pilots will point out that even a small superiority makes a big difference.

The Hun continues developing his fighter force by every possible means, dispersing his factories, using shadow-plants, whatever. Our graph of his fighter inventory shows the line moving sharply upward and his total single- and twin-engine fighters is again in the thousands. There will be days when that non-com at the plotting map table in the control room will need more Swastika blocks to keep track of them.

Maybe it's just as well. Our Mustang pilots like a good fight.

Fallen Eagle

by Samuel L. Sox, Jr.

On 21 July 1989 common bees did to Col. William T Whisner what no German or Korean pilot could accomplish in five years of aerial combat, inflicting the blow—by way of an allergic reaction to their sting—that ended the life of one of only seven flyers with the distinction of becoming an ace during two conflicts.

While he served with the 352nd Fighter Group during WWII, he was credited with 15.5 aerial and three ground victories. He became an "ace in a day" when, on 21 November 1944, he claimed six victories in a single mission. (This was later reduced to five.) As a member of the 487th Fighter Squadron, he flew the P-51B and D models, "Princess Elizabeth" and "Moonbeam McSwine," respectively.

Bill loved to fly, and thus, when the Korean conflict reared its ugly head, he served with the famous 51st Fighter Group, home of WWII's highest scoring ETO ace Col. Francis "Gabby" Gabreski. Flying the P-80 Shooting Star and later the F-86 Sabre, Whisner scored another five aerial victories over the North Korean and Red Chinese pilots, bringing his total career victory score to 21.5. Prior to his retirement in 1972, he served as an advisor in Vietnam.

Early in 1989, Vlado Lenoch, of Joliet, Illinois, purchased a P-51D which had been restored in the markings of Whisner's Mustang named Moonbeam McSwine. He wanted very much to meet Whisner, take him for a ride and have him autograph the rudder of his aircraft and a date was set. Sadly, three days before Vlado was to finally meet his hero, he received a call advising him that Whisner had died.

But William T. Whisner would fly once again in a "Moonbeam McSwine." Lenoch would fly his beautiful Mustang to Alexandria, Louisiana, and there, at Whisner's widow's request, take to the air for one last sortie. Over Lake Ponchartrain in Bill's beloved state of Louisiana, Lenoch slid the canopy sporting the name of "Capt. Wm. T. Whisner" back and released to the blue skies and the lake below the ashes of his fallen hero, launching his friend into a flight eternal. God speed, Bill Whisner, and blue skies forever.

Adapted from an article originally published in the February, 1990 issue of Air Classics magazine.

Thunderbolt Liberator

by Allen Matthews with Joe Noah

Forty years later, a B-24 pilot discovers the Mustang pilot who saved his life.

It was 22 December 1943 and I was flying co-pilot on my first combat mission, a harrowing mission I'll never forget. It was to Osnabruck, a railroad town in northwest Germany. Actually, I was substituting for a pilot who had burned himself on our pot-bellied stove the night before this mission. So this was not my regular crew.

Records show that 574 bombers from the Eighth Air Force were dispatched on this mission, and that heavy clouds disrupted some of the formations. I was a member of the 445th Bomb Group and our ship was assigned to the low element of the lead formation. The Liberators of the 445th formed up at 3000 feet and then began climbing to the assigned altitude of 22,000 feet. As our B-24 reached that point we leveled out and immediately lost power in our number two engine when its supercharger malfunctioned. We feathered the prop and increased manifold pressure to 50 inches and rpm to 2400 on the three remaining engines but were still unable to maintain formation.

Our pilot, Lt. Glenn Jorgenson, restarted number two and attempted to gain power by jockeying the supercharger controls back and forth. This procedure helped and it enabled us to stay in formation during the bomb run. But once the bombs were dropped our problems returned. The air at 22,000 feet was just too thin for our number two engine to operate effectively without its supercharger and we could not keep up with the other B-24s.

As the 445th BG withdrew across Holland, our Liberator, along with two others, fell further and further behind the formation. We were sure to attract the attention of the Luftwaffe. Being a straggler over enemy territory is not good for ones' health, and we knew it! We began to dive for the overcast. As we did, we spotted a group of about 15 enemy fighters, Me-109s, attacking one of the B-24s which was lagging behind the main formation and immediately stopped our dive in an attempt to avoid those fighters. A few seconds later one of the straggling Liberators burst into flames and spun into the overcast. The enemy fighters then swarmed on the other straggler and after about ten passes sent it crashing to earth.

Then the enemy fighters saw us, and one of our gunners warned us of an attack coming in from the rear. We started taking evasive action, diving, climbing, turning, slipping, any maneuver to thwart the enemy's aim. The gunners reported eight Me-210s coming in at between five and seven o'clock. On their first pass our number one engine began to overspeed. It was hit by 20mm cannon shells and ran up to 3300 rpm. The prop shaft turned red hot. On the same pass, a 20mm hit our number three gas tank leaving a four-inch hole, but no fire! Incredible!

The oil gauge on number one engine had dropped to zero immediately after being hit. We were too late trying to feather the prop and all the oil was gone. So we cut off the mag switches. gas supply and mixture control and let the prop windmill. It was the best we could do in that situation. Unfortunately, the prop was windmilling very slowly and causing a huge drag on that outboard engine.

Now the enemy fighters were making their second pass, more successful than the first. Two or three shells exploded in the bomb bay throwing parts of the bomb racks into the radio compartment and breaking the gas gauges and damaging the radios. The hydraulic system was also knocked out, leaving the tail turret inoperative. Although the gunner, Sgt. Frank Socco, was unable to rotate the turret, he continued firing in hopes he could at least scare the attacking enemy fighters.

Another 20mm went into the nose compartment exploding in the side of our navigator, Lt. Arthur E. Barks, where his flak jacket was open. He was killed instantly. That explosion also set our ship on fire filing the cockpit with smoke and drove fragments into the legs of the bombardier, Lt. Roy Stahl. It ripped off his connections to the oxygen system and the radio. Stahl went for the fire extinguisher. By the time he made his way through the smoke and found the extinguisher, the fire was out. He did this at 20,000 feet without oxygen or gloves. Two or three more shells exploded in the waist, slightly injuring both waist gunners, Sgt. Schaffer and Sgt. Dodson.

Even so, both kept firing at the enemy fighters as they made passes on "Lizzie," our Ford-built B-24H. Then a Ju-88 fired a rocket that went through both rudders leaving a hole about the size of a basketball in each.

Meanwhile, the engineer in the top turret, Sgt. Charles Jones, scored a direct hit on one of the fighters; it burst into flames as it spun down through the overcast. The right waist gunner got off a few good bursts which hit another fighter, leaving him smoking. After that pass the fighters suddenly and for no apparent reason left us to cope with our 300-mile trip back home on the remaining two overworked engines: number two, which began delivering more power as we descended, and number four engine.

As we approached the Continental coastline and the North Sea, we started throwing out everything that would come loose, steel helmets, flak suits, radios, ammunition and guns. Since every pound less could mean the difference between our surviving or not, the body of the navigator was thrown out as well.

We were forced to increase power settings on number two and four engines to about 60 inches and 2500 rpm. To reach such high power settings the engineer used the screwdriver blade on his pocket knife to remove the stops on the throttle

Major George E. Preddy, Jr., top Mustang ace, and high scorer (aerial) for the 352nd FG.
(Courtesy Troy White, Stardust Studios)

quadrant. These settings were greater than takeoff power (later our local Pratt & Whitney representative challenged these figures saying that at that power setting the engines should last only three to five minutes). These two engines were better than the book gave them credit for being. We dropped about eight degrees of flap to maintain 140 mph and a descent rate of 250 fpm estimated to get us home.

With all radio equipment either abandoned or inoperative, and the gyros tumbled, we continued our flight across the North Sea, descending through a 2000 foot overcast. We were flying by the seat of our pants, the airspeed indicator and the needle and ball.

Just before the navigator was hit, he had given us a heading of 250 degrees so we kept that heading as best we could. When we reached 3000 feet our number four engine cut out and the B-24 swerved and started side-slipping down despite full opposite rudder. We were pumping fuel from the number one engine all the way across to number four and the booster pump could not keep up with our consumption. After losing about 1000 feet the engine cut back in and we were able to regain control. About a minute later it cut out again, and this time we lost about 600 feet before it cut back in. We saw that the fuel pressure was low, so we reduced power on number four and kept the booster pump on. The engine ran smoothly but with lower power settings than we were using on the number two, which was doing fine now since we were at a low enough altitude not to need the supercharger.

At 1500 feet now and still losing altitude, we prepared to ditch even though ditching in the North Sea in December was a last resort since no B-24 had ditched successfully and the freezing waters of the North Sea were prohibitive. We tried restarting number three and it worked, even with no oil and a hole in he fuel tank, and we began gaining altitude until it got red hot and we had to cut it off to avoid a fire, a fate worse than ditching.

With 5000 feet between us and the water, we continued on but we continued losing altitude. Fifteen minutes later we saw the Cliffs of Dover, a truly beautiful sight. We spotted a long runway which turned out to be Manston and prepared to land, but only the right main gear would go down and lock. We had no choice. We slipped her out over the grass and made a very smooth landing on one wheel and one wing tip. When the plane finally came to a violent stop with the left wing tip acting as a landing gear, all remaining crewmen jumped out safely. Lizzie couldn't burn; we were out of fuel.

Through the years, remembering that mission, I had always wondered why the enemy fighters left us so suddenly for surely we were easy prey. My daughter, doing some research for the Greensboro, North Carolina museum, read a copy of biography of Major George Preddy written by the co-author of this story. She thought I might enjoy reading it and sent me a copy.

As I read about Preddy's mission of 22 December 1943, chills went up my spine. I had discovered without a doubt that it was Major George Preddy who drove the enemy away for us. Leading Crown Prince Blue Flight of the 487th Fighter Squadron, Preddy had become separated from his wingman, Lt. Richard Grow. So his Blue Three, Lt. John R. Bennett, and Preddy spotted our plane and saw the predicament we were in, according to his encounter report. He immediately dived his P-47 into the enemy attackers, leaving Bennett up for top cover. Bennett was chased into the clouds by Me-109s, and the gaggle of Me-210s Preddy went after dispersed as soon as they saw him coming. We never did see Preddy or his wingman, but Preddy picked one of the Me-210s out and closed on him and gave him a short burst before the 210 escaped into cloud cover.

Preddy then pulled back up and attacked another Me-210 which was attacking a crippled B-24. That had to be us! Preddy started firing at 60 degrees deflection from 400 yards, closed to 200 yards continuing to fire and the enemy aircraft began to disintegrate with large pieces flying off as he went down into the clouds in flames. Preddy broke back up and he heard his missing wingman call on the radio and warn that an Me-109 was on his tail. Preddy reported that he threw the stick in left corner and saw the Me-109 behind him but was out of range. Lt. Grow then called that an enemy aircraft was on his tail but Preddy couldn't find him to go to his aid. Now low on fuel, Preddy went into the clouds and set course for Bodney, his home base in Norfolk, England.

So, that's how I found out some 40 years later that a Thunderbolt pilot named George Preddy was the one who rescued me and my crew from a fate similar to that which claimed our comrades in the other two stragglers. For unusual courage demonstrated that day, George Freddy was recommended for the Distinguished Service Cross but the award he received was the Silver Star. The citation reads in part:

***George E. Preddy Jr** 0-430846. Captain, Army Air Force for gallantry in action while escorting bombers withdrawing from a mission over Germany 22 December 1943. Captain Preddy observed a lone crippled bomber being a tracked by a large number of enemy fighters. Though outnumbered six to one, he unhesitatingly led his flight in an attack on the enemy and pressed it home with such viciousness that the enemy planes were scattered and forced to cease their attacks on the bomber Captain Preddy personally destroyed one of the enemy aircraft…*

As it turned out, I was not the last B-24 crewman to be saved by George Preddy. On 12 June 1944, when he was leading the 352nd Fighter Group on a mission giving area support to bombers on various targets in the vicinity of Rennes, France, a group of about 18 B-24s were last out and withdrawing. A dozen Me-109s made a quarter stern attack on them from out of the sun. Preddy led his group straight for them; they were at the same altitude! He reported that some of the 109s turned into them, and others broke for the deck. George selected one from the pack and followed him down, firing from various ranges and angles. He finally got in a few hits on the 109 which promptly lost some of its speed, causing Preddy to overshoot him. George pulled above him to start another attack when the pilot bailed out.

But the enemy fighters had been successful in knocking down one B-24 from the 448th Bomb Group. Ben C. Isgrig. Jr. found out in 1970 that the Mustang pilot who saved his life was George Preddy. He was the bombardier who had bailed out of his stricken Liberator, and on the way down an Me-109 made a pass at him while he was helplessly dangling in his chute. Isgrig said, 'Just as the Me-109 passed me, a P-51 came in on his tail and shot him down.' The German pilot bailed out. I have always been extremely grateful to the P-51 pilot who very possibly saved my life, but never made any effort to find out who it was. Recently, I read *The Mighty Eighth*, and in it found that members of the 352nd Fighter Group reported shooting down three Me-109s on 12 June over Rennes. In trying to find out the individual pilot that got the Me-109, I wrote the Air Force and received word that Major Preddy was the only pilot who reported that his adversary bailed out. He was the one!"

The most significant mission of Preddy's career was the one of 6 August 1944 when he was leading the 352nd Fighter Group. Preddy was about to rendezvous with the bombers his group was assigned to escort when he spotted more than 30 Me-109s with similar intentions. George drove right into the pack of enemy aircraft and shot down five, starting with tail-end Charlie. He made his initial attack at 28,000 feet. He got on the tail of his sixth victim and chased him all the way down to 5000 feet where, after a tough dogfight, Preddy came out victorious. He was recommended for the Medal of Honor and awarded the Distin guished Service Cross, our second highest medal, for this feat.

Ironically, Major Preddy was shot down by Allied ground forces on Christmas Day 1944 while he was chasing a Focke-Wulf 190 near Liege, Belgium. When George spotted the enemy aircraft, it was near the deck. Preddy dived his P-51 Mustang, Cripes A' Mighty, in hot pursuit. As he closed on the enemy fighter, American antiaircraft guns opened fire on the enemy aircraft not knowing it was being pursued by a friendly aircraft. They missed the 190 and got Preddy; he was fatally wounded by one of the rounds from the AA guns and crashed to earth. The late Major George E. Preddy, Jr. is the top Mustang ace and the eighth ranking American ace of all time and the sixth ranking American Air Force ace of all our wars with 26.83 aerial victories and five destroyed on the ground by strafing. His brother, Lt. William R. Preddy, was killed less than four months later while strafing enemy air fields near Prague, Czechoslovakia. His combat record included two aerial victories while flying Mustangs with the 339th Fighter Group. He had only been in the ETO for three months when he, too, was killed in action.

A Letter from Fritz Koal, JG/27, GAF

(This letter from a German pilot who was shot down by one of our most famous 352nd aces, was sent to the late Mindy White through a German interpreter when she was doing WWII research for her aviation artist husband, Troy White. The original translation has been edited.

"It is very interesting for me to see how you are doing research in personal activities during the awful war some 50 years after the end of WW2. For God's sake I did forget some of the things I've seen. But flying of course has its distinctive events one can hardly forget.

"At the age of 20 I was a "Flugzeugfuher" (pilot) attached to 6.Staffel JG 27. That was in 1944. (Koal was born on Dec 23 1924). The unit was based at Feis am Wagratn, Austria. After JG/27 (suffered) very bloody combat, it was in the process of getting new mounts and fresh personnel. Beginning in July 1944 we started intercepting and engaging allied bomber and fighter units. The missions were flown in the Alps, Austria, Hungary and all through Romania and finally in the skies of the Deutsches Reich. At this time we were attached to the "Reichsverteidigung" (Defense of the Reich).

"In November 1944 the unit transferred to Hopsten (a small airfield near what is now Rheine AFB). We were flying high altitude missions against allied bomber and fighter groups. In the early days of December, 1944, a short time before the Battle of the Bulge. We were told that a major offensive strike would take place in the West. On Dec. 16 the offensive started and we flew several missions daily in order to cover the ground forces. We engaged the bombers or went on "Freie Jagd" (free chase) looking for enemy a/c to attack. The casualties were very high and the "Einsatzstarke" (operational strength) went down rapidly.

"Our Gruppe consisted of four Staffein, each of them sporting an operational strength of 16 Me-109 aircraft totaling up to 64 aircraft. Coming back what took place on Dec 25, 1944 I recall that we had nearly reached the operational strength. The 6./JG27 was able to start with 16 planes.

"We were to stop allied bombers from attacking German units in the Battle of the Bulge area. The 5./JG27, 7 and 8./JG27 were to engage the bombers while 6./JG27 was to cover them from above "Hoenschutz" (high altitude security or screen). Since the Staffelkapitan (commanding officer leading a Staffel in combat) was not able to fly, I was appointed to take over the command. At this time I was senior "Schwarmfuhrer" (Koal was in fact the pilot with the most experience still alive at that time so he was appointed not only to lead the Scwarm (a flight of four a/c) but the entire 6th Staffel).

"December 25, 1944 was among the most wonderful days in December 1944. The cold skies were blue and we could hardly spot a single cloud! At about lunchtime the Gruppe was called to take off. Approximately 60 Me-109s took off led by Hauptmann Kutschka, the new Gruppen-Konmiandeur of II./JG 27. The planes waited until everybody was airborne, then all of us climbed to an altitude of about 9,000 metres. My Staffel climbed up to 10,500 metres. When thinking back, I recall a Lightning that cruised about 2,000 metres above our unit tracking our movements. In order not to disrupt our unit, I did not attack her.

"While heading towards the target area, we lost the Lightning. When we reached the target area we saw bunches of Mustangs attacking the Staffelin below us. I called for immediate action, and after checking whether everyone was diving or not, I turned my aircraft and speeded downward into the battle. My very high speed brought me down there faster than I would have expected. I was in a perfect position and opened fire on a Mustang. I could clearly view several hits on the aircraft. At that moment my ship was hit hard by enemy fire and I could not see what happened to the Mustang I was chasing.

"I was attacked and surprised by this attack. My wingman was supposed to keep me safe when I dove down. My aircraft dropped into a spin and I was not able to get out. I was unable to get the ship back under control and blew off the canopy. Due to negative forces I was unable to get out of the plane at first. After a couple of seconds I somehow left the plane and baled into the unknown. When my chute opened, I was unable to see anything. I thought:, Gee, you are blind. But why? Did the "Kehlkofmikrofn" (radio transmitter) cut off your nerves? Feeling some sort of liquid in my eyes, I figured that some sort of oil or cooling water burnt me. I did not know. For an awful long time I was hanging under my chute, breathing the cold, clean air and was not aware where I was drifting to. At about 500 metres altitude I spotted the ground with my right eye. The landing was not much of a problem.

"Since I had no idea whether I had landed in Allied or German territory, I ran off and tried to hide in a bush. A little later German soldiers appeared and aided me. I then returned to the Staffel where our doctor sent me to a hospital at Osnabruck. Due to my injured eyes I was sent to my home address in order to recover and afterwards had a vacation in Bad Wiessee in order to gain strength to return to my unit.

"In March 1945 I returned and flew several missions over Berlin until April 30, 1945. (Koal shot down three more American planes in April, ending the war with six confirmed victories). On this specific date the last six Me-1 09s of 6/JG 27 surrendered at Leek AFB, a small base in northern Germany near the border to Denmark. WWII ended for me.

"I feel sorry that Major G.E. Preddy died when shot down by friendly fire. This is a sad ending for an ace, but was not so uncommon as some people think. I recall these accidents in the Luftwaffe as well.

"When thinking back I have to feel thankful that Major Preddy scoring the victory in our encounter. I got out of the war for a while and survived. Who knows whether I would have survived without this break?"

Editor's note: Fritz Koal was piloting one of the two German planes that Major Preddy shot down in his last wartime engagement just minutes before he was so tragically killed by friendly fire.

Preddy holds up six fingers to Asst Chief McVay.

"Second to None"

Continuing to tally more kills as the war raged on, Preddy scored an incredible six aerial victories on 6 August 1944. Following a 30-day leave soon thereafter, during which he had the opportunity to fly mock combat with younger brother Bill, Major Preddy returned to the 352nd FG to take command of the 328th squadron. There he would score two more victories flying from Bodney, this time in his brand new P-51D 44-14906 "Cripes A' Mighty." His first victory came during the epic air battle November 2, 1944 while he led the 328th to a new 8th Air Force record for enemy aircraft destroyed in a single mission. During this escort mission to Merseberg, his squadron destroyed 25 aircraft in the air in about 20 minutes. In "Second To None" Major Preddy is seen as he prepares to shoot down an Me 109 from JG 27. According to George, "He went down leaving a long column of white smoke. A P-51 from another group went in behind him and the enemy aircraft continued on down crashing into the ground…" (Courtesy Troy White, Stardust Studios)

Author/editor "Punchy" Powell and wingman
strafe a German airfield in April 1944. Painting by aviation artist Robert Bailey.
Reprinted courtesy of the artist.

Lt. Ray Mitchell downs an Me-109 on Christmas Day, 1944. Flying on Lt. Jim Lambright's wing,
Mitchell downed the Messerschmidt when it committed to Lambright during the battle over Koblenz, Germany.
Courtesy of artist Charles Taylor

MISS LACE'S FIRST ENCOUNTER
Courtesy of Troy White

Captain Don Bryan of the 328th FS, 352nd FG rolls out behind an Arado 234 "Blitz" bomber flown by Hauptman Hirshberger of 6./KG 76. (Courtesy Troy White, Stardust Studios)

Captain Ed Heller of the 486th closes on the tail of an Fw-190D from II./JG 301 on March 2, 1945 in his Mustang "HELL-ER BUST." (Courtesy Troy White, Stardust Studios)

"FULL HOUSE-ACES HIGH"

Major George Preddy, Lt. Col. John C. Meyers, and Capt. Bill Whisner, in "Cripes A' Mighty," "Petie 3rd," and "Moonbeam McSwine," respectively, accompany two B-17Gs. "Wiggling" their wings, they signal the bomber pilots that they will break off and head for Bodney. Courtesy of aviation artist John Doughty. www.highironillustrations.com

"Twilight Victory"

Captain Walter "Wally" Starck blows past a German Me 109 that he has just damaged on July 1, 1944. Starck was leading the 352nd FG when controllers vectored him to an air battle that was already in progress. Upon arriving at the scene, Starck quickly dispatched one of the enemy 109s and damaged this second aircraft before it ducked into a cloud. His wingman, Lt. Sheldon Heyer, narrowly missed colliding with Starck's first victim as the Nazi pilot bailed out. Heyer noted that the second Me 109 started smoking before it entered the clouds.
(Courtesy Troy White, Stardust Studios)

Painting of General J.C. Meyer at the Pentagon
Photograph Courtesy of USSTRATCOM Public Affairs

"The Battle For New Year's Day"
Limited edition print by world-renowned aviation artist Nicolas Trudgian.
(Printed courtesy of the artist)

The New Year's Day Legend
The Y-29 "Bluenosers" Speak
By Marc Hamel and Robert "Punchy" Powell

The facts have been known for 57 years about the fantastic "white wash" 23 to 0 victory by the 487th Fighter Squadron over the Luftwaffe on January 1st, 1945. The story of this aerial melee over Asch was penned at the end of the war, and entitled "The Legend of Y-29." However the entire story, from the mouths of the men who participated, has never been put in print. This is their story, with a surprise recent addition to the 23 victories thought to be achieved.

The 352nd Fighter Group (the famed "Bluenosed Bastards of Bodney") was notified in December of 1944 to prepare for a move to the Continent. As Axis airfields and terrain for new airstrips were overrun, the Allied Air Forces began shifting fighter units from England into Europe proper. This had several benefits. The tactical fighter groups, which bombed and strafed in support of ground actions, could stay close to the ground troops. They could therefore provide faster response and increased attack missions per day. The escort fighter units, such as the Bluenosers under 8th Fighter Command, could fly farther into Axis territory, stay in the target area longer, as well as escort tactical fighters. These forward bases also allowed the RAF's shorter-legged fighters (such as the Spitfires) to get into combat over the gradually shrinking Axis territory. These rapidly proliferating Allied fields proved to be too tempting a target for the beleaguered Luftwaffe, much like a steak before a hungry dog. So the stage was set.

"Operation Bodenplatte" was hatched by the Luftwaffe as a do-or-die attempt to wipe these hated Allied aircraft from the Continent, and prevent them from harassing the German troops retreating from the "Bulge." The ten Jagdgeschwadern (JG) assigned to undertake this task could conceivably commit 100 aircraft each (1,000 total) to the mission to strafe the 17 targeted Allied advance airfields. Numbers regarding the mission vary widely, but it is generally agreed that the JG's were only able to get somewhat less than this total number into the air, perhaps as few as 800. Of this total, many pilots were true "greenhorns," with little flying time under their belts.

The plan was to attack these 17 RAF and USAAF bases on New Year's Day at 9:20 am, flying west from Germany at lower than 300 feet to avoid detection. To the German's dismay, roughly 100 of their fighters were downed by "friendly fire" on this round trip to and from Germany. The Luftwaffe's JG 11 (including I, II, and III/JG 11), commanded by 32 victory ace Guenther Specht, sped towards Asch after they rendezvoused near Koblenz on New Year's morning...

An advanced skeleton unit of the 352nd FG was temporarily attached to the Tactical Air Force and moved towards Y-29 at Asch, Belgium two days before Christmas of 1944 (the field was located just north of the current town of Zutendaal). Here they would share the field with the P-47's of the 366th FG (9th AAF) and attempt to make the best of things.

Pilot Ray Littge of the 487th FS wrote after the war, "We arrived on the Continent on the afternoon of December 23rd. After completing an escort mission to a target in the Ruhr Valley we landed at A-84 near Brussels towards evening, having been unable to find Y-29 (our base-to-be) due to weather. We were never to forget this date. It started a period of time, elapsing over several months, which was punctuated with thrilling air flights, casualties that hurt us deeply, and a score of aerial combat victories which put our squadron in the lead for victories in the Eighth Air Force."

328th FS ace Don Bryan recalls this stop at A-84 , "On December 22nd, after we had finished an 8th AF mission, we were told we were deploying to Asch (Y-29). We were told that it was to be a deployment and we should take all of our clothes and supplies we'd need to stay for an extended time. Since we were told to take our clothing, we all dressed up in our Class A uniforms. I even had my Bancroft hat in the cockpit.

"We all loaded up and started engines, but got a call from the tower to shut down. Squadron CO Earl Abbott went to get codes and found we were going on a mission before landing at Y-29. Fine. We got the the signal to start engines again. Everything was okay, except Earl forgot to turn on his generator. He aborted about 5 min. after we all got airborne. I took over the squadron but didn't have a clue as to what we were to do or who to talk to. I slowed up and forced one of the other squadrons to take lead. Again everything okay except we didn't have the call sign to get any directions from the 9th AF radar. They would respond and then shut up.

Major Don Bryan confers with his crew chief. (Berlow)

Dick DeBruin cuts up on the 328th Squadron Jeep. (Dick DeBruin)

"We milled around over Belgium for an hour or so doing nothing. About that time Hank White, Blue Lead, called that two enemy aircraft were coming in on me. He said to hold and he would give me the 'break' call. Boy is that 'fun,' not being able to see them and hoping Hank would give me the break in time. He did. We all dumped tanks and broke right. The first thing I saw was the enemy aircraft in my windscreen. I kicked rudder and fired about 5 rounds per gun. I hit him and he popped out of the cockpit like a jack-in-the-box. I asked were the other German was and my Number Three said, "I got him." I called for a form-up and looked out at my squadron in perfect spread formation. Then the fun started.

"The other squadrons were in front of us and higher. They also dropped tanks, and we had a real thrill dodging them. We milled around for some time, not knowing where Y-29 was. The damn radar stations wouldn't help us, so we found an airfield with P-47's on it and landed there. Now remember a few things. One, we were all in class A uniforms. Two, our aircraft were all

(con'd on p. 282)

Wartime Letters Home
by Nelson Jesup

JUNE 9, 1944 (D-DAY + 3)

Well, I finally got a bath and a shave and I sure feel better.

This is the first evening I haven't flown and I'm going to bed as soon as I finish this although it is real early...Went out on a mission early this morning and we dive-bombed a Nazi troop and truck convoy and then strafed them. We took off in the semi-darkness in the rain with a 500 lb. bomb under each wing. Try it sometime—some fun! We did a good job and left a lot of them burning...

JUNE 11, 1944 (D-DAY + 5)

Up at 3 am. Sunday, too. Was on a five-hour escort and strafing mission and had quite an amazing experience—one of the damnedest things I ever heard of. There were several layers of overcast and one down very low. Three of us hit the deck over France to come back. There we were barreling along as we approached a small bridge and, just as we got over it, bombs dropped by B-17s above the overcast started hitting the bridge . . . we went all through the flame and smoke and debris and our planes were tossed around by the terrific explosions under us. I thought I'd 'had it' as we say, but I got out of it with a beat-up airplane. Thought I would have to bailout for awhile but I managed to get it back. There was one hole in the leading edge of my right wing that you could put your head in. The other two guys got back OK, too. We had no idea the B-17s were up there over us as we had been escorting them before. Oh, well, it adds variety nd they say that is the spice of life—I wonder!

Also saw my first Jerry today. Almost forgot to tell you. My element leader and I took after him but he escaped into the cloud layer.

JUNE 26, 1944

Back to war again...after a two-day pass in London! Dive-bombed some marshaling yards and then strafed. I got a couple barges in a canal, too—cut them in half. Some Fw-190s jumped us when we were on the deck and got one of our boys, then escaped us by ducking into the clouds... Got 19 letters today dated from May 11th to June 17th. Boy, am I happy!

JUNE 28, 1944

On yesterday's mission we dive-bombed and strafed again. I got two trains but good. I blew one up by strafing the locomotive and set fire to some freight cars on another. We put our bombs on some marshaling yards. Unfortunately, we lost the boy who bunks next to me. He was an old experienced hand at the game . . .

JULY 9TH, 1944

As for the mission yesterday morning I came within a hair of bailing out over the Channel. I developed a leak in a gas line and gas was spraying into the cockpit. I could wipe it off the canopy. It got so bad that even with my oxygen mask on (I was at 22,000 ft), I was getting sick and having difficulty breathing. I was almost to the French coast when I turned back toward England and I had decided to bailout after sticking it out until I got closer to the English coast where Air-Sea-Rescue boats could pick me up. I unfastened my safety belt, shoulder straps, oxygen equipment and switched over to the emergency channel on my radio and called in so they could get a fix on me. I was about to climb out when for some unknown reason the gas spray stopped and I stayed with it and got home—with a beautiful headache.

Nelson Jesup taking photos in Belgium. Note 487th Squadron sign at back left, very similar to ones at Bodney.

NOVEMBER 2, 1944

I'm tired but happy tonight. Today was a big day. I got my first victory, destroying a Fw-190. We really had a big fight today and our Group set an 8th AF record for the most kills in one day. Our squadron only got seven since we weren't in the hottest part of the fight. I was element leader in my flight and my wingman had to abort the mission. I saw a couple 190's coming down. I broke into them after calling my flight leader and about 15 of them went past heading for the Forts. I rolled over on my back to follow them and climbed right up one's tail and let him have short squirts. With each squirt I observed strikes and pieces flying off the e/a. He rolled over and bailed out. By this time I had two Jerries on my tail but I managed to finally outrun them. I was alone and had to head home as my engine was running rough from the long period it had been operating at high speed. I think I might have gotten another victory or two if I had still had my wingman with me. It's one helluva feeling to be all alone in a swarm of enemy planes in the middle of a dogfight. It sure felt good when I came over the field and saw the boys who weren't flying lined up outside the hut. They knew by the way I buzzed the field for landing that I had gotten something and they jumped all over my plane banging me on the back . . .

NOVEMBER 21, 1944

We went to Merseburg again yesterday and it was a pretty good trip for me—I got a double-kill—both Fw-190s. Bill Whisner, my flight leader, got six destroyed (later confirmed as five) to tie the record for a day's kills. Our squadron got 12 altogether and one of the other squadrons got nine. Our squadron was operating in two sections with Colonel Meyer leading. The bombers were starting home and we climbed up through the overcast and saw a big bunch of contrails so we climbed up and attacked. We were heavily outnumbered but we piled in after them. The first one I clobbered bailed out and the second one was killed. The whole fight lasted about 20 minutes, and that's long for a fight. I can't remember all that happened but in emotions—fear aplenty, exultation and only God knows what. I should have gotten some more but my shooting wasn't so hot. We were at 30,000 feet and I was in and out of contrails and fighting propwash and I wasted a lot of ammunition. At one point I ended up only feet away from another's wing. I could have counted his whiskers. He looked at me and I looked at him and he rolled over dove down to the deck. One Jerry and I were making head-on passes at each other but neither of us did any harm . . .

Jesup's P-51D "Peanuts" shivers in the snow in the winter of '44-'45.

JANUARY 1, 1945 (at Y-29)

Happy New Year! We are in Belgium close to the front lines and beating Hell out of the Jerries. We are living in tents and never take our clothes off, not even for bed. It is terribly cold here. Living is damn rough but the food is good and we are busy. Jerry pays us visits most every night and the sky is full of tracers. I got a Jerry this morning—my fourth altogether. The Jerries attacked our field just as we took off and in the ensuing battle twelve of us shot down 23 of them without a loss to ourselves. The attacking planes were both 190s and 109s and plenty of them. I should have gotten more but I didn't. Our boys on the ground saw it all and were strafed by a couple of 190s during the battle. So many things happened that it would take pages to write and I am too tired and the light is so poor for writing much. It's like a fiction story.

Our beds are cots and we have everything imaginable piled on top of us as it really gets cold at night. For baths and shaves we take a truck into a small town nearby and bathe in a coal miners' shower room about once a week . . .

Lt. Dean Huston running up "The Hawk-Eye-Owan" at Y 29.

(con'd from p. 279)

polished and waxed (you could use my Mustang 'Little One III' as a mirror). Three, we didn't have a clue as to where we were—just on a 9th AF base. Four, I had just shot down an enemy aircraft and was felling kind of high.

"They parked me along side a P-47 with the other 328th men to my right. I filled out the Form 1 (a Form 1 on a mission), popped my Bancroft Flighter on my head, and walked up to two pilots that were standing along side of the P-47. This P-47 was about as cruddy as anything I've seen with dirty exhaust stains all over the side, muddy and just crappie looking. The pilots looked worse if anything. Dressed in OD's with those combat boots. They looked like they hadn't shaved in a week or changed their clothes either. They might have had a shower a month or so ago.

"Anyway, I hopped out of 'Little One III,' walked up to them, and said 'Hi. Where am I?' The 9th AF pilot looked at me, looked at 'Little One III,' looked at each other and said, 'G_ _ D_ _ 8th AF' and walked away. Took me a while to see the humor in it, but I did, with my plane shining like a mirror."

Littge continues, "Around noon the next day, we flew to our newly assigned base, Y-29. By coincidence, the base we left was to be our permanent base two months later. At the moment it didn't impress us much, as it was empty and the P-47 outfit that had been there was taking away the last of its equipment. We were eager to get out of there, and to the new base near the front lines. Rumors were that we were to fly with General Quesada's 9th Air Force. This was great news. This outfit had been meeting the Hun often, had inflicted heavy losses, and was supporting our troops in dive bombing missions. All of us were sweating out our first mission here!"

Though their base at Bodney, England was Spartan living, the conditions at Y-29 airfield were brutally primitive despite the best intentions of the 852nd Engineer Aviation Battallion who constructed the base. In the roughest winter in anyone's memory, the officers and enlisted men were housed in tents, with virtually no protection from the pervading cold and deep snow other than their heavily layered clothing. This is the same appalling weather experienced by the ground troops during the famous Battle of the Bulge. Littge adds, "We lived in tents hidden partly in the woods on the south side of the field, and it was extremely cold in those tents at night. Most of us were frozen stiff the first night, when we hit the sack without adequate cover, not knowing how cold it got there after dark. The morning found covers frosted, and the ground inside the tent covered with a layer of white frost. After the first night we went to bed in our heavy flying suits and wrapped ourselves in ten or twelve blankets. The last one to turn in would fill the stove with fuel until it was red hot and then retire in the same manner."

The skeleton force of men from the 352nd included three squadrons of pilots (328th, 486th, and 487th Squadrons) along with a representative ground support team (totaling 75 pilots, 6 ground officers and 100 airmen). Rather than a ground crew team being assigned to a single plane, the hard-pressed selected crews found themselves individually maintaining numerous planes with no shelter.

Throughout the last week of December 1944 the 352nd was in the thick of the action over enemy territory, as the skies finally opened over Europe and allowed air support for the embattled troops in the Bulge. The Bluenosers claimed fifty enemy aircraft destroyed by the three squadrons combined in these actions. To the Bluenoser's horror, they lost their beloved 328th CO George Preddy to friendly fire over Langerwehe, Germany, on Christmas Day. Preddy was the top Mustang ace for the USAAF with 26.83 aerial and 5 ground strafing victories.

Richard "Dick" DeBruin (Supply and Transportation Officer, 352nd FG Headquarters)

"The German Luftwaffe planned 'Operation Bodenplatte' for January 1st, 1945. Their theory was that the Allies would celebrate New Year's Eve to the point of being hung over from too much libation, and everyone would be sleeping on New Year's morning. The Luftwaffe would have all of their fighters in the air early for the purpose of strafing Allied airfields and destroying all grounded aircraft. At this point I want to state that the Germans were not aware of the depth of Lt. Col. J.C. Meyer's thinking and planning. He was a strong advocate of the theory 'think like the enemy' to defeat the enemy.

"The plan for a New Year's Day mission formed in J.C. Meyer's mind the afternoon of 31 December, 1944. Our Group C.O., Colonel Jim Mayden, had returned to Bodney that day and J.C. Meyer was acting C.O. He tried to get 9th Air Force approval *(from General "Pete" Quesada – author)* to take an early mission on New Year's Day. The early mission was denied, but approval for a later mission was granted. Meyer advised the pilots involved and the ground crews to not celebrate but to 'hit the sack' early.

"The 366th Fighter Group of the 9th Air Force had been stationed at Y-29 for a while flying P-47's. Among the luxuries, if any could be called that at Y-29, was a small tent from which drinks (some intoxicating) were dispensed. We had permission from the 366th FG C.O. to stop at the tent on occasion for a drink (drinks to be paid for of course). After our evening meal on New Year's Eve, Meyer said, 'Let's go over to the 366th area for one drink and then to bed.' I drove one of the Jeeps, and a group of us (including Meyer) went for one drink. He appeared to be very calm that evening and one could imagine his thoughts were about the next morning.

"New Year's morning, the four us in our tent (Cy Reap, Russ Cookingham, Doug Laird, and myself) arose at 530 hours *(5:30 am – author)* for a quick breakfast and then to the flight line. We would therefore be ready when the enlisted crews preflighted and got the P-51's ready for takeoff. In the meantime, J.C. Meyer continued to be persistent in telephoning the 9th Air Force for permission to fly an early mission. He planned to lead a flight of twelve pilots, the eleven others being from the 487th Squadron. Meyer had commanded the 487th Squadron before being promoted to 352nd Fighter Group Deputy C.O. Permission was finally granted for an 0930 hours takeoff. Eight P-47's of the 390th FS, 366th FG across the field were scheduled for takeoff at 0915 hours.

Lt. Col. William Halton (Berlow)

"At about 0900 hours, anticipating a routine takeoff in one half hour, I went into the Operations and Intelligence tent to write my daily report to Ray Wesley. I had about one half page written when I heard our P-51's start engines. I left the tent and positioned myself near the edge of the plank steel runway. J.C. Meyer was on my left ready to taxi for his takeoff spot.

"Meyer was gaining speed as he passed me, and I turned 90 degrees right to watch his takeoff. He started climbing and his wheels were not yet fully retracted. To my astonishment and disbelief, I saw an enemy fighter plane 'on the deck' flying directly across the path of his takeoff..."

Al Giesting (C-Flight Chief, 487th FS)

"I started New Year's off early by being awakened at about 5:30 am after New Year's Eve and my birthday. We had gone to bed early, but were sent to the foxholes in front of our tents by "Bed Check Charlie" *(usually a Ju-88 night-bombing for harassment – author)* who made a couple of visits every night. We went ahead and got up and went through the pine trees to the mess hall about 100 yards north of the tents. After eating and straightening up our tent, we walked the ¼ mile down to the flight line. It was still dark when we started to pre-flighted the Mustangs. They were very cold, and after uncovering them and plugging in the battery charger, we checked the coolant and started the engines with maximum effort. All of my C-Flight started without too much trouble.

"As I had only three crew chiefs for seven planes, I also doubled as a crew chief (I was C-Flight Chief at the time). It was then that I checked the mags (magnetos) on one of the Mustangs with the belt around the stick. This belt held the stick back and kept the tail down while the engine was run up. The Mustang's guns fired, and started New Year's off with a bang. After quickly shutting down the engine I looked out and saw guys lying in the snow. One by one they got up. The last pilot had forgotten to turn off his gun switch the day before, and my hand on the stick triggered the guns when I was running up the engine. I expected to make a trip in to see Colonel Meyer, but he never said a word about it. I guess he figured we were doing the best we could—and we sure were. The armorers were a little mad as they had to get out and clean the guns and add about 30 rounds per gun.

Alex Sears (right) with his crew at Y29

J.C. Meyer points out newly applied victory markings painted by Conkey.

"We'd been alerted that there would be a patrol mission before the 352nd FG's escort mission for that day. When the pilots came out for this patrol they didn't seem to be in a hurry as they were waiting for the fog and frosty haze to burn off. About nine o'clock they started. The P-47's across the field took off first and went toward the northeast in the direction of the slag piles. About this time our '51's were just getting to the western end of the runway. I was standing on the wing of one of C-Flight's planes to watch the take-off. I heard a big explosion and looked towards where the P-47's were. They were dropping their bombs. Then I noticed a lot of planes coming our way. About this time the 487th pilots must have seen the same thing and started to take off – all bunched up. If the front one had cracked up the whole bunch would have had it as they were almost prop-to-tail. They all got off okay and the battle over Y-29 started.

"At this time I saw '109's and '190's over the eastern end of the field, and I was still on the wing of the plane and looking for a place to go. I went for the roadway. There was a foxhole at the gun emplacement, but it was already full. About this time a '109 went by at about 50 feet away and 30 feet off the ground. He'd come across the western end of the runway and was firing at an old Mosquito that was lying on the field there. When I saw his bullets kicking up snow I headed for the ditch. The '109 didn't have anyone chasing him and I guess he got away for the moment. I then saw several German planes with '51's on their tails,

Our Worst Enemy
by Bob Powell

One morning, "Stormy," our appropriately nicknamed weather officer, briefed us for a mission telling us that the heavy overcast we were under would be marginal for takeoff but we would break out on top at Angels 8 (8,000 feet) and find it clear by the time we got over the Continent. He was never so wrong!

As usual, we took off in tight formation by squadrons in our flights of four with each flight following another about every 15 or 20 seconds, thus getting 48 airplanes plus a couple of "spares" in the air in less than four minutes. All three squadrons took up the same course, but each was separated from the other by some distance as we climbed up into the soup at about 170 mph in tight—very tight—formation. Actually, in the overcast you were flying only about 15 or 20 feet apart and the airplane you were flying formation on was only a shadowy, ghost-like image.

In these instances, only the squadron leader was flying instruments. All the others are flying tight formation. This can get very hairy over a long period of time, as we would find out on this mission. We were still in that stuff at 8,000...at 10,000...at 15,000...and 18,000 feet. By this time we had crossed the North Sea and were over the Continent, we were three squadrons, separated, on the same heading, and still in the soup. How did we know we were over the Continent? Easy. We were starting to hear the high-altitude radar flak explode and as bunched up squadrons we made easier targets for the German gun crews.

The tension was mounting by the minute. Since we were maintaining radio silence, there was no radio chatter, at least until we started hearing the flak. We continued climbing, now at 22,000 feet. I remember feeling cold sweat running down from my armpits. Think about this: If only one pilot had failed to switch his fuel tanks (we did this on our wing tanks about every thirty minutes for stability) and his engine had cut out, even for that one second or so it took to switch tanks, his sudden hiccup in speed could drag him back into the planes behind and below him. Not a good thought! The "pucker factor" was coming into play as we continued to climb.

Suddenly, radio silence was broken. One of the guys got on the R/T popped in saying, "I'd eat a sack of crap to be out of this stuff." and another pilot came on with this, "You can start now, brother, my pants are full of it." Then, Colonel Meyer, who was leading the Group, toned in said, "Knock it off, boys," and we had radio silence again as we continued toward our assigned altitude of 27,000 feet and our rendezvous with the bombers. A few minutes later Colonel Meyer came on the R/T again, advising that the bombers were aborting the mission and returning to base because of the weather and we would be doing the same. Then he gave us our turn-around procedure.

"Turndown Squadron (328th) will make a 180 now and start a letdown on a course for base. Angus Squadron (486th) will make a starboard 180 in about three minutes and letdown. The 487th will continue for five minutes and do the same. Squadron leaders acknowledge."

I think Earl Abbott was leading my squadron, the 328th and Earl opted to take us down, still in tight formation, to see if we could get below the overcast and to speed up our return to base. We descended in a shallow dive after making our turn, and it seemed that we went down forever, particularly after being in tight formation for so a long as we had been.

We broke out below the cloud layers clocking about 300 mph and I will never the immediate reaction of the pilots in being in the clear and seeing the ground again. Sixteen planes, tightly bunched, suddenly, as though a spring had been released, simultaneously popped into our normal combat formation without a signal. It was a beautiful to see and be a part of, a spontaneous exhibition of combat discipline by experienced combat pilots. And, I am sure that other pilots, as I did, expelled simultaneous "Whews!" as well.

We stayed just below the cloud level and returned to Bodney, an uneventful mission fought against our worst enemy—weather. I doubt that there was a dry flight suit hanging in our huts that night.

That mission made a great impression on me. Maintaining close formation under those conditions for that long said a lot about our training, our discipline, our flying skill, and our leaders. Had there been a "weakest link," there would surely have been some tragic mishaps. It greatly increased my confidence in my fellow pilots. Living on "the edge of the envelope" can make men out of boys very quickly.

and one went straight in southeast of the field and another crashed off the eastern end of the runway. I could see others already burning off the field.

"Another German came across the center of the field with a '51 on him. He pulled up to the left and that was all she wrote. I saw another guy bail out but his chute didn't open or it had burned before I caught sight of him. There was so much going on that one guy couldn't possibly keep track of everything."

John C. "J.C." Meyer (Pilot and Deputy Group Commander, 352nd FG)
Tally - 2 Fw-190's destroyed

From his Encounter Report: "Leading twelve ships of the 487th Squadron taking off to the northeast from Y-29, Belgium. Immediately upon getting my wheels up I spotted fifteen plus '190's headed towards the field from the east. I attacked one getting a good two-second burst at 300 yards, 30 degree deflection, getting good hits on the wing roots. The enemy aircraft half rolled, crashing into the ground. I then selected another '190 chasing it into vic northwest of Liege. On my first attack I got good hits at 10 degrees, 250 yards. The enemy aircraft half rolled and recovered just on top of trees. I attacked but periodically had to break off because of intense friendly ground fire. At least on three occasions I got good hits on the '190, and on the last attack the enemy aircraft started smoking profusely and then crashed into the ground. Out of ammunition I returned to the field but could not land as the field was under attack. I proceeded west and was bounced twice by '109's and was able to evade by diving and speed. The squadron was engaged before it ever formed; immediately after takeoff. Claim: two Fw-190's Destroyed."

Author's Note: According to research by historian John Beaman, Meyer's first victory was most probably Gefreiter Bohm, flying Fw-190 Weisse 3 (White 3) of III./JG 11.

Alex F. Sears (pilot, 487th FS) Tally – 1 Me-109 Destroyed

From his Encounter Report: "I was flying White 2 on Colonel Meyer's wing. We had just taken off when we were bounced by forty or fifty Me-109's and Fw-190's. One Me-109 came at me from head-on and we made several passes at each other, both of us firing. On the third pass I got some strikes on his engine and shot part of the tail section away. He started burning, went down in a lazy spiral and crashed."

Raymond H. "Ray" Littge (pilot, 487th FS)
Tally – 2 Fw-190's Destroyed

From his Encounter Report: "Flying White Three, we had not yet formed up after takeoff when we spotted 50 plus '109's and '190's northwest of our field. I picked out a '190 flying at 3000 feet, and made one turn with it to the right. I got strikes on the wing roots and cockpit. He snap-stalled to the right, recovered, then snapped again, and hit the ground and exploded.

"I then picked out another '190 and started chasing it on the deck. I got strikes on the wings and engine, and he started throwing black smoke. After this, he used violent evasive action, and I used up all of my ammo but got no more strikes. He had taken a westerly heading, and since I probably could not have found the flight again, I stayed with him. He took me almost to Paris, when he started climbing, and at approximately 3000 feet he bailed out in Allied territory."

William J. "Bill" Halton (pilot, Yellow Flight Leader 487th FS) Tally – 1 Fw-190 Destroyed

From his Encounter Report: "I was leading Yellow Flight. As Colonel Meyer, Squadron Leader took off, he ran into a large number of enemy aircraft northeast of the field. I took off and climbed to 3000 feet and bounced three '190's on the deck headed northeast. I dove on one and he crashed and blew up as a result of Yellow 3's (Lt. Moats) fire whose victory I confirm. I fired at one at 0 degree deflection and he spun in and exploded in some woods. I fired at the lead ship and used all the rest of my ammo but observed no hits. I broke off my attack and returned to the field. I was attacked three times by '109's but managed to evade them. The enemy pilots were very eager. The '109's gave me plenty of trouble."

Henry W. "Stew" Stewart II (pilot, 487th FS)
Tally - 3 Me-109s Destroyed

"On New Year's Eve J.C. Meyer went through the 487th and picked 11 pilots to fly with him on January 1st. I was happy that I was one of the ones selected for the flight. We were told the night before, 'No partying. Get to bed early. Be ready to go.' I was assigned to fly as William Halton's wingman. We were lined up in two's as opposed to our regular Bodney takeoff pattern (of four's) because the steel strip at Asch was narrow. We would be the third pair to go. J.C. pulled off the strip and I watched. As his wheels were retracting he was shooting at a '190 and it was going down in flames. We had to wait for the two in front of us to take off but after the sight of the '109's and '190's

Ray Mitchell flashes a smile from "Carol" at A-84, Chievres.

we took off into the prop wash without waiting very long. I stayed with Bill as we left the runway but at that point the radio was alive with 'Every man for himself' and 'Get the Germans.'

"I turned to the east away from the field at about 2000 feet. All of a sudden I saw a P-51 chasing a '109 about 1000 feet above me. The '109 did a Split-S *(1/2 of a loop, from top downward – author)* and came down in front of me about 100 feet off the ground. You could actually see the wings of the '109 shudder and shake. I wondered how that plane could do it. I didn't spend much time thinking about it because I closed on him and opened fire. He crashed into the trees. He didn't have much of a chance to gloat over his perfect Split-S.

"I turned to the south and caught up with another '109 that was headed toward the airfield. I know he didn't see me because he made no evasive moves at all and I just caught him with a few bursts of gun fire.

"The next one to me was the toughest. I saw a '109 headed west from our airfield and followed him. I caught up near the coal mines. We started making turns around the slag piles at the mine. We were about 100 feet off the ground going around 175 to 200 mph on the first turn. I started to overrun him on the second turn when we were down to 120 to 130 mph. I found I could turn with him by using the flaps instead of throttle. Each time he tightened his turn I did the same with the stick and flap handle. After the third turn around the slag pile he took off and I stayed with him closing for a shot. I pulled the trigger but nothing happened. I looked down at the switch panel and saw that my knee had pressed the metal guard over the gun switch and shut off the guns. I tried to move the guard over but it wouldn't work so I ripped off the metal guard. I turned on the switch and got the '109. I saw him hit the trees.

"I looked up and saw another '109 about 1000 feet above me. He was headed east (and for home I am sure). I climbed up after him and got him in my sights. I pulled the trigger and found I was out of ammunition. I closed on him and dropped below looking up at the tail assembly of the '109. I wondered if I could use my prop to cut his tail off. I thought that it would not work, and so I broke off and started for the base.

"Later one of the crew chiefs called me over to a plane to look at the metal guard over the gun switch. We all tried to pull it off and found we could not move it.

"We assembled at the Operations tent and spent a lot of time making out our reports. The great event of the day was when the command car arrived with all of the Generals. I remember watching them get out of the car and walking towards Operations. Everyone stood and stared—no one called us to attention or anything. Then the Generals were being introduced to the pilots and shook our hands. It was unreal. We all agreed we would not wash our hands for quite a while."

A nice promo photo of armorer Jack Rider, pilot Sandy Moats and armorer Jim Bleidner. (Berlow)

Raymond "Ray" Mitchell (pilot, 328th FS)

"We 328th Squadron pilots were on our way to our airplanes when the field at Y-29 came under attack, and several of us were caught out in the open. We scurried into a ditch at the edge of the field as the first '109's and '190's came across the field. I remember they were on the deck and bullets flew through the trees in back of us. We all emptied our Colt .45s at the low flying aircraft, which probably was useless but exciting anyway. I also remember a P-51 chasing a '109 around a distant slag pile (Henry Stewart, see last month's Part I - author) and aircraft all over the place, some burning or smoking. As I recall, they shot up an old disabled B-17 that had been abandoned at Y-29."

Sanford Moats (pilot, 487th FS)
Tally - 4 Fw-190's Destroyed

From Encounter Report: "I was flying Yellow Three in Major Halton's Flight. As I took off I spotted about 15 plus Fw-190's at 100 feet coming from 3 o'clock on their way to make a pass at the airstrip north of our field. At the same time I noticed approximately 15 Me-109's flying top cover at 3500 feet just below a thin cloud cover. Two '190's broke into my wingman Lt. Huston and myself, and we entered a Lufbery *(defensive circular-turning maneuver – author)* to the left under intense light friendly (?) flak. I closed on the first '190 and looked back to see a '190 closing on the tail of my wingman. I called him to break as the '190 started shooting at him. I then fired a short burst at 300 yards and 30 degree deflection at the '190 ahead of me, observing strikes in the cockpit area and left wing root. He burst into flames and I saw him crash and explode as I continued the turn. The pilot did not get out. Approximately 50 enemy aircraft were in the vicinity and the entire area was full of friendly flak.

"I chased a '190 that was strafing a marshalling yard as my second target. He broke left and started to climb. I fired a short burst at 200 yards and 20 degrees deflection, observing a concentration of strikes on both wing roots. Both wings folded up over the canopy of the enemy aircraft and he dropped straight in. The pilot did not get out.

"I continued my left turn and rolled out slightly above and behind another '190 which broke left. I fired a short burst and observed hits on the left side of the fuselage, canopy , and left wing root. He burst into flames, the canopy came off, and he crashed. The pilot did not get out.

"I then broke into several '109's and '190's who were coming at me from the rear, heading towards Germany. They split up and I picked one '190 who broke into me. We made several head on passes and I pulled up and came down on his tail, firing a two second burst, observing strikes from wingtip to wingtip. He leveled off and hit the deck. I closed and fired several bursts from dead astern, observing strikes all over the tail and wing sections. As we passed over Maastrich, I fired a short burst that exploded his belly tank and my aircraft was hit by 40mm ground fire. At this time I only had one gun firing and the enemy aircraft kept taking violent evasive action on the deck as we crossed the front line. I fired a burst at him every time I had a chance and observed many strikes in his tail section. I climbed above the enemy aircraft and made an attack from above and to the right. I fired and observed a few strikes around the right wing root. The enemy aircraft broke left, pulled up slightly, and dove into the ground. The pilot did not get out.

"I returned to the airdrome and was fired at by ground batteries. A lone '109 came across the field and I made a pass at him. He broke up into me and I fired my remaining few rounds of ammo at 90 degrees deflection and 100 yards range. He hit the deck and I chased him into Germany before returning to our airdrome to give it top cover. Another '109 came by and I followed him through a couple of barrel rolls but could not shoot as I was out of ammo. He went straight up chopped throttle, and tried to get on my tail. He couldn't, stalled out, recovered, and Split-S'ed at 1500 feet, barely pulling out above the trees. Another P-51 came into the area at this time and shot the enemy aircraft down four miles from the field. I later found out that the pilot of the P-51 was Captain Stewart. All enemy aircraft were very aggressive and the last '109 was absolutely hot. We were all handicapped by fuel in our fuselage tanks and by ground fire, and I feel very proud of our 12 ships destroying 23 bandits on takeoff without a loss. Claim: 4 Fw-190's destroyed."

Lt Jack Diamond's "Twyla Sue"

Dean Huston (pilot, 487th FS)
Tally – One Fw-190

From his Encounter Report: "I was flying Yellow 4 and managed to join White 5, Lt. Moats, just after takeoff even though the fight was already going on, almost at the edge of the field. I was about 300 yards astern and above Moats when an Fw-190 started to crawl up behind him. Lt. Moats was already engaged with another ship and I told him to break left. He did and the Fw-190 followed but not with enough lead. I broke down and under him, coming up in a turn about 100 to 200 yards behind him, firing about a 2 to 3 second burst. Hits were observed all over the wings and fuselage, with flames resulting immediately. He went into a dive from about 2000 feet and crashed into the ground. I saw no chute."

Today he relates, "I remember the famous January 1st melee, when I was in the group of 12 that took off the Belgian strip only to fly right into the Luftwaffe at the east end of the field. I got credit for a '190 as I recall, but the most noteworthy part of it was that I got shot down by our own anti-aircraft ground positions. I was on the tail of a '109 at low level, right at the field, when I sensed my '51 was hit by gunfire (later confirmed to be that of our own anti-aircraft batteries firing at the Germans). I looked around and nothing was on my tail so I continued for a few seconds. I then happened to glance at the oil pressure gauge and it read zero. I immediately wheeled around for an emergency landing, as I was still fairly close to the base. As I touched down, the engine stopped (froze) while I was rolling down the runway. I pulled over onto the grass as I slowed down, so the runway would be clear for others landing.

"Talk about a close call! I was too low to have bailed out. If I hadn't been close to the field I'd have had to belly in on the best possible area. Once again, I was so low I had little choice on that. I guess I have the notoriety of being the only 352nd shot down by our own gunfire at Y-29. All I know is that I was one lucky hombre that day. I must have had the right four-leaf clover in my flight suit.

"Later that day a group of us walked or rode out to the area just east of the strip where a German fighter had 'augured in.' The body of the pilot was still there and I remember feeling that he was barely out of high school. I will never forget it....nor the day."

Walter G. "Jack" Diamond (pilot, 487th FS)
Tally – 1 Fw-190 Destroyed, 1 Fw-190 damaged

"On that morning I was assigned to fly wingman to Captain Whisner and I took off following him. After takeoff when we would normally form up I saw no P-51's. I saw aerial activity around an airfield a few miles away and I thought it must be an enemy attack. It was soon every man for himself. I kept searching the sky for planes and called Whisner three times to let him know I was not with him. He did not reply so I took off on my own looking for German planes.

Bill Whisner being congratulated by Nelson Jesup after 5 in one mission (Berlow)

A cold 328th flight line shot at Y-29 Asch, Belgium.

"I soon spotted a lone Fw-190 at a low altitude and headed for it. Apparently the pilot did not see me and I closed in and started firing short bursts. There were a few strikes on the wings and fuselage and as I maneuvered closer and kept firing there were more hits. The '190 started a gradual descent but took no other evasive action. As I followed he flew into the ground at a low angle and pieces of the plane flew off everywhere. I looked back as I zoomed over but saw no explosion.

"I climbed back up a few thousand feet and continued to look for planes. When I spotted another Fw-190 I closed in on his tail firing short bursts as I approached. Again, this pilot took no evasive action except for slight turns. I moved in so close that my plane was bouncing up and down violently from his prop wash. I kept firing every time my plane lined up with his. There were numerous strikes on his wings, but none to a vital spot that would bring him down. I remember thinking I could chop his tail off with my prop, but was not ready to trade even – plane for plane, and especially not pilot for pilot. He was obviously a very new pilot and took no evasive action with me on his tail and unable to get a decisive hit. I was soon out of ammunition in four guns and decided to break it off while I still had ammunition left in two guns in case I was attacked. I was also very concerned that there was no one watching out for me (no wingman) and I was very vulnerable to attack while engrossed with this close-range '190.

"It turned out that I saw no more German planes. I flew around a little while looking and waiting for things to quiet down at our field. One thing I did not want was to be attacked while going in to land."

William "Whiz" Whisner (pilot, Red Flight Leader 487th FS) Tally – 2 Me-109's & 2 Fw-190's Destroyed

From his Encounter Report: "I was leading Red Flight. As we were taxiing out to the strip I saw some air activity east of the field. The squadron, consisting of three four-ship flights, was taking off singly. As I started down the strip Colonel Meyer called the Controller and inquired about bandits in the vic. As I pulled my wheels up the Controller said there were bandits east of the field. We didn't take time to form up but set a course, wide open, straight for the bandits. There were a few P-47's mixing it up with the bandits as I arrived. I ran into about thirty '190's at 1500 feet. There were many '109's above them. I picked out a '190 and pressed the trigger. Nothing happened. I reached down and turned on my gun switch and gave him a couple of good bursts. As I watched him hit the ground and explode, I felt myself being hit. I broke sharply to the right and up. A '190 was about 50 yards behind me firing away. As I was turning with him, another '51 attacked him and he broke off his attack on me. I then saw that I had several 20 mm holes in each wing, and another hit in my oil tank. My left aileron was also out, I was losing oil, but my pressure and temperature were steady. Being over friendly territory, I could see no reason for landing immediately, so I turned toward a big dogfight and shortly had another '190 in my sights. After hitting him several times, he attempted to bail out, but I gave him a burst as he raised up and he went in with his plane which exploded and burned. There were several '109's in the vic so I engaged one of them. We fought for five or ten minutes and I finally managed to get behind him. I hit him good and the pilot bailed out at 200 feet. I clobbered him as he bailed out and he tumbled to the ground. At this time I saw 15 or 20 fires from crashed planes. Bandits were reported strafing the field, so I headed for the strip. I saw a '109 strafe the northeast end of the strip. I started after him and he turned into me. We made two head-on passes, and on the second I hit him in the nose and wings. He crashed and burned east of the strip. I chased several more bandits but they evaded in the clouds. I had oil on my windshield and canopy, so I came back to the strip and landed. All of the enemy aircraft were very aggressive, and extremely good pilots. I am very happy that we were able to shoot down 23 with a loss of none. We were outnumbered five to one, with full fuselage tanks. The P-47's on this field did a fine job and helped us considerably. The cooperation among our fighters was extremely good, and we did the job as a team."

Nelson R. Jesup (pilot, 487th FS)
Tally – 1 Fw-190 Destroyed

"I was flying Red Three in Captain Whisner's Flight. As soon as I had taken off I saw P-47's, P-51's, and Me-109's and Fw-190's in a dogfight off to the left and I headed into the fight. After pulling up sharply to avoid a mid-air collision with an Me-109, I made a 180 degree turn and chased another Me-109 up into the clouds and lost him. I dove below the cloud layer and started after a couple of enemy aircraft but they were shot down before I could close up. I saw an Fw-190 heading up through the clouds and went after him at full throttle. He leveled off on the top of the clouds at 4500 feet and headed due east. I chased him through the cloud tops for several minutes and slowly closed on him as he started a left turn. At about 600 yards I fired 2 bursts from about 10 degrees deflection but did not hit him. At 400 yards I fired another burst and observed a cluster of strikes in the region of the cockpit. He slowly rolled off to the left and went down through a hole in the clouds in a spiral and went into the ground. I did not see the pilot get out. After this I climbed up through the clouds again and was attacked by a Typhoon, so I dove down through the clouds and returned to base."

Jesup adds, "I recall that I needed all of the speed I could get at takeoff, and so shoved everything to the firewall. Once airborne, I yanked the wheels up and stayed right on the deck. As I was flying down the runway, a blue-spinnered Me-109 appeared coming right at me. He was low, but I was lower, and he passed about 20 feet above me and to my left. By the time I got to what I thought was the end of the field (visibilty being poor due to fog and blown snow) I pulled up sharply. I later overheard a couple of pilots from another squadron saying, 'Will you ever forget that next to last plane taking off right on the deck like a bat-out-of-hell and then pulling straight up?"

Earl Akard (486th FS, Ground Crew)

"For the next 15 minutes, more or less, I could look in almost any direction and enemy airplanes were being engaged in battle. I could see planes going down and huge billows of smoke rising from the ground, and in the air, too. All of this happened while I was standing on the wing of an airplane. Straight ahead and to my right planes were taking off and immediately they were being engaged in battle. To my knowledge, it was always a German plane going down. We did not lose a single plane.

"A large sergeant friend of mine from ordnance, for lack of a foxhole, fell down along side one of our bombs on the ground (not sure of the size) during the heat of the battle. Afterwards, I kidded him about doing so. He merely said, 'That was kinda dumb of me.'"

Richard "Dick" DeBruin (Supply and Transportation Officer, 352nd FG Headquarters)

"I recall some instances taking place on the afternoon of January 1, 1945. I drove a Jeep carrying Army Graves Registration personnel looking for bodies of German pilots. I have a clear recollection of locating a crashed German aircraft a few miles east of Y-29. The pilot's body was thrown clear of the wreckage and pieces of his body (such as both legs) were severed. It was a gruesome sight and I was glad it was not my job to locate bodies for identification and burial. During the afternoon a group of Air Force Generals arrived at the field to congratulate J.C. Meyer and the pilots. Two of the Generals were Carl Spaatz and Hoyt Vandenberg. I was quite impressed and stood at the outer fringe of the group admiring our pilots, who provided me with the most exciting day of my military career."

Al Rigby and Dean Huston indicate victories at Y-29. (Rigby)

America's Newest Ace, Bluenoser Alden "Al" Rigby (pilot, 487th FS) Tally – *2 Fw-190's Destroyed, *½ (Shared) Me-109 Destroyed, *½ (Shared) Me-109 Destroyed

***Note: Later adjusted to 2 Fw-190's and 2 Me-109's Destroyed**

Potentially the most interesting revelations regarding the New Year's Day battle are associated with pilot Al Rigby. Regarding his New Year's Day exploits he details, "If I could adequately describe the morning hours of January 1, 1945, I could write for a good living. I will try to describe the events, and my feelings as my records and memory permits. Actually the morning was filled with such history making excitement that it would be difficult to forget.

"The day started for me about 7AM. The weather was the same dark, damp, cold foggy feeling. The fog had lifted a little, and was being replaced with haze, and a cloud cover at about 1500 ft. I checked my plane before breakfast, and found the crew getting the ice from the wings, and the frost from the canopy and windshield. I went to the briefing tent and learned from Col. Meyer that he had requested a short patrol mission before the Berlin run. At about 9AM the fog and haze had thinned to a point of being able to see the trees at the end of the runway to the east. General Queseda had just given the 'ok' for a short mission, using only part of our planes. Start engines at 9:00, takeoff at 9:20, and be back on the ground at 10:15. This would give us time to refuel, and meet the bombers overhead at noon. The briefing was the bare essentials, since we did not expect more than a look at the 'Bulge.' I would be in Colonel Meyer's flight as White 4. This was New Year's Day, and we had not seen the 'Hun' aircraft for 2 days. The German pilots could be celebrating a little also. WRONG! Little did we know of their plans for exactly 9:20AM at Asch, and 16 other Allied bases.

"I kicked the tires, and climbed aboard at 9:00. The plane had been warmed up, and the tanks topped off. The cockpit was warm, and I was ready for a comfortable ride, as I rolled into position behind the Col. The P47s had taken off a few minutes earlier, and headed straight for the front lines below the clouds. We had just gotten the green light from the makeshift tower, when we noticed anti-aircraft fire just East of the field. Surprise, and even shock, would be an understatement. We next saw what looked like at least 50 German fighter aircraft about to make their first pass on our field. We could not have been in a worse position, unless loaded with external fuel or bombs. We were sitting ducks, and our chances were slim and none. It was not a difficult decision to take off, since that was the slim chance. The next 30 minutes were filled with action and anxiety, that perhaps had not been seen, or felt before or since. I had turned on my gun heater switch earlier, and now had the presence of mind (and prompting) to turn the main switch on.

"The takeoff roll was very close, rapid, and somewhat organized. We did not wait for help from the tower, or our own departure Control Officer. We just went. I am certain there were a few short prayers to just get off the ground. I had my own sort of set prayer, consisting of 6 words that had been used many times. Being caught on the ground was simply a fighter pilot's nightmare. We had made the situation even worse by having our fuselage tanks filled. This would make a big difference in our maneuverability, until about 50 gallons could be burned off. This would be my first takeoff ever with the gun sight illuminated on the windshield. Things were happening too fast to even be afraid—that could come later. There was no training to cover such a situation, instinct simply had to take over, and it would have to be an individual effort.

"Getting off the ground was extremely difficult. I was fighting Meyer's prop wash, so I had to keep the plane on the steel mat a little longer to establish better control. It was of some comfort to just get airborne. Our ground gunners were firing a lot of rounds at the enemy, and in all of the confusion, were firing at us as well. This would have been their first test in anything near such conditions, so they were not hitting anyone, but it was a little disturbing.

"My landing gear had just snapped into the up position, when I opened fire on an Fw190 which was on Littge's tail. I told him on the radio to 'break left,' this put the '190 right in my sight. I could see strikes from the tail up through the nose. The plane rolled over from about 300 ft., and went straight in. I then picked out another Fw190 headed east. It appeared that he was headed for 'The Fatherland.' I dropped down on his tail and opened fire at a greater distance than was necessary, since I had the speed advantage. During the chase my gun sight failed. The bulb had burned out, and I did not have the time to change it, had I even known where the spare was. I expended even more ammunition before enough hits brought the smoke and crash in the trees. I was now in a very difficult position, no gun sight, low on ammunition, and high on fuel. I had my tracers loaded to show only when I had fired down to 300 rounds. I was now into that short supply, with still a lot of fighting to be done. I knew that mine would have to be at very close range without the sight.

"There did not seem to be any overexcitement, or even caution. It was not just another day at the office, but more of a day that all of the training had led up to. The odds were getting better with each minute, and I did have reason to be even a little optimistic. Considering getting off the ground in the first place, and being over friendly territory was much more than could be hoped for a few minutes earlier. The friendly territory added another dimension, since bailing out (if necessary) meant friends on the ground for a change.

"I did not have any trouble finding the field after the lengthy chase on the second '190. The flak was still there, though not nearly as heavy, and I could see at least 2 dog fights. I could see a few fires on the ground, and wondered if any could be 'ours?' I could see a P47 in a turn with an Me109 at about 1000 ft. I knew that the 'Jug' could not turn with the German at the low altitude, which left me with a bit of a problem. I really needed what ammo I had left for self-preservation, but when the '109 had the advantage, I did not have a choice. As the P47 mushed to the outside, I came up from beneath, and from very close range fired enough rounds to see hits on the left wing, through the cockpit, and right wing. The '109 went in from about 500 ft. Before joining the fight, I reasoned that only I would know of my ammo shortage, and gun sight problem. I thought perhaps sheer numbers would count for something. I have reason to believe that this P-47 pilot whose 'bacon I saved' was actually a pilot named Jack Kennedy of the 366th FG.

"The fuselage tank would now permit reasonable maneuverability near the ground, and I would very soon need that. I knew that I was now down to what could be my last burst, even if all 6 guns were working.

"My last fight was with the best German pilot I had seen at any time. He could well have been their Group Commander. I would be the 2nd or 3rd P51 pilot to try for a reasonable shot. He put the '109 through maneuvers that had us mostly watching, i.e., a Split-S from about 1000 ft. I recall seeing the aircraft shudder, then pull wing tip streamers as his prop wash shook the tree tops. He was then back in the fight and very aggressive. I was glad to have another P51 in the vicinity, since my firepower could only be a bluff as far as I knew. I recall being very impressed by the way the '109 was being flown, and hoped that I could in some way get in a reasonable firing position. I knew that I would only have one chance, (if any) because of his ability, and my limited ammo. After about 5 minutes, I did not see any more firing from the German. It could have been that his situation was as bad as mine. His maneuvers now seemed to be on the defensive side. It was what seemed like 10 minutes (but was probably less) before the other P51 turned the '109 in my direction, where he turned broad side to me from something less than 3040 yards. It was close enough for me to see the pilot clearly, and what proved to be the last of my ammunition score a few hits on the left wing, the engine, and then shatter the canopy and cockpit. I had again guessed right for the very close proximity, high deflection angle firing without the gun sight. Some might think in terms of being "lucky." That could well have been, but I am convinced of other factors being involved (help from above for one).

"The fight was over, but I was not at all anxious to land, though I knew the fighting had to be over. I would take my chances without ammo in the air rather than be in any hurry to get back on the ground at Y29, or any field to the west. I could see several fires burning near the field, and what looked like 2 or 3 on the field, but the runway looked good. I could see the rows of P51s and P47s, and could not believe the field could have gotten by with so little visible damage.

"A check with the tower was not all that reassuring about the condition of the field. After about 15 minutes of looking things over, I decided it would be safe to get back on the ground. I had clearance to land, and would follow the P51 on what was to be his break on the 360 degree overhead pattern. Instead, he came in on the deck and pulled up in the frequently done victory roll over the runway, with anti-aircraft fire following him. The ground gunners were still on edge. I had thought of giving the ground troops a little thrill also, but suddenly changed my mind. They had probably had enough for one day anyway. The frost had melted on the steel mats, and the landing was a bit slippery. I was just happy to be back where it all started in one piece.

"Landing to the west left only a short taxi to my parking place, and the fox hole used some during this mission by the crew. As I cut the engine, there was some emotion that I had not given any thought to. Sgt. Gillette knew something of what had happened, but of course did not know the numbers, my gun sight problem, or my ammo predicament. He was almost in tears as I made my account to him. I assured him that it was most probable that I had done better without the sight, because of the low altitude, and very close range. We had always had a close relationship, but the events of this day, and our visible emotions about what had happened, left us with even more common bond.

"I do not recall any celebrations. There was a lot of excitement, but nothing that was not rather subdued, or even 'matter of fact.'

"I describe this event as being historic, since it was the only time in World War II that American pilots had been in such a situation, at least in the European Theater. It was the only time that American Pilots had fought over friendly territory in Europe, and certainly the only time anywhere under such conditions. There could not be any manuals written, or even instructions given to cover the emergency we found ourselves in. At least 2 years of training, and considerable combat experience suggested (demanded) that we get airborne at any price. The timing of our takeoff, however risky, had probably saved lives, and certainly saved the near 100 parked aircraft on the field. Another miracle, 9 of us had shot down 23 of the German fighters, without losing a plane or pilot. This encounter has been referred to as 'The Legend of Y29.' I would also add the word 'Miracle' in that title. I have relived that day many times over the years since. It had to be a once in a lifetime experience for any involved. We were in the right place at almost the wrong time. One minute, or even 30 seconds later, and the day would have been a total disaster. I would probably have been history, instead of writing it. My mind still sees twelve sitting ducks (our 487th flight) on the west end of the steel matting at Y-29, and anti-aircraft fire and the enemy on the east. Perhaps another viewpoint would be close-range, with moderate ground fire and some low cloud cover. The troops on the ground had a better view, but not the feel of it."

In October 2000 the American Fighter Aces Association (AFAA) awarded Rigby two ½ victories, adding to his wartime total of four confirmed destroyed. This additional one total victory bestows upon Al the unofficial but highly visible designation of "ace." It also indicates that the Bluenosers official victory tally for January 1st, 1945 should be revised to read "Twenty-four Destroyed, None Lost."

Rigby's curiosity over who "shared" those two ½ victories led him to research the matter in the last several years. Amazingly, there was no other claim for either of Al's ½ credits, so the AFAA reviewed the case and decided they should both be full credits on Rigby's score sheet. A delighted Al Rigby adds, "During the War the 8th AF counted ground victories the same as air, I was then an Ace, so no big deal on the once-shared kills. I did not realize that the 'Ace' part would not extend beyond the 8th AF. It has been a source of a little irritation for these many years, since I knew in the same few days following the big encounter that I was the only one sharing. Why the powers that be did not resolve the shared part, has been a part of the irritation. The other part has been my passive/satisfied attitude at the time. It was perhaps largely my fault for not following through for a resolution. Since 1993 there have been books and magazine articles that have confirmed the need for some action on my part. Finding the American Fighter Aces Association (AFAA) address on the 'Net, I decided that with nothing to lose, and a victory to gain to 'go for it.' The Executive Director of the Association, Col. J. Ward Boyce, has been most helpful. In October of 2000, I was advised by the President of the Association (Com. Dean Laird) that I was to receive full credit for the two previously-shared victories. I had sent the AFAA copies of articles, pages from our 352nd F.G. History, pages from 2 books printed only in England, pages from the book "The Mighty Eighth, The Silver Star award, etc. "

Insight Into "The Battle"

Some interesting conclusions can be drawn from the descriptions of the pilots who participated in the action. First, it is evident from these narratives that the Bluenosers encountered veteran Luftwaffe pilots as well as "greenhorns" with little combat experience. Obviously Whisner and Moats tangled with veterans (and quite possibly unit leaders), while Lt. Diamond apparently latched onto a couple of inexperienced pilots.

Second, on New Year's Day, it appears likely that the Focke-Wulf Fw-190's were tasked with the ground strafing duties with the Messerschmitt Me-109's flying top cover for them. This is logical, as a similar approach was used while attacking heavy bombers. As the war progressed, the Focke-Wulfs were found better suited to modifications aimed at adding fire power and armor. This enabled them to better attack and survive the daunting USAAF bomber boxes, but left them weighty and sadly vulnerable to roving P-38's, P-47's, P-51's and other Allied fighters. For this reason, they were often escorted by Me-109's as cover, as can be seen at the attack on Asch.

We should not overlook the role of the 9th Air Force men across the field in New Year's Day. Their 390th squadron's pilots lifted off the field, disrupted, and occupied the attacking Germans just long enough for the Bluenosers to get into the air. *(See "The Key Block" for more details.)*

As mentioned earlier, the Luftwaffe committed roughly 800 to 1,000 aircraft to this mission, losing a total of around 300 to all causes New Year's Day. Over 200 Luftwaffe pilots were killed-in-action, with 30 to 60 being unit commanders. As miserable as any of the losses were for the Germans, this last figure was the one they could least afford. Few experienced Luftwaffe pilots had survived the war's meat-grinder against the Russians hordes, or the Western Front's bomber boxes and roving fighters. Bodenplatte removed the majority of these survivors, spelling the virtual death-knell of the German's aerial fighting ability.

The numbers lost on the Allied side are harder to obtain. Due to the apparently lack-of-preparedness by the Allied commanders involved, their loss numbers were immediately downplayed. The initial published figure of around 145 is preposterously low, considering that Eindhoven (the hardest hit) lost nearly 120 combat ready aircraft on the ground. The Germans claimed between 350 and 500 Allied planes destroyed (Galland used the figure of 400), not illogical considering the success of the attack. Recent insights into the numbers say that a realistic total lies somewhere near 200 destroyed total from the 17 airfields.

Rather than diminishing over the last 57 years, interest in the legendary battle over Y-29 (and Operation Bodenplatte in general) seems to be ever increasing. It is hoped that this documentation of the events by the men who were there will add to the understanding of the winter air war.

Necessity: The Mother of Invention
by Ray Mitchell

Living conditions at Y-29 were a far-cry from those the pilots and airmen of the 352nd had experienced at Bodney and particularly during one of Europe's coldest winters, but Americans seem to have great ingenuity and find themselves able to adapt to most anything.

When we arrived at the airfield in the woods at Asch, Belgium, only about 15 miles from the battlefront, we were told to dig a fox hole using those short G.I. shovels the first night. We really didn't think it was necessary until 'Bed-Check Charley,' the nightly German plane came over and dropped anti-personnel bombs on the field. After that we found those shovels very useful even in the frozen ground.

Our first night there we gathered pine bows to lay on the flor of the tents and covered them with blankets to sleep on. Fortunately, a C-47 flew in the next day with a load of parkas and we slept in these for the next few days until things could get better organized.

Living in the tents in that frigid weather we soon found a way to get some heat. We took the oxygen bottles from the wrecked B-17 on the field along with the copper tubing to fashion a gas supply in a nearby pine tree. The gas was piped down into a small pot-bellied stove in the tent and a 'drip-system' was rigged to allow the 100 octane gas to drip into the stove. That stove would turn red hot and no one seemed concerned that it might blow up. We may have had the warmest tent on the field despite the near zero temperatures.

The latrine was about as primitive as it gets. It was a hole in the ground with a piece of canvas on posts to give some semblance of privacy and the seat was a 2 x 6" board, maybe a little wider. Whatever it was, it was COLD!

On the day of the "Legend of Y-29," the pilots of the 328th Squadron were still on the ground preparing for a mission when the 109s and 190s started strafing the field where we were. The bullets were flying over our heads and cracking in the pine trees behind us as they targeted some of the parked aircraft. We jumped into a drainage ditch as they flew over us at less than treetop level. Some of us emptied our .45 pistols at them even knowing it would be useless, but it gave us an outlet for our frustration.

Later, a P-47 from the 366th FG across the field which we had been escorting on a bomber mission landed behind me with a 500 lb. bomb under his wing that he had been unable to shake loose. When he hit the runway it fell off and exploded. As I looked back in my rearview mirror I saw the front half of the plane fall over on the prop and the pilot scurried out of the cockpit and down over the prop and ran to the side of the runway—unhurt. The bomb had blown away all of the fuselage from behind the pilot.

It just wasn't his day.

Earl Abbott's "PE-P" and Mac McCarthy's "PE-B" running up in Belgium.

The 328th's flight line at Y-29, Belgium.

The "Key Block"
THE 390TH FIGHTER SQUADRON OVER ASCH
by Tom Ivie

During the early morning hours of January 1, 1945 a sight which had not been seen in Germany for four years was taking shape. The sky was darkened and the ground reverberated t;rom the sound of some 800 low-flying German fighters and fighter-bombers headed westward on an offensive mission. This was the beginning of OPERATION BODENPLATTE, the Luftwaffe's mighty offensive designed to annihilate the Allied air forces based on the continent. Their targets were sixteen air fields located in the low countries and in northeastern France. Luftwaffe planners believed that if they could destroy these bases they could end Allied fighter superiority, which in turn would allow German fighter forces to more effectively defend German cities against bombing attacks.

In spite of their high hopes the dreams of the Luftwaffe were shattered by the events of the day. Their pilots were able to do considerable damage, but at a cost the Luftwaffe could ill-afford. (They claimed between 350-500 planes destroyed. Allied figures place the number at about 200.) German losses came to about 300 planes and 232 pilots. Most importantly this failed mission cost them over 30 of their most experienced commanders. German losses can be attributed to several factors: accidents, anti-aircraft fire (their own and Allied) and swift and decisive actions by Allied fighter pilots. Without a doubt the most decisive air battle of the day took place over Y-29, the American airfield at Asch, Belgium, home of the 366th Fighter Group, 9th Air Force and, temporarily, the 352nd Fighter Group, 8th Air Force.

The air battle over Asch was also one of the best examples of training and teamwork demonstrated by American fighter pilots in World War II. What made it so remarkable was that the reaction to the German attack was unplanned, but it was carried off in a magnificent style. Part of this story is well known, but the whole story has not been told in detail. The best analogy of what took place is like a crucial play of an exciting football game. Many times a game has been won because a player grabbed a punt or intercepted a pass and ran 90 yards for a touchdown. In most cases the runner's success was due to a "key block" thrown by one of his teammates during the run, but few people, if any, ever remember who threw that block.

The 487th Fighter Squadron, 352nd Fighter Group, flying Mustangs, is well remembered for its incredible victory that day and deservedly so, but most accounts of that day fail to mention the "key block" thrown by the 390th Fighter Squadon, 366th Fighter Group, flying Thunderbolts, which cleared the ways. Finally, here is the story of the eight pilots who threw that block.

In spite of the bitter cold of that New Year morning and the celebrations held the night before, the 390th was able to get the eight Thunderbolts of Red and Yellow flights airborne at 0915 hours As we shall later see, this time of takeoff, which was fifteen minutes early, was a critical factor in the day's events. The mission, led by Captain Lowell B. Smith, was to be an armed reconnaisance over the "Bulge" area. Flying with Captain Smith in Red Flight were Lts. John Kennedy, Melvyn Paisley and F/O Dave Johnson. Yellow Flight was composed of Lts. John Feeney, Robert Brulle, Currie Davis, and Joe Lackey.

Lowell Smith described how the day began: "We got up in the dark and ate our powdered eggs, then I briefed the men on our mission and we headed to our planes which our crew chiefs had warmed up for us and we taxied out for a sunrise takeoff. The days were short at that time of year. We circled the field once and joined up before setting our course to the front (only about 15 miles from Asch). As we joined up, my wingman, Jack Kennedy, reported flak bursts just NE of our field so I led the squadron to investigate. As we approached the flak we sighted enemy aircraft strafing Y-32, a British base located about seven miles northeast of Y-29, and another force was seen headed toward Y-29."

The anti-aircraft fire had announced the arrival of JG 11 commanded by Lt Colonel Guenther Specht. Specht, a German ace with 32 victories, had taken his mixed force of 50 or 60 Me-109s and Fw-190s to the target area, arriving at 0920 hours, right on schedule. Unfortunately for Specht, the 390th Fighter Squadron had taken off 15 minutes early and by doing so destroyed any hopes the Germans had for success.

Lowell Smith again describes the action: "After seeing the enemy aircraft I called, "Tallyho" and the fight was on. The German aircraft had penetrated on the deck, which accounted for the heavy flak over the front and my squadron was climbing and was well above the Germans. When the order was given to jettison bombs and arm guns the squadron moved into battle formation, opening up the formation for freedom of maneuvering. We were in a perfect position for the attack even though we estimated the German strength to be over fifty aircraft. Our attack went smoothly as we peeled off and came in on the tail of our targets, who apparently were concentrating on hitting our airfield and destroying the fighters on the ground. We hit the lead wave and I am sure we accounted for some of their leaders since shortly after we shot down the first of our victims the thrust at our airfield was dulled when some of the Germans turned to fight. In an aerial battle when you are taken by surprise, as these Germans were, you can never tell how many aircraft you are fighting against and there was obviously 'confusion in their ranks.'"

It's difficult to tell who scored the first victories over the Germans, but quite possibly it was Lt. Melvyn Paisley. His encounter report stated, "We were taking off on another dive-bombing mission when one of our pilots saw flak bursts in the vicinity of our base. I then spotted the enemy aircraft and was ordered to take over and lead the flight to them. On the way in I jumped an Me-109. Instead of using my guns, I chose to initiate my attack with the rockets I was carrying. I missed him with the first two but got him with my third." He then followed up this unusual victory with two quick victories over Fw-190s.

Even as Paisley's victims were crashing to earth his squadron mates were adding to the score. Two of those victories were scored by Captain Smith and Lt Bob Brulle. The first was scored by Captain Smith who slid in behind a 190 and laid a well-placed burst into the 190's cockpit. The enemy aircraft caught fire and headed downward as Smith turned and pursued a second 190. Smith's victory was confirmed seconds later by Lt. Kennedy and Flight Officer Johnson, who witnessed the crash.

Lt. Brulle's kill came almost simultaneously with Smith's. He reported it as follows: "I jettisoned my bombs and did a steep wingover getting into position on the tail of an Fw-190. I started firing and closed to 100 yards, observing many strikes. He exploded when he hit the ground. I then got on the tail of another

Fw-190 and used up the rest of my ammo while firing at him. When I broke off I saw him trailing smoke but never saw him hit the ground."

Several other pilots made victory claims during this first pass, but those records were not available to the author. It is believed though that the victories claimed by Lts Davis and Feeney and FLO Lackey were scored at this time. Most importantly this sudden and deadly attack by the eight pilots of the 390th momentarily diverted the enemy's attention, buying time for Colonel Meyer's 487th "Hunhunters" time to get rolling at Y-29.

The 390th's "key block" had allowed the twelve Mustangs to fight their way into the air and join the battle. Now the odds were a little more even and the 352nd FG's most dramatic air battle of the war was launched.

Lowell Smith recalled this stage of the battle: "As is always the case in a dogfight, the battles tend to boil down to individual combats where it is the fighting skills of the pilots that makes the difference. I tangled with another Fw-190 and saw him crash into the ground and then pursued another German as he retreated toward his homeland. Meanwhile my wingman and I had become separated and I feared the worst."

Captain Smith's wingman did get involved and did make it back to base, but only after a few of his tailfeathers had been shot away. Jack Kennedy gave this account of his experience: "I don't know how many German planes there were, but I guess there were 50 or 60. We were in the midst of them in seconds. I don't know if I lost Captain Smith or he lost me. With all the aircraft just off the deck, who could tell? I got on the tail of a 109 which I thought at first was a Spitfire. This 109 filled my gunsight and I got off a few rounds when I was suddenly hit in the tailend. I found two 109s on my tail. I broke to the right and of course those guys followed me still firing away. I got hit again in the right wing and a hell of a fire broke out. I thought about jumping but changed my mind because I didn't want to risk jumping through the fire. I broke to the left and flew past a slagpile with the 109s still firing away. Finally two P-51s from the 352nd drove them off and I headed for a low cloud; by that time the fire was out and I headed in for a landing. Since my hydraulic fluid had burned away I had no flaps, so I came in hot. I hit the strip at over 100 MPH and then I found I didn't have any brakes either. I ran out of runway when the plane was still doing about 30 mph, but it finally came to a halt and I climbed out. When I got back to my tent I truly realized what a long night it had been. I had flown the mission with my flight suit pulled on over my pajamas."

During the period that Lt. Kennedy was having his troubles several of his squadron mates were able to claim more victories. Lt. Paisley's fourth kill of the day was described in his encounter report: "An enemy plane got on the tail of my wingman, (F/O Johnson). I shot at it, noticed several strikes and shook him off. I then pulled an Fw-190 off the tail of a Mustang, knocking him down in one straight pass from above."

While Paisley was downing his third Focke-Wulf of the day, his wingman was attacking two others. In the ensuing dogfight he shot down both of them but in doing so his own plane was heavily damaged and Johnson had to bail out. He came down near the site where one of the enemy planes had crashed. After borrowing a bike from a farmer, Johnson rode to the scene of the crash and removed the German officer's papers and pistol, then pedaled back to Y-29 to announce his survival and his successes. Afterwards he showed the documents to his friend, Sandy Ross, and Ross recalls that the German held the rank of Lt. Colonel. After cross-referencing several accounts of the battle of Y-29 it appears that F/O Johnson may have downed the Wing Commander of the attacking German force, Lt. Co. Guenther Specht.

On the ground at Y-29 the personnel of the 366th and 352nd Fighter Groups climbed out of their foxholes and slit trenches to watch the drama unfolding above them. It was a show they had never before witnessed and they risked injury or death to watch their pilots in action. However, from time to time a low-flying enemy plane would cross the field headed in their direction and these "cheerleaders" would scurry for their slit trenches. Luckily for the spectators, the enemy fire was directed toward the parked aircraft and only one casualty was suffered.

It was a rough day for the AA gunners at Asch. With the sky filled with both enemy and friendly planes it was difficult for them to fire for fear of hitting their own aircraft. The situation was also unpleasant for the pilots who needed to land. Bob Brulle was one of these pilots, here describing this ordeal: "When I realized I was out of ammo and broke off I hightailed it west to get out of the area. It happened so fast that I flew right over our field and had all our AA gunners shooting at me. I rocked my wings violently and climbed to 3,000 feet, just below the cloud cover. I felt safe there knowing I could duck into the clouds if any enemy plane came after me. From that vantage point I circled around and had a front seat for the rest of the battle below. While circling I noticed a 109 heading back toward Germany pursued by two P-51 s about 1000 yards back. As the 109 started to go below my wing I started to roll over so I could keep an eye on him. He must have seen me and thinking I was going to dive on him, veered away. This allowed one of the P51s to turn inside him and get close enough to shoot him down.

"When there was a lull in the battle I went in and landed. Just as I was turning off our runway two Me-109s came over. I remember seeing them come right towards me. I shut off my engine and got out of my airplane and was running away from it as they flew over. They strafed an aircraft on the other side of the field. The two 109s were being pursued by a Mustang. He had them in his sights but was not firing since his fire would have raked across our field. The AA fire missed the two 109s but hit the P-51 and the pilot dropped his gear and landed. I don't know who he was, but I'll bet he was mad!!"

Within forty-five minutes the air battle was over. The German attack had been completely broken up and at least 42 of their number lay in smoldering wrecks on or near the airfield at Asch. The eight pilots of the 390th Fighter Squadron who played such an important role in the battle claimed 12 destroyed, 2 probably destroyed, 6 damaged. They lost one plane in aerial combat (F/O Johnson's) and two others were damaged.

Lt. Melvyn Paisley had emerged from the fighting as the squadron's top gun with claims of 4-0-3. Captain Smith and his wingman Dave Johnson were second with two each, The remaining scoring was as follows: Lt. Robert V. Brulle 1-1-0; Lt Currie Davis 1-0-0; Flight Officer Joe Lackey 1-0-1; Lt John Feeney 1-0-1; and Lt. Jack Kennedy 0-1-0.

In addition to the victories scored by the 390th Fighter Squadron, the 487th Fighter Squadron claimed 23 destroyed and one damaged, and the anti-aircraft gunners downed seven.

Operation Bodenplatte had been a complete disaster for the Luftwaffe and it truly sounded the death knell for German air power. The air battle at Asch, above all others, drove that point home. So spectacular was that victory that the 487th Fighter Squadron named their account of the battle "THE LEGEND OF Y-29." That title is quite fitting and this article is dedicated to ALL of the men who contributed to that Legend.

Under Attack at Y-29
by Russell Liebfarth

On January 1st, the German Air Force initiated an all out attack on Allied fighter bases in Belgium and France with their Operation Bodenplatte, when practically every Me-109 and Fw-190 was used to destroy the Allied fighters harassing their retreat from the Battle of the Bulge. Ju-88s led the way, dropping smoke bombs on the designated airfields to guide the attacking fighters.

"Some 50-plus Luftwaffe fighters hit our airfield at Asch, Belgium, just 15 miles from the front lines in a surprise attack just as 12 Mustangs from the 487th Squadron were preparing to takeoff on a patrol mission led by Col. John C. Meyer. Under attack, these 12 pilots roared into the air.

"Barely off the ground with his wheels still not fully retracted, Colonel Meyer scored the first victory of the day when a 109 crossed in front of him strafing aircraft on one side of the field.

"Our ground support personnel and pilots of the other two squadrons not involved on the mission suddenly became spectators to a dramatic air battle right over their heads. They also found themselves 'in harms way' and rushed to find somewhere to hide from the strafing enemy fighters, some choosing strange hiding places such as behind stacks of external fuel tanks and bombs, behind military vehicles and running to reach the slit trenches which had been bulldozed for their protection.

"A yellow-nosed 109 was heading right for Lieutenant Vitek and me as we ran like Hell for a trench. I jumped in first and he piled in right on top of me. He told me to get down so he could get all the way in but I couldn't. Somebody was already under me and the 109 went right over us firing at one of the P-51s parked on the ground, but only doing only minor damage. It was back in service a couple days later.

"Anyway, Vitek started yelling that he was hit and I got out of the trench to see if I could help him with his wounds. But there were no wounds. He had been hit in the butt by one of the free-falling expended 20mm shell casings from the strafing German fighter that went over us on its strafing run. (No Purple Heart for Vitek). Another 109 strafed a scrapped B-17 parked on the field. Boy, how the Jerries hated those B-17s.

"There was one Me-109 pilot who put on one heck of a show, out-maneuvering four P-51 pilots. He finally broke away executing a 'split-S' at an extremely low altitude for that maneuver and we kept waiting to hear the explosion of his crash, but there was none. Suddenly he came across our strip and the 51's caught him and shot him down. Later we learned that he was a very young lad, but he sure could fly that 109.

"Later, we watched a P-47 from the Ninth Air Force Group sharing the field with us who was returning from a mission and making his landing approach with gear down when a Me-109 flew up within about 200 yards of his six. Major Jackson and I were yelling for him to break but of course, he couldn't hear us. The 109 never fired a shot, most likely because he was out of ammunition due to his strafing runs on other Allied airfields that morning. I've often wondered if that pilot ever knew how close he came to being blown out of the air.

"Of course the tension was extremely high during and after the attack on our field and around noon some low-flying British Typhoons crossed our strip apparently planning to land. They were being flown by Royal Australian pilots. One of our pilots, up patrolling over our field, dived down on them thinking this was another enemy attack and shot one of them down. Jackson and I had tried to call him to tell him they were 'friendly' but he was gripping his mike button on the throttle so tightly that he never heard us and we could hear his guns firing in our earphones. Sadly, the Aussies lost one of their own to friendly fire.

"That was one New Year's Day that can never be forgotten."

Target Munich
by Ray Mitchell

A public relations photo of pilot Ray Mitchell of the 328th Squadron. (Berlow)

War is an ugly experience and many innocent people lose their lives, but freedom is costly. Today, these many years after my wartime experience as a fighter pilot I think about some of the missions we flew.

Our February 25, 1945 mission to Munich left an indelible impression of excitement and an air of sadness that has lingered in my mind over these years, even as my wife and I visited the beautiful rebuilt Munich just a few years ago. I think of the young people destroyed on that day. It was a five hour and fifteen minute flight with some 750 heavy bombers and 400 P-P51s involved. The weather was clear and crisp as we flew at 28,000 feet and the temperature at that altitude was about 50 degrees below zero and the heater in the Mustang cockpit did not provide enough heat to keep our fingers and toes from getting numb even with our gloves and heavy socks.

As we proceeded over France and into Austria I couldn't help but see the contrast between the beautiful snow capped Alps on our right forming a serene landscape leading us on to the destruction of a great city because a terrible dictator wanted to rule the world and take away our freedoms.

We were assigned at the briefing to a particular box of bombers with an insignia that identified the Bomb Group we were escorting. Each Bomb Group normally had a triangle or square design on the vertical tail section with an alphabetical letter identifying it. We left our base in Chieves Belgium to rendezvous with the bombers just prior to getting into enemy territory. After meeting the bomber force we positioned our fighter group off to each side of the bombers far enough not to encroach on their formations and course, but close enough to keep any German fighters at bay. Being faster than the bombers it was necessary to weave to keep our position with them.

A critical time was when the Bombers made their bomb run from the I.P. to the target. It was at this time when they had to fly straight and level for maximum bombing accuracy and they were subjected to intense and accurate flak concentrations. The enemy fighters usually hit them before and after this period to avoid their own flak.

On this day there were no enemy fighters in the immediate area and flak was not as heavy as it normally was when they were bombing oil refineries and other industrial targets where bomber losses were catastrophic. On this day the bombers did a saturation bombing on one end of the city. It was like a virtual giant walking block by block over the whole city. It was an awesome sight! It left the main body of the city burning from one side to the other amd the smoke was visible as far away as France as we could see while flying back to our base.

The John C. Meyer Story

by Tom Ivie

The story of General John C. Meyer's military career is full of action, courage, leadership and flying skill. As a member of the 352nd Fighter Group during WWII, he became the highest-scoring ace in the Eighth Air Force with 24 aerial victories and thirteen enemy aircraft destroyed on the ground. His superb flying abilities and his tally of 37 destroyed underscores his tenacity and deadly marksmanship.

Not only was John Meyers' individual scoring record exceptional, but the true test of a leader is how well the men under his command perform. His men did well. His 487th Squadron became the highest-scoring squadron in the 352nd FG, claiming 514.33 enemy aircraft destroyed, air and ground, 236.5 as aerial victories. His 487th Squadron had 14 aces and was the only fighter squadron in the Eighth Air Force to be awarded a Distinguished Unit Citation.

John Meyer stayed in the service after WWII and when the Korean War broke out, he returned to combat, commanding the famous 4th Fighter Wing. He flew thirty-one missions in F-86 Sabres and downed two MIG fighters in aerial combat, boosting his career total to 26 aerial victories.

After the end of the Korean War, Meyer began a series of command assignments of increasing responsibility and rose to the rank of a four-star general. He served as Vice Chief of Staff of the U.S. Air Force and Commanding General of the Strategic Air Force before retiring in 1975.

The irony of this story is that, in spite of all of his accomplishments, Meyer has been one of the least publicized of America's top fighter aces. On the other hand, his P-51D named "Petie 2" is quite well known. Photos of this fighter have appeared in many aviation books, magazines and action paintings and continue to be popular with aviation modelers worldwide. Ironically, he only scored two of his 37 victories while flying this particular aircraft.

With these ironies and his wartime record, the question is "Why has John Meyer been overlooked for so long?" The answer is just as elusive, but it is likely that wartime correspondents just never got the opportunity to know him as well as they knew many of the other high-scoring aces. The reason, according to former members of the 352nd FG, was that "JC" was a serious-minded, no-nonsense officer who gave his full time and effort to carrying out his command and combat duties and kept his public relations efforts to a minimum.

What kind of man was Colonel John Meyer during WWII? The officers and non-coms who served under his command describe him as a leader who demanded excellence from all of his men, but was always fair in his treatment of them. These policies gained their complete respect and his squadron performed in an exceptional manner.

He was also a strong disciplinarian. On one occasion, when he observed one of his pilots "landing long" at Bodney, he met him when he climbed out of his plane, handed him a yardstick and ordered him to use it to measure the length of the field. All of his pilots got the message.

They also describe him as a leader of great courage and a real "student of aerial warfare." He learned the strategy of aerial combat quickly and with each lesson he perfected the tactics and honed the skills of his 487th pilots. His premise was that to be successful in combat you must be able to "think like the enemy," so as to anticipate his moves and defeat him.

His armorer, James Bleidner, recalls that Meyer had the traits of a superstitious athlete. His superstitions included a particular pair of 'long Johns,' his letter sweater from Dartmouth, and a certain silk scarf. These items of apparel were worn on every combat mission since he felt that flying missions without them would bring him bad luck.

Meyer's military career began in 1939 when he decided to drop out of Dartmouth after his sophomore year and join the Army Air Corps. He entered the program in October 1939 and received his wings commission as a 2nd Lieutenant in July, 1940. For the first ten months of his career he served as a flight instructor at Randolph Field, Texas and Gunter Field, Alabama. His introduction to pursuit aviation came in May, 1941 when he was assigned to the 33rd Pursuit Sqdn of the 8th Pursuit Group at Mitchel Field, NY.

In July, 1941 the Group was transferred to Iceland and he was part of the first American Expeditionary Force of WWII. His tour in Iceland lasted until September, 1942 flying P-40s on defensive and convoy patrols. While there he served as a Flight Leader and was promoted to 1st Lieutenant. .

Shortly after his return to the U.S., Lieutenant Meyer was assigned to the 352nd FG and on arrival at Westover Field was given command of the 34th FS. In January, 1943 he was promoted to Captain. At that time the 352nd FG consisted of the 21st FS, the 34th FS and the 328th FS. In May, 1943 the 21st FS was redesignated as the 486th FS and the 34th FS became the 487th FS.

The Group received the new P-47 Thunderbolts in March, 1943 and trained in them until June when they received orders to go to England, arriving there July 4, 1943 and assigned to the 67 Fw of 8th Fighter Command. Their new home was the former RAF field at Bodney, and the Group spent their first few weeks organizing – and preparing the base for combat operations. John Meyer described those pre-combat days in England in an essay he wrote after the war.

"As we accepted and modified airplanes, stacked bombs and stored pile after pile of ammo, a bond of premonition and excitement encompassed our polyglot group and drew us into a closer camaraderie. This strange community, particularly to an aura of danger, dropped its cosmopolitan values and substituted valor, life and discretion."

By Sept 4th, the 352 had completed its preparations and was declared combat ready. The Group flew its first combat mission on 9 September, 1943 but it would not encounter the Luftwaffe until its 27th mission on 26 November '43 and it was John Meyer's 487th Sqdn that fought the opening round.

The 352nd mission that day was to provide withdrawal support for bombers returning from Bremen. The rendezvous was made at 27,000 feet in the vicinity of Strucklingen and the Group positioned around the bomber formation and withdrew along the bomber route. The first two boxes of B-17s were holding good formation, but the last two boxes of B-24s were spread out, inviting an attack by enemy fighters. That attack was not long in coming. Six Bf-109s zoomed in and attacked two straggling Liberators and were, in turn, bounced by John Meyer and Yellow Flight of the 487th FS. In the ensuing air battle, three of the Messerschmitts were knocked out of the sky. The first enemy fighter destroyed by the 352 FG went down under a hail of gunfire from John Meyer's P-47 named "Lambie." Moments later he claimed a second 109 as a "probable." This action is described in this extract from Meyer's encounter report:

"As the e/a started to chandelle upwards for another pass at the B-24, I closed to within 300 yards of him and fired a burst using about 10 degrees deflection, The e/a, which I identified as a Bf-109G with rocket guns slung under the wings, exploded with a large burst of flame and disintegrated in the air. Fragments from it flew past me in my turn and struck and damaged Yellow 4, flown by Major Therriault. At this moment another e/a came into my sights. He was turning away from me but I had no difficulty turning inside of him.

I took a wide two-ring shot at him using about 45 degree deflection which placed him out of my sight below the fuselage of my ship, so I saw nothing that followed. Major Ross, flying Yellow 2, saw this aircraft start to give off large quantities of smoke from the right side of its cowling. At this moment I was attacked from above and to the right by a third e/a so I continued my turn very steeply and found myself right head-on to him. I fired a short burst head-on at point blank range and saw hits on him.

"After my third encounter I turned quickly and saw this same e/a attacked by Yellow 3, Captain Dilling. Dilling was still firing and the e/a was twisting straight down obviously out of control and giving off large quantities of black smoke that obscured the rear portion of his aircraft."

In addition to the victories by Major Meyer and Captain Dilling, Lieutenants Bennett and Berkshire claimed a third 109.

The Group's next encounter with the Luftwaffe took place on 1 December and, although Meyer himself did not score, his squadron accounted for three enemy fighters. Three days later, on 4 December, the Group headed toward enemy territory on a fighter sweep and engaged enemy fighters as soon as their Thunderbolts crossed into Holland. The first encounter involved the 328th FS and ended in a scoreless draw, but moments later it was the 487th's turn and they blasted three of the Bf-109s from the sky. One of these kills was claimed by John Meyer, who exploded a 109 in the air with only one burst. His next victory was not quite as easy.

On 22 December, the 8th Air Force had dispatched its bombers to strike at targets in the Munster and Osnabruck areas, and it was the 352nd's responsibility to escort the lst Task Force home. Just as the 487th was assuming its position alongside the bombers, the enemy struck. Lieutenant McMahan and an unknown other drove off the first 109 and just seconds later, Major Meyer and Lieutenant Riley bounced two more Messerschmitts which were tormenting another Fortress.

Meyer went after the German leader, who turned out to be a talented and experienced pilot, and found himself locked into a real dogfight. After several passes Major Meyer was able to zero in on him and flamed the 109 with two long bursts firing from 100 yards down to 25 yards. As he swerved to miss the flaming 109, Meyer narrowly missed colliding with the German's wingman. Freeing himself from this entanglement Meyer formed up with Lt Riley and headed home. With his third aerial victory, John Meyer was now tied with Captain Dilling and Lieutenant Meroney for first place in 352nd scoring.

The next three months gave the 352nd few opportunities as aerial combat declined drastically and the Group only scored forty-three victories. For John Meyer personally it was a period of no victories; however, when the Group converted to the NA P-51 Mustangs in April, 1944, his score increased rapidly.

John Meyer's scoring surge began on 10 April, 1944, when the 352nd escorted bombers to their targets in Belgium. The Group was engaged in several dogfights during the mission and claimed 7-0-1 aerial victories. The battles began

J.C. Meyer by "Petie 2nd" HO-M, in flight gear.

just as the 352nd made its rendezvous with the B-I 7s over France. Major Meyer was the first to see the enemy fighters, and pushed his new P51B, "Lambie II," over into a dive after them. His first target was the leader of two Fw-190s. He hit the 190 with three bursts and it dropped one landing gear and began smoking heavily. At this point Lieutenant Carl Luksic flew in and finished it off. No sooner had Meyer broken off from his battle with the 190 than a Bf-109 tried to sneak in behind him. Meyer out-turned him and now the German hunter became the hunted. After about a two-mile chase "Lambie II" overtook the 109 and Major Meyer sent it crashing into the woods below.

During the next several missions J.C. Meyer could find no aerial targets, so he went after them on the ground and flamed six enemy planes. He got the first two on an airfield near Berlin, the next three on 13 April near Stuttgart, and the remaining victory was scored on 22 April. With his seven and one-half victories in April, Meyer increased his total to 10 1/2 and leaped into acedom. Along with the title of "ace" he also received his promotion to the rank of Lt. Colonel.

The flurry of activity in April that enabled Meyer to increase his score so rapidly also kept his ground crew jumping. Asst C/C W.H. Kohlhas tells the story of the end of one of those missions.

"Colonel Meyer had just landed his P-51 after a mission in which he scored some victories and told the S/Sergeant and me to get the plane ready for the next one and to be sure to get those additional swastikas painted on it. Sergeant Conkey spoke up and said, 'Colonel Meyer, we have a lot to do to get this plane ready for the next mission and for me those swastikas rank very low in my priority.' Colonel Meyer, a master of dry wit, replied, 'No, Sergeant Conkey, in my book it ranks high and here's why. On one of my early missions two German fighters covered in victory markings flew alongside me and looked me over—scared the hell out of me. Now that I have the opportunity to do the same, if I can't shoot them down, I might scare them down.' Needless to say, we darn well painted the swastikas on his plane without further discussion."

The scoring momentum that Lt. Col. Meyer built up during April continued right into May. The mission of 8 May turned out to be most memorable. The 352nd was to escort the lst Task Force over the target city of Brunswick. As they arrived at the point of rendezvous, an enormous air battle was already in progress.

Almost immediately the "Bluenosers" encountered a German force of 100 to 150 fighters and, in a series of engagements that lasted nearly an hour, the 352nd claimed 27-2-6 victories. The 487th led the way with fifteen of those kills, three of which were claimed by John Meyer. After an unsuccessful dogfight with Fw-190s which went from 24,000 feet down to the deck, Colonel Meyer and two other 352nd pilots intercepted a flight of fifteen Me-109s as they climbed back to the bomber formation. Before the Germans knew what hit them, Meyer had ripped three of their number out of the formation and sent them crashing down to earth. Two of these 109s went down in flames, but the pilot of the third Messerschrnitt jumped when he saw the bluenosed P-51 bearing in on him. Not a shot was fired and history does not record if it was Meyer's swastika theory that brought him this kill or not. For his outstanding performance on this mission John Meyer was awarded the Distinguished Service Cross, our nation's second highest award for bravery.

Four days later L/C Meyer added to his score with an aerial victory and a strafing victory but this time found the going was a bit rougher. The story is detailed in his encounter report.

"Rendezvous with the first box of bombers was northwest of Frankfurt and we saw a combat wing of B-17s about 20 miles south of course. There were several explosions among these bombers and no flak. I led an eight-ship section toward them and as we approached we saw 20 ro 25 parachutes floating below at various altitudes. We then saw 15 plus 109s and Fw-190s headed toward the bombers and started to attack, lost them in the haze against the ground. We spied another group of 15 plus bogies we believed to be enemy. They headed for the deck and we started after them and again lost them in the thick haze. Proceeding toward the bomber box, about ten-plus 109s crossed directly over us at 22,000 feet and started turning in behind us. As we turned into them they headed for the deck and disappeared into the haze, all but one, which I followed. I caught him at 5,000 feet and fired a short burst at 200 yds and 10 degrees deflection. The e/a caught fire and crashed into an airdrome after the pilot had successfully bailed out.

"The airdrome had about ten-plus He-177s and 20 various types of aircraft, mostly twin-engined. I attacked one of the He-177s and it caught fire, burning profusely. As I pulled up I saw a Bf-109 on my tail at 250 yards and firing. He was painted robin-egg blue on the bottom and sides and either black or dark brown on top. In about a turn and a half I was on his tail, then he dropped some flaps and I was unable to get sufficient deflection on him to fire. Unlike other Huns who in similar situations have broken for the deck and set themselves up, this Jerry continued his tight turn and seemed very willing to continue the fight. I tried dropping 10-degrees of flaps and then 20-degrees and although this helped momentarily to decrease the radius of the turn, my airspeed dropped off so much that I think nothing was gained. At this time I was receiving ground fire from the field directly below, this combat taking place about 3000 feet. Just then another 109 joined in the fight climbing up and dropping down behind me, climbing back up and dropping down behind me as I completed the next orbit. I then broke for the deck flying as low as possible and headed south into the sun, using the valleys and hills for evasion. The two Huns followed me for awhile, but always out of range. I pulled 67 inches for about 30 minutes and at some point finally lost them."

These two victories turned out to be the last of his first combat tour. With a score of 15.5 and nearly ten months of combat flying to his credit, he was sent back to the U.S. for a well-deserved leave.

Returning to Bodney on 2 August 44, he resumed his duties as commanding officer of the 487th. Within a matter of days he was ready to return to combat flying and for that occasion his ground crew had his new mount readied for him. This plane was a P51D named Petie 2 and it would replace "Petie," his P51B which was lost with another pilot at the controls. (Lambie II was to be renamed "Petie" when he returned from his leave).

One of his first missions was a fighter sweep to France, and the German flak gunners tried to make it his last. The date was 13 August and the 487th was patrolling near Fauville when the air was suddenly filled with light flak. Most of the Mustangs escaped without damage but Lieutenant Barnes was shot down and seconds later Petie II shuddered from hits. Flak punctured the fuselage and wounded Col Meyer in the thigh, but he was able to return to Bodney without further incident.

Twelve days later L/C Meyer had the chance to return the favor and did so in a strafing attack on Neubrandenburg airfield. In two passes across the field, Meyer flamed a Ju-52 and a He-177 and damaged two other unidentified twin-engined aircraft. In addition to Lt. Meyer's victories, the 487th destroyed another five e/a and one tank car on the airdrome. These victories, incidentally, were the only two that John Meyer scored in the well-known "Petie II."

During the first few days of September, 1944, the 352nd FG participated in a number of interdiction and bombing raids into France and had no real opportunity to engage the Luftwaffe. These raids ended with the mission of 10 September when the Bluenosers escorted bombers to targets deep in Germany. Although the escort itself was uneventful, Meyer was able to find some targets on the ground. Leaving the 328th and 486th with the bombers, Meyer led his 487th down to the deck and scored heavily in a strafing atttack on Wertheim airdrome. Flying George Preddy's Cripes 3rd, Meyer made eight passes across the field and flamed four enemy planes. His squadron mates added another seventeen planes to the list for a squadron total of 21.

The success of the 10 September mission carried right over to the next day when the 487th downed twelve enemy planes in a series of air battles over Germany. Leading the 487th in scoring was John Meyer with four aerial victories. In his most successful combat to date Meyer downed three Bf-l09s and an Fw-190. The remarkable thing about this feat is that he only used 587 rounds to down all four e/a. His fourth kill was scored with only his right wing guns firing. This was exceptional shooting and especially so since he was using a borrowed plane on the mission.

With eight victories in two days, "JC" had raised his total to 25¼ and he was now closing in on George Preddy, who led the Group with 27. The event was treated with great fanfare by the Group PRO, Sheldon Berlow, and resulted in the last known photos of "Petie." The plane was decked out with 26 swastikas representing Meyer's 25 confirmed kills and photographed for posterity.

Soon after the missions of 10 and 11 September the entire Group went on a scoring drought which lasted until 2 November when the Group claimed 38-1-4. However, when the next opportunity for a real "knock-down, drag-out" dogfight with the Luftwaffe presented itself, John Meyer was there. The date was 21 November and it was an escort mission to Germany.

Leading "B" Group of the 352nd in his brand new P-51D, "Petie 3," Meyer observed and engaged a large group of enemy fighters near Merseburg. The gaggle of forty German fighters was heading for a bomber formation as Meyer, with his formation of 12 planes fell in behind them.

Meyer reported, "We gave chase to the bandit force which was heading on an interception

J.C. Meyer in his Jeep bundled up against the English cold. (Jim Bleidner)

course for our bombers on their withdrawal. Since the enemy force was flying just above lO/lOths cloud cover, surprise was essential to prevent their escape. To achieve this and also keep our own force from becoming too strung out in the long chase, we 'essed' and attempted to imitate the enemy top cover actions as much as possible. We were evidently successful in as we closed on the high cover of the enemy after a ten mm chase without discovery."

As his force closed in on the German formation Meyer assigned each of them targets and then gave the word to attack. For himself, Meyer chose the German flight leader and with three bursts sent the 190 down in flames. As the Fw began its downward spiral, Meyer picked out another 190 and fell right in behind it. The German pilot, apparently oblivious to the trailing Mustang, flew straight and level and it took only one burst to send him down in flames. No sooner had this Fw been dispatched than Meyer spotted a third 190 and opened fire. This time the German fighter exploded in mid-air. After scoring his third victory Meyer saw that the enemy force had been completely routed and led his pilots back to Bodney. As a result of his leadership in this battle, his pilots were able to claim a total of 22½ victories and suffered no losses. For this action, Meyer was awarded his second DSC.

John Meyer's last victory while the Group was based at Bodney came on 27 November. He was leading a strafing mission against an airfield near Munster when three 109s attacked. The unfortunate 109 pilots missed their targets and soon found themselves on the receiving end of the gunfire. In quick order Meyer shot one of the 109s down and his wingman, Captain Hamilton, nailed the other two.

Within a few days of this air battle the Euro weather worsened and flying was severely hampered. During this period encounters with the Luftwaffe dropped off considerably and the weather became the American fighter pilots most dangerous enemy. The severe weather also served as a natural camouflage for the forces that Field Marshal von Runstedt formed in the Ardennes Forest in prep for the counterattack which became known as the Battle of the Bulge. Under cover of the horrible weather the Germans struck with a fury and drove the Allied forces back. Requests for air support went virtually unheeded for the first few days of the enemy offensive, and our forces were continually swept aside by columns of German panzer forces. Finally, after Pattons Third Army was pushed into the line on 20 December, the Germans momentum was slowed. It was about this time that Patton uttered his famous prayer for good weather, and on 23 December it was answered. The skies cleared and Americans now filled the skies and struck back at the Wehrmacht.

In an effort to build up the tactical air forces covering the front, the 352nd was dispatched to

Asch, an advanced base in Belgium just 15 miles from the front lines. From here they would operate under the command of the 9th Tactical Air Force. The Bluenosers didn't have to wait long for action. It came on Christmas Day and the 328th, now under command of the 487th high-scoring ace, Major George E. Preddy, claimed 11 enemy fighters, but at the cost of Major Preddy's life, unfortunately from American groundfire.

On 26 December the 352nd picked right up where it had left off the day before and destroyed another 13 1/2 enemy fighters. This time it was the 487th that did the damage. L/C Meyer was leading sixteen Mustangs on a freelance patrol and at 0945 hours the controller directed them to bogies near Ollheim. As they arrived in the area, P-47s were dive-bombing targets on the ground and the tactical pilots had not noticed the two flights of enemy fighters preparing to attack them. J.C. Meyer led his White and Red flights in a bounce on the lead group of c/a and sent his Yellow and Blue flights against the German's high cover. In the attack Meyer claimed 2 1/2 Me-109s and his squadron accounted for another eleven.

On 27 December, the 8th Air Force dispatched its fleet of heavy bombers to strike the marshaling yards and rail lines that were supporting the German offensive. In an effort to defend these făcilities the Luftwaffe mustered a large force of fighters and sent them up after the B-17s. These gaggles of enemy fighters were quickly picked up by our radar and the controllers vectored the 352nd into the battle. The melee began when the 486th encountered a large force of enemy fighters southwest of Bonn and downed seven of them. That encounter had just broken up when Meyer and his 487 arrived on the scene. Just then a gaggle of 190s was seen approaching them from the south and a second formation was observed to the north. Meyer split his force and sent the Mustangs after both German formations. He led White and Red flights in a diving attack against the FWs and in a brief skirmish downed two of them. In addition to Meyers's personal kills, the pilots of the 487th claimed another thirteen for a squadron total of fifteen.

Four days later on 31 December John Meyer led the 328th on an area patrol and had his first encounter with one of Germany's new jets, the Me-262s. This encounter took place near Viviers after the controller had picked up some activity and vectored them there. Just as the squadron dropped through some clouds they spotted an Arado 234 and it was attacked by Capt Donald S. Bryan. Colonel Meyer recorded the engagement as follows:

"I saw Captain Bryan get some good hits on the starboard engine of the e/a. Just then I saw another e/a of the same type pull up underneath Captain Bryan at 6 oclock to him. Calling for him to break, I gave chase to the e/a firing one burst at 1,500 yds before the e/a pulled up into the overcast. I climbed up through the overcast and spottted the target at 11 oclock to me and heading 60 degrees. Since I was between target a/c and the Rhine, I continued the chase. I seemed to be neither gaining or losing ground pulling 67 inches and 3000 rpm. Just west of Bonn the e/a went back into the clouds and I went under also, losing sight of him, I continued this heading and again sighted him in a port turn at 5,000 feet above the overcast at 3,000 feet. Circling the area I saw another e/a of the same type headed toward Cologne. I gave chase firing sevral long bursts at extreme ranges observing no hits. Being low on gas, White 2 and I turned west five miles north of a large landing strip east of Euskirchen when we sighted 12+ 109s with belly tanks or bombs in three flights, one at 5,000 feet and two at higher altitudes as high as 10,000 feet. We flew right through the enemy formation nearly head-on and headed directly for base. We did not engage the e/a who obviously saw us but did not give chase. "I claim one Ar-234 destroyed (air). The enemy a/c was same size and general fuselage shape as an A-26. No canopy bulge on top of fuselage, glass enclosed nose, fuselage generally conforms with that of Ar-234 diagrams. However, the nose seems more bulbous similar to the Ju-88. Plan view showed wings well swept back in similar plan form of Ju-290. Engine nacelles were very close to fuselage. Believe good shot of plan view will appear in combat film."

On 1 January 45, the German High Command launched another offensive. This time it was an aerial offensive designed to destroy TAF's forces based in Holland and Belgium. The Germans believed that if these planes could ge destroyed their Ardennes offensive could be salvaged. This

487th CO J.C. Meyer

offensive was called Operation Bodenpaitte and it began in the early morning hours of 1 January with the dispatch of nearly 800 aircraft. Their attack was quite successful against some of the Allied airfields and nearly 150 planes were destroyed. Heaviest hit was the RAF which lost 120 planes. The Luftwaffe didn't get off easily though. Nearly 100 of their dwindling supply of planes and pilots were lost in the attack. Nearly one fourth of their planes went down at or near Asch, Belgium, where the 352nd teamed up with the 9 AF's 366 FG to smash the attack.

In the battle, now known to its veterans as the "Legend of Y-29," the 352nd destroyed 23 German fighters and fighter-bombers without a loss. It was the 487th, again led by Colonel John C. Meyer, that supplied all the punch, gaining the 487th Squadron a unique honor—the only award of the DUC to a fighter squadron of the 8th Air Force in WWII. This battle also earned Colonel Meyer his third DSC for extraordinary heroism.

Veterans of the 352nd cite this battle above all others as one that exemplifies John Meyer's gift for leadership. Armorer Jim Bleidner stated it this way, "While it is true that he showed extraordinary heroism in that battle and many others, it is also true that his planning ahead and ability to anticipate the enemy's actions played a very important role. In this case, he was convinced that the Germans would believe the forward airfields would be vulnerable on New Years' Day because the pilots and crews would be suffering hangovers from celebrating the night before. Meyer called his pilots together on New Years Eve and said, "No parties until the following night." As it happened, he was correct in his analysis.

The Group had not been scheduled for an early morning mission but because of Meyer's intuitive feelings, he prevailed upon the 9th AF commander, General Quesada, to allow him to put up a patrol. With permission granted, Colonel Meyer led twelve Mustangs of the 487th down the runway and into the battle under fire. As his wheels left the ground Meyer spotted the enemy fighters headed toward his airfield. His encounter report tells the story:

"Immediately upon pulling up my wheels, I spotted fifteen-plus 190s headed toward the field from the east. I attacked one, getting a two-second burst at 300 yards and 30-degrees deflection and got good hits on the fuselage and wing roots. The e/a half-rolled, crashing into the ground. I then selected another 190 and chased it to the vicinity of northwest Liege. On my first attack I got good hits at 10 degrees, 250 yds. The e/a half-rolled and recovered on top of the trees. I attacked but periodically had to break off because of friendly ground fire. At last on three occasions I got good hits on him and on the last attack the e/a started smoking profusely and then crashed into the ground. Out of ammo I returned to the base but could not land because it was under attack. I proceeded west and was bounced twice by 109s but was able to evade by diving and using my speed. The squadron was engaged before it was ever formed on takeoff." These two victories were John Meyer's 23rd and 24th aerial victories and with his thirteen ground victories, he had now achieved a total of 37 kills, the highest in the ETO.

Following JC's example, the eleven other members of "Meyer's Maulers" lifted their Mustangs into the air and tore into the German attackers. Within minutes, these 487th pilots sent twenty-one more enemy planes crashing to the earth—with no losses. It was a magnificent victory for the 487th and "JC" later wrote these words of praise for his men:

"For the first time in my experience in the European air war the American fighters had neither the advantage of superior tactical position, numbers or equipment. For the first time there was no measurement involved in the final determination other than the relative skill and initiative of the pilots. It was no time for leadership or organization. It was man against man."

In that same document he noted the important part the ground crews played in the successful outcome of this and all their air battles. He stated, "This day was particularly important to the crew chiefs of the 352nd FG. In a year and a half of sweat and tears their only knowledge of the war was the stories their pilots told, the missing faces, torn aircraft and sometimes bloody cockpits. Until now these men had lived on the fringe of the crusade in Europe, with nothing to show for their contribution but a solitary European 'theater ribbon' on their jackets. Until now they had even been denied the self-satisfying fruition of the tortured mind and aching hearts behind their skillful hands.

"On 1 Jan they stood out 100 strong, each man at the edge of his foxhole, many not even venturing to get in it, in order to see his skillful pilot employing the airplane which he had inspired with caressing hands of a lover the night before, actually destroying the enemy in flaming cauldrons well within the reach of their eyes. Fifteen of the enemy airplanes destroyed came down within a mile of the field."

John C. Meyer's last combat mission of his second tour was flown during that same afternoon when he was alerted for another assignment. However, fate stepped in and he was never to report. In early January a tragic traffic accident accomplished what no German pilot or gunner was able to do. Meyer was traveling the snow-covered roads of Belgium in an ammunition carrier when an accident happened and he was sent to the hospital with a leg injury which almost cost him his leg. From there he was returned to the U.S. to rest and recuperate.

For Col. John Meyer, WWII had now ended, but the leadership and personal bravery he had demonstrated during combat would serve as the platform from which he would build his long and successful Air Force career. General John C. Meyer retired in 1974 after over thirty-five years of service to his country. He died on December 2, 1975.

Editor and 352nd historian "Punchy" Powell, when recalled to active duty for the Korean War, had an opportunity to again be with J. C. Meyer. While serving as a pilot and PRO at Wright-Patterson AFB, Powell was responsible for a series of TV programs being featured each week on one of the Dayton TV stations. One of these was to simulate an F-86 mission in Korea. Knowing "JC" had just returned from a combat tour in Korea, he contacted then Brigadier General Meyer and asked him if he would participate in the program. So, B/G Meyer, to accommodate one of his old WWII pilot friends, flew into Dayton to take the lead role in the TV program.

Since he had never met "Punchy's" wife, Betty, he visited the Powells in their small home and Betty prepared breakfast for him. Afterwards, while enjoying his cigar, he remarked, "Well, Punchy, you picked yourself a mighty fine wife. She makes a damned good cup of coffee." Punchy had long before learned of "JC's" sarcastic sense of humor.

Years later, after General Meyer had retired, Powell called him again and persuaded him to be the speaker for the first reunion of the newly organized Eighth Air Force Historical Society in Florida and he accommodated Punchy again. Sadly, only a few months later he died of a heart attack after his daily jogging, an exercise he had done even during WWII.

The author would like to thank the following individuals for their assistance in making this article possible—Sheldon Berlow, James Bleidner, Albert Giesting, William Kohlhas, Joseph "Red" McVay, J. Griffin Murphey III, Robert H. "Punchy" Powell and Dwayne Tabatt.

Perfidia

By Marc Hamel and Troy White

Percentage wise, few Allied fighter pilots glimpsed a German jet during World War II. Ace Donald S. 'Don' Bryan caught four with his exceptional eyesight while flying with the 352nd Fighter Group. Never one to pass up a good opportunity, Bryan damaged the third of these, and destroyed the fourth. The riveting "Bluenoser" painting of the same name by Troy White, portrays the pivotal moment in Bryan's ultimate test as a fighter pilot. The fourth time was indeed the charm. ***(See painting pg 273.)***

Stationed with the 328th Squadron of the 352nd, the "Bluenosed Bastards of Bodney," Don Bryan ultimately achieved 13.33 aerial victories, including five in one day on November 2, 1944. On March 14, 1945 with 12.33 victories to his credit, he was itching to have another crack at a German jet before the war ended.

When the German Arado 234 "Blitz" bombers became operational in the fall of 1944, they created quite a stir among Allied pilots. With roughly a 50 mph speed advantage over our best fighters, they

Arado 234 being rolled out from its hangar.

proved a worthy adversary. Additionally, none of our pilots was really sure at the time whether these jets carried offensive air-to-air armament. The only way to get a crack at one of these speedsters was to either dive upon them, cut them off in a turn or wait for them to return to their lair. The predominant unit flying these jet bombers was KG 76, which became operational in late November.

In his first Arado encounter in December 1944, Bryan mistook an Arado for an A-26 Invader flying overhead and called it in as a "Big Friend." His number four man replied with something unintelligible. Upon returnimg to base that day he queried Bryan, "Did you see that A-26? It had German crosses on it!"

The Group Intelligence office was soon abuzz. When asked what the jet looked like Bryan allowed, "If you ever see an A-26 with German crosses, no props, and going like Hell, that's what it looks like!" A couple of weeks later he encountered another. By the time he set up his usual rear-end approach it was far enough away that Bryan notes, "I didn't even try to shoot at him."

As a Christmas present to himself on December 21, 1944, Bryan finally got a piece of one of these elusive German jets. He mentally estimated where the German was headed, cut across the predicted turn, and lay in wait for a shot. One flaw. "I was hanging it all in the wind, as the German would have an easy shot at me out in front of him." Bryan certainly didn't know whether the Arado had forward firing guns at this point. In any event, Don took a long shot estimating deflection after the jet bomber rushed by, and saw at least one good hit on a wing. Credit one damaged.

On March 14, 1945 near Vic Blsaffthal, Germany, Bryan was leading a flight of four P-51s escorting 9th Air Force A-26 Invaders. Don was flying George Middleton's "Worra Bird 3rd Baser" Mustang coded PE-J crewed by Jim Shenk since his famous "Little One III" had been badly damaged in the German attack on Y-29 New Year's Day.

This escort mission took them near the famous Ludendorf Bridge at Remagen which had just been captured a week before. The Germans were desperate to drop the bridge and stymie the Allied push across the Rhine. KG 76 had commenced tactical attacks on the bridge on the 9th, following up again on the 11th and 13th.

Bryan's encounter report narrates: "I observed an Ar-234 cross in front of us headed for the Remagen bridgehead. I dropped tanks and started after the enemy aircraft. He was traveling about 50 mph faster than we were and crossed the Rhine south of the bridge going west, then turned north and made a very shallow dive run on the bridge, but he did not drop his bombs. I saw several P-47's to the northwest, so I headed in a northeasterly direction. I could not catch the e/a in a straight run, and thought he might turn east to avoid combat with the P-47's. When he saw the Thunderbolts he did turn east, passing directly under me. I turned east and when he passed under me I dove down on him and opened fire at about 250 yards."

Don Bryan in a serious PR pose. (Sheldon Berlow)

Arado 234 as seen from above

Committing early, Bryan rolled right into a dive to assure he would be on top of the jet as it passed. As the jet whooshed by him, Don then rolled wings-vertical left directly behind it and prepared to fire. This is the moment of truth depicted by White's painting.

Don says today, "I really had to hang my butt out in front of that guy to get the jet because if he was armed he could have had me. I just got ahead of him and let him fly by me. I rolled in behind the 234 and fired when all g's were neutral."

His encounter report continues, "I hit him with the first burst and apparently knocked his right jet out. He made a shallow turn to the right and started very mild evasive maneuvers. They consisted of shallow turns and a few shallow dives and climbs." There was no jet or prop wash or anything, so I squirted him. I fired almost all of my ammo at him and before I had finished I had knocked his left jet out. I had to stay close to him after I knocked out his jets and he started to slow down. The other P-51s of the squadron and those P-47's were lining up for a shot at him and I wasn't about to give them the chance!" The encounter report concludes, "The e/a was emitting much white smoke but I don't believe it caught fire. At about the time I finished firing he rolled over on his back and dived straight into the ground and exploded. Just before hitting the ground the pilot jettisoned his canopy, but did not get out. I claim one (1) Arado 234 destroyed (air). The fourth time's the charm."

Don adds, "How close did I stay to him on the way down? My "g" meter registered just a fraction short of eight g's when I pulled out of my dive following him down. It was either that or hit the ground!"

The Arado was one of eleven from 6/KG76 tasked to attack the bridge that day. Four of these downed. Bryan's Ar-234 was piloted by Hauptmann Hirschberger, flying his first and only mission with this unit. As the encounter report details, he attempted to exit the aircraft, but did not make it.

The genesis for this article was the completion of the painting "Perfidia" by noted artist Troy White featured on page 273. White's fanatical attention to detail resulted in much of the research material presented in this chapter.

"... All The Way"

Edwin L. Heller, Fighter Ace

by Marc Hamel from interviews with Edwin L. Heller

"I was leading 'A2' Group which consisted of six of us. I headed north from the target as there were more bombers up there. As I approached them, I noticed 15 plus bogies coming in from the right. Just then they nosed over and went through a box of bombers. I dropped tanks and told my squadron we would attack. I tried to bounce the last of them, but the leading flight turned into me, so I turned into them. Then a Lufbery (circular aerial battle) ensued of all planes. The fight was underway only a few minutes when I noticed at least six Fw- 190's (Focke-Wulf fighters) burning and spinning down. At one time during the scrap, lining up on one E/A (enemy aircraft), I glanced back to clear my tail and saw a 190 queuing up on me. At that instant he burst into flame, as my wingman Lt. Paulsen got him in his sights before the Hun could get a shot at me. I made some hits on my 190, but as it was a 90 degree deflection shot I only damaged him. There were more 190's in the sky so I let him go and got on another Hun's tail. This one was pulling up on another box of bombers when I closed on him. I got many hits on the cockpit and a lot of pieces came off. I am sure I killed him, as the plane went down out of control. I looked around for another fight but there weren't any Jerry planes left."

These March 2nd, 1945 combat Encounter Report lines describe Heller's graduation into the unofficial, but highly recognized, ranks of American fighter aces. To his 5.5 aerial and an astounding 16.5 ground victories during World War Two, he went 'All the Way' adding 3.5 Mig kills over Korea.

While playing high school football, young Ed Heller was told by coach Doug Crate, "Don't give up Ed, play all the way to the very end!" In looking back over his career Heller relates, "I never forgot that. Go all the way." Whether this great American fighter ace was clawing Luftwaffe and Korean aircraft from the skies, or a POW resisting barbaric Chinese attempts to turn his loyalty from our great country, those words were his personal motto. "All the Way." All the way indeed.

See Troy White's painting, pg. 273.

Early Days

The youngest of six children, Heller was born into relative luxury on December 5, 1918, in Philadelphia. His father, Clyde, was the affluent president of a western gold mining firm, while his mother developed a school on their property where Ed began his studies. When he was around six years old, the Hellers moved to Beverly Hills, California for three years, with Ed attending military school.

Soon after, the stock market crashed and the Hellers lost virtually everything. Ed began working summers for a grounds keeper to stay in school, though he was more interested in sports than studies. Buckling down and studying however, he graduated Friends Central College Preparatory School.

After graduation, Heller worked for a wire rope company before trying his luck in California. Staying with his sister, he worked in sign painting for a year before turning back for home. With little money, Heller began a spellbinding odyssey across the country hitch-hiking and hopping trains. "I'm glad I did it, as it was a wonderful experience and I'm proud of accomplishing it," Heller allows. "But I would never do it again."

On return to Pennsylvania, Heller tried slaughter house work for a few months, deciding, "It was a sweatshop. It was work, just work." Obtaining a posting to the Pennsylvania State Motor Police through his uncle's connections, Ed states, "I was elated to find myself out of the labor force, and elevated to a little higher rung in life." Completing training he was posted to a substation near Scranton. Attending to the usual variety of Motor Police tasks, Heller found this duty to his liking. "One day as I was on desk duty there was an interruption on the radio. It was President Roosevelt, and he told us the Japanese had bombed Pearl Harbor and we were at war. I was very excited, and told the others all about it. I soon heard that the army was reducing their requirements for aviation cadet training eligibility, so on my next day off I drove to Philadelphia and signed up." Heller had found his life's work.

Flight Training

Heller's military training began at Maxwell Field, Alabama, on May 9th, 1942, which included Pre-flight, and brief course work at Craig Field. Unlike some, Heller thrived on his military training and excelled. Interestingly he noted in a letter home "Men over 5'-10" and 180 go into bombers, so I guess I will fly a bomber if I make the grade." What a mistake the Army Air Force would have made!

Primary training at Dorr Field, Florida was where he actually began flying in the Stearman PT-17 "Caydet" biplane. Describing this flying as "a heck of a lot of fun and a big thrill." He took to flying like a fish to water. Basic flight training was next at Shaw Field, South Carolina flying the BT-13 trainers. He mastered the syllabus, and was assigned to twin-engine training (indicating bomber duty). Inexplicably, the powers-that-be suddenly reversed their decision and sent him into single-engine training. "I think I will like single engine," noted Heller during the war. "After all, now that I am in this war I would like to get some fun out of it. The pursuit ships will give it to me." On December 18th, he arrived at Spence Field, Georgia for Advanced training in North American AT-6's, and moved to Eglin Field, Florida for gunnery training in January. He graduated from flight training with 227.55 hours in Class 43-B on February 15th, 1943. He was now 2nd Lieutenant Edwin L. Heller.

He was assigned to the 322nd Squadron, 326th Fighter Group at Westover Field, Mass. This was an Operational Training Unit (OTU) transitioning pilots into the awesome Republic P-47 Thunderbolt fighter. He wrote, "What a ship! Twice as big as the AT, and about twice as many gadgets on it." After completing OTU late in March, Ed was assigned to the "Fighting" 21st Fighter Squadron, 352nd Fighter Group where he would ultimately achieve his greatest success.

Ed is not looking too sure about the name "Happy" bestowed upon his P-47 PZ-Hbar by flight leader Frank Greene. (Ted Fahrenwald)

This unit was to take the P-47 into combat against the Luftwaffe over Hitler's Fortress Europe.

In May of '43, the 21st FS's official designation was changed to the 486th FS. The unit trained hard in combat tactics with their big 'Bolts, and boarded the Queen Elizabeth with the other two squadrons of the 352nd at the end of June for the journey to England.

Combat, European Theater of Operation

The 352nd was stationed at Bodney Airfield in East Anglia, an area which would come to be known as "Little America" by the British because of the numerous 8th Air Force bomber and fighter bases located there. The 352nd became operational in September, 1943.

Under command of "that pilot's pilot," Major Luther H. Richmond, Heller and the other 486th pilots were learning quickly and itching to test their skills. Becoming a First Lieutenant on December 1, 1943, Heller soon shared in the sinking of a German E-boat in February of '44. In early March the 486th converted to the longer-range P-51 B/C Mustang. His combat P-47 had been named "Happy" at the insistence of his flight leader Franklin Greene, who wanted a "Snow White & the Seven Dwarves" flight. Heller grumbles, "Happy was no damn name for a combat aircraft!" He hit his stride ground-strafing targets with his destruction of six locomotives and shared a Junkers Ju-52 transport victory in April. This set the stage for a brilliant demonstration of the fighter pilot's art. His April 24th, 1944 Encounter Report reads:

"Blue Flight consisted of Captain MacKean and myself. We had permission to strafe a field, in the approximate area of Ingolstadt, and we made a west to east pass. We each picked out a line of three planes. I was behind Captain MacKean and saw his strikes on all three. We circled left on the deck (near the ground) and came back for a north to south pass. We each picked out two planes in the SE corner of the field. I damaged one and destroyed the other as Captain MacKean destroyed both of his. We then made a 180 degree turn on the deck, keeping low. I was still behind Captain MacKean and I saw him hit the ground in a plowed field, raising a cloud of dust. He said over the radio that he was hit and was going to bail out, but couldn't get the altitude however and decided to crash land. I thought he was going to make it fine when he suddenly dropped in from about 50 feet just on the edge of the field. His plane broke up and burst into flames. I circled two or three times at 100 feet but could not see him get out.

Ed Heller and his crew chief Clio Moralez with their F-86 "HELL-ER-BUST X" (Clio Moralez)

"After Captain MacKean crashed, I climbed to 10,000 feet and started home by myself. After about 15 minutes I saw an airfield (near Bokingen) with about 35 plus planes on it. I circled it, planned my approach, and picked targets. I hit the deck at about two miles from the field and approached the field on the tree tops. As I saw the field before me, I zoomed up to 100 feet and came down on three Messerschmitt Me-110's in a row. I could see strikes on the trio. I kept low over the field and after I was away zoomed up. I looked back and saw two planes smoking and in flames. I then climbed to 10,000 feet and started home again. After about 10 minutes, I saw another field (near Crailsham). Using the same tactics I approached the field by a big hangar. I sprayed it and jumped over it. There was a mess of planes right in front of me. I hit three of them good, but only set one on fire. I was out in the middle of the field then and saw an Me-110 at the end of the runway at about 45 degrees angle from me. I skidded around, gave it a long burst, and saw it burst into flames. I also killed one of the ground crew as he first stood there and looked at me. When I looked back, two of the planes were smoking and burning. I climbed back to 10,000 feet and proceeded home without any other excitement." Heller was flying his olive green P-51B "HELL-ER-BUST" coded PZ-H. Named by its ground crew (at Heller's request) he laughs, "Everybody likes that plane because of its name. It's more famous than I am!"

Distinguished Service Cross

Setting a strafing record for the 8th FC, this mission earned Heller our nation's second highest honor, The Distinguished Service Cross (DSC). The award write-up in part read, "For extraordinary heroism in action against the enemy, 24 April 1944..." During these attacks Heller destroyed seven enemy aircraft, damaged five others, and inflicted damage on buildings and other installations. The courage, skill, and determination to destroy the enemy displayed by Lieutenant Heller reflect highest credit upon himself and the Armed Forces of the United States." Coach Crate must have been proud.

On May 8th again flying his "HELL-ER-BUST," Ed chalked up wins in the aerial victory column. "I was flying Blue Three when I saw about 20 plus bandits. Captain Frank Cutler who was leading my flight attacked a 109 (Messerschmitt Me-109 fighter). I started to follow but saw another 109 nearby (who was queuing up on Cutler) and went after him. I got behind him and opened fire at about 150 yards, no deflection. He blew up on my first burst. By this time I was fairly low. Looking up I noticed several bandits coming down through the bomber formation. They seemed to hit the bombers and continue down to the deck. I picked one out and went after him. He was apparently damaged by my first burst for he took violent evasive action. By this time we were both on the deck. I got a few more strikes. We got into a Lufbery at 130 mph and I had 10 degrees of flaps. He finally straightened out and I noticed he was preparing to crash land, as his engine was smoking badly. I gave him a last burst, just as he hit the top of a bridge and went into the river. I took a picture of his plane in the river, then returned by myself."

Ed adds to this report with, "This was the hairiest combat I was ever in. On this day Cutler was probably smiling in anticipation as the two of us turned to attack 20 plus 109's, but I was scared shitless. Anyway, when I got that second 109 down on the deck it was pretty hairy as he would turn around buildings so I couldn't pull a lead on him. Finally we got into a just-above-stalling Lufbery Circle and it was him or me just a few feet above the ground. I was very thankful when he broke out and I got the final burst in. As I dove in to take a picture of him, my coolant boiled over and came right back on the windscreen. I pulled up and, in reflex, automatically jammed the coolant door open. I climbed to 10,000 feet, heading home, and made a blind call on the radio that I would most likely bail out, but that I got two 109's. We had been way south of Berlin, and I chose 10,000 feet as it was too high for the light stuff and too low for the big guns. I was real concerned that my engine would quit over the Channel, but she kept going and I got home with it. The North American tech rep couldn't believe I'd flown for two hours after popping my coolant. He surmised that I'd saved a fraction of coolant and it turned to steam, providing the necessary cooling to keep the engine from seizing up."

After three separate successful train-strafing missions (destroying eight locomotives) in the week prior, Heller's May 28th Encounter Report read: "I was flying Blue Three in a three-ship flight when contrails over the target Dessau were seen. We started climbing up to meet them and reached 30,000 feet where they came down at us. I saw Blue Leader get on one of them and another came near me so I got on his tail and we split-essed toward the deck. I finally got him right over Dessau at 20,000 feet where he bailed out. I never saw him open his chute. I was by myself and joined up with three red-nosed P-51s with "QP" markings (Joseph L. Lang, 334th FS, 4th Group).

One of these and myself saw another 109 at the same time, but he got there first so I stayed back to cover him. He got a few hits and overshot so I came up and finished him off. I had to break away to avoid a collision with the P-51, but saw the 109 going down smoking at 3,000 feet. From his actions I knew he could not pull out in time." Heller was flying Captain Woodrow Anderson's Mustang "Texas Bluebonnet" (PZ-A). This half credit victory was confirmed by Lang and Michael McPharlin of the 4th FG.

In support of the Invasion of Europe near St. Lo, Heller again notched up strafing scores in "HELL-ER-BUST" on June 10th with one truck destroyed solo, and three trucks shared with fellow 486th pilot Joe Gerst.

The Russia Shuttle

The 486th participated in the famed three-way England/Russia/Italy "Frantic" Shuttle Mission in late June of 1944. The June 21st Encounter Report reads: "I was flying Yellow three on the first leg of a bomber escort mission to Russia. When the Jerries came in to attack the bombers from 11 o'clock high, we broke up their bounce by making our pass at a four ship flight of them from 3 o'clock. As Yellow leader was closing in on the tail of a 109, I observed another Jerry queuing up on his tail. I immediately chased the Hun that was pursuing Yellow Leader and forced him to break for the deck. I followed him down to about 8,000 feet where I finished him off with a short burst from 200 yards dead astern. The Me-109 poured out black smoke and I last saw the E/A when it went into the ground (near Brest Litovsk)." Heller's victory is considered to be a record; the farthest away from home base at approximately 1,000 miles.

Heller expands on this today with: "After I had shot down the 109, I started back to the bombers. At about Angels 20 (20,000 feet) I saw below me three 109's in a fight with one P-51, so I peeled over to help him. Well, when I got close I saw they were actually four 109's, three being camouflaged. I took a burst at one and zoomed full-throttle with all of them on my tail. I climbed in a spiral so steep I was just above stall. It seemed that when a 109 shot its guns, it stalled. Three of them stalled out, and the fourth got a burst in on me and hit my rudder. It almost caused my knee to hit my chin. He then stalled out. I took off and finally found the bomber string. I surmise that if I hadn't been so scared, I could have winged over and gotten that last 109 after he stalled out."

After repairing the rudder fabric on landing at Piryatin, Russia, Heller followed the rest of the 486th to Kharkov and thence to the cratered field at Chugiev. When his engine failed on takeoff from this rough field, Heller jammed on the brakes and stopped just short of a ditch. Again attempting takeoff after checks, the engine failed after the "point of no return" and "HELL-ER-BUST'"ended up in a ditch, damaging a prop blade. The remainder of the 486th left on the next leg of the Shuttle (to Italy), leaving Heller to his own devices. Proving his determination to go "All the Way," he had the prop repaired with a bastard blade from a B-17. After these repairs were completed over a couple of days, the Mustang was again lined up for takeoff after ground runup. A few seconds after becoming airborne, a connecting rod came sailing through the cowling, and Heller jammed the P-51 back onto the ground and stopped without additional damage. He was forced to wait two weeks for a replacement Merlin engine to be brought in and installed in the Mustang. The rest of the Group was now long gone, so Ed decided to fly home via the southern Air Transport Command (ATC) route over the Middle East. After a short hop over to the large bomber base at Poltava, Heller bid goodbye to his newfound Russian interpreter friend and was presented with a bottle of wine for the long trip ahead. After checks, Ed lifted off for Teheran, and lost his hydraulics in the process. After a hot, no-flaps-or-brakes landing at Teheran, he waited (with dysentery) for another two weeks for repairs. After seeing Cairo, Benghazi, the Gulf of Sidra, Tunisia, and the Atlas mountains, he came to roost in Casablanca. Here, the resident Commanding Officer fancied the Mustang, confiscated it for himself, and grounded Heller. This resulted in Ed arriving back in England via ATC after all. "What a trip!" he recalls today.

Headed home in August, Ed relaxed on a well earned 30-day leave. He had amassed over

Ed in the winter of '44-'45 with his P-51D "HELL-ER BUST" crewed by Charles Agee. Photo prior to victory tally being transferred to fuselage side as red Iron Crosses. (Ed Heller)

275 flying hours in more than 75 missions, including two tour extensions. On returning to duty in mid- September, he was notified that he was now Captain Edwin Heller. Rejoining the 486th in England in October, he received a new P-51D, promptly named "HELL-ER-BUST." Again coded PZ*H, this natural aluminum Mustang was also crewed by S/Sgt. Charles Agee.

A few days before Christmas, the 486th received word that they would move to a forward base in Belgium to support the push into Germany. Ed soon wrote, "We are about seven miles behind the lines and flying three missions a day. We are supporting areas and doing a good job. We live in tents in zero degree weather and it's pretty rugged." A cold February saw Ed move up the ladder to Squadron Assistant Operations Officer.

Ace!

Regarding the March 2nd, 1945 report that leads off this article, Heller allows, "This was truly a hairy fight. A hell of a mess. Planes were flying in all directions, and several times I barely averted a head-on collision." The value of a truly proficient wingman was graphically depicted here by Gene Paulsen (flying his "Sweet Bet," PZ-F), Paulsen finished the day with a "probable" and a "destroyed" Fw-190 to his credit, either of which could have ended Heller's career at 445 total hours. These Fw-190's were from II/JG 301, according to careful research completed by aviation artist Troy White.

After returning to Bodney with the 486th on April 15th, the Heller & "HELL-ER-BUST" team duplicated the earlier seven victory feat during a fight over Ganaker Airfield: "I was leading Red Flight on a strafing mission. We found this field with 100 E/A and Colonel Jackson made a dive pass to determine flak. There was quite a bit, so my flight and another busted it. I was told to make a pass, so I hit the deck and went across the field. We then set up a traffic pattern for strafing. I made about 10 or 12 passes usually destroying one on each pass. Twice I got two on one pass. The E/A's were undoubtedly filled with gas as they either blew up or burned readily with a few hits. After about a half hour Colonel Jackson signaled everyone home. There were 40 plus fires burning by then. My plane and others received battle damage from small arms fire. We returned home uneventfully."

Recently he added, "It's easy to say there was quite a bit of flak, and my flight helped bust it, but that doesn't tell the story. When Willie O. Jackson took his flight down they were shot at from all over the field. I noted where a lot of guns positions were stationed, so when we went in we could bust them up. Strafing is very dangerous when the target is heavily defended, as you don't have the altitude to bail out and you can't always know where the firing is coming from. When I got home that day, Crew Chief Charles Agee showed me a bullet hole right through the canopy, six inches behind my head."

Heller finished out the war in Europe with 520 combat hours and 22 victories (5.5 in the air), ranking him fourth overall in total combined victories (and tops in strafing) for his fighter group. On June 12th Heller was awarded the Silver Star by General Anderson for his April 16th strafing mission. This increased his wartime awards tally to the Distinguished Service Cross (DSC), Silver Star, Distinguished Flying Cross (DFC) with 5 Oak Leaf Clusters (OLC's), Air Medal with 14 OLC's, ETO ribbon with six stars, French Croix de Guerre w/ Etoile d'Argent, and the Unit Citation (generally referred to as the Presidential Unit Citation).

In August of '45 Ed was promoted to Major, and in November returned to the States and married his sweetheart, Judy Rae. After a brief assignment post-war in Florida, Heller heard from his friend and former Crew Chief, Charles Agee, about a posting with the West Virginia Air National Guard as Guard Instructor reporting to the Adjutant General. He spent an enjoyable three years there, and with Charles Agee's help convinced the powers-that-be to paint the WV/ANG Mustang's with sweptback blue noses just like those flown by the 352nd Fighter Group during WWII.

Korea

Heller's next posting was as Commanding Officer, 62nd Fighter Interceptor Squadron at Selfridge AFB. His friend Al Kelley commanded the 61st FIS at the same time. "When the Korean War began, I volunteered to go. The 16th FIS of the 51st FIW was deployed and I got in through my buddy Kelley and arrived in Korea in September 1952. After transition I was made squadron CO of the 16th." The unit was flying the F-86 Sabre, Heller's was logically named "HELL-ER-BUST X" and sported a shark's mouth.

Soon after, Heller claimed his first Russian Mig fighter victory. "Near the Yalu a flight of four Migs cut through my formation. I positioned my flight to attack, and as I closed, they broke into elements; one zoomed up right and one broke left. Considering the slower climb of my '86, I broke left and got behind the number four guy. I got a burst into him, and he broke away and just bailed out. I told my element to keep an eye on the Mig zooming up, and I went back for a gun camera shot of the guy in his chute. Boy, you should have seen him climbing his risers, as he thought I was going to shoot him! Soon after that mission I shared a victory with my wingman. On a routine patrol we bounced a Mig, but I had a tank I couldn't shake off. The '86 kept skidding, and I couldn't center the ball even standing on the rudder. I got a few good bursts in, but couldn't finish him due to that tank. I flew wing while my wingman finished him off, and the Mig went in."

Clio Moralez, Crew Chief for "HELL-ER-BUST X" relates, "I met Major Heller the day I was assigned to be his crew chief. He struck me as a no-nonsense type of man, and I certainly did not want to fail, especially as he was the Squadron Commander. I had been his crew chief for little more than a week, when one morning he met me at the plane saying 'Good morning Clio. Is the plane ready to go?' As always, I answered 'Yes

Sir.' He then proceeded to climb into the plane and strap himself in. Surprised, I asked him if he wasn't going to go through his preflight. He answered by telling me that he never had a crew chief that he could not depend on completely. He followed with, 'When I ask you if everything is all right and you say yes, I believe everything is all right because you know what you are doing.' I quickly re-checked my preflight preparations. I was amazed, but at the same time felt good about the confidence he had in me. Heller must have also shown the other men the same sort of confidence he had in me. He was that kind of Commander. He had confidence in himself and those around him."

On return to base from a mission on January 19, 1953, Heller was notified of his promotion to Lieutenant Colonel. Recently, he related his thoughts about a significant mission the next day:

"My next Korean war victory was over a lone Mig which was hightailing it near the border. I pursued, closed, and fired a burst from a couple of hundred feet. When he went out, he and his seat just missed my left wing. If it had hit me, I would have been 'walking on air' too. As we were ending the patrol I looked across the border and saw a flight of Migs milling around like they were on a training mission. I closed on one who zoomed, but not fast, so I closed to about 200 feet and got him good. He winged over and went right into the ground. I then realized that I had less than a 1000 pounds of fuel, my wingman had less, and we were still in China. This was not good, and we were pretty low too. We went to 100%, climbed up over 40 thousand feet and started coasting home. We barely made it back to base, and landed on fumes." It seemed as though Heller was slashing through the Korean Air Force to become a two-war ace.

January 21, 1953

"I guess my most interesting mission in Korea was not a victory for me but just the opposite; I was the victim. I still have dreams about it. The day before I was shot down, we had crossed the river and I had gotten the two Migs. So this day I was going to go across again. I briefed my flight that this was what we were going to do, and they went along with it. They were all experienced men; Erickson, Herrick and Dolph Overton (the latter a fast rising Sabre ace who remained Ed's friend until Ed died). We took off and I was leading the flight on a normal patrol, and went up and down the river for a couple of minutes. Then we went under another flight of F-86's, and on signal we cut our IFF (an electronic transmitter that allowed identification of friend or foe) off and went across the river.

"It wasn't long before we ran into a couple of flights of Migs. I think I counted about six or eight of them. (Dolph Overton recalls being separated from Ed's element at the first break, as his element leader could not follow Heller's turn.) I was in perfect position to bounce one Mig, in a peeling off maneuver as I was above him. As I always did when I was about to make a maneuver like that, I checked my six o'clock. Then I glanced straight down (being in a wing-over to the right) and saw a Mig coming up at me from below, zooming up. I really wasn't concerned this time as my maneuver to make a pass at the first Mig was a perfect defensive move to evade the one that was coming up against me. I gave him a large angle-off shot of about a half a second, so I wasn't concerned. But in a half a second I got concerned because he hit me—right in the cockpit with a cannon shell. If it wasn't for the armor plate he'd have shot the heck out of me. It broke my right arm, which felt like a huge hammer blow, and cut most of the stick. I was stunned for a few seconds, then felt my arm. It was ripped with three holes, from my elbow clear up to my shoulder. This concerned me greatly, and I lifted my right hand from the stick and placed it on my lap. I was greatly relieved to find that I could still move my fingers.

"I then started taking stock of the cockpit. The whole instrument panel was shot out except one gauge, and the console on my right was shot away. There was red fluid, either my blood or hydraulic fluid, on the remains of the console.

"Then I realized I was going straight down, with the engine still wide open. I tried to pull out of the dive with my still-working left hand, and the stick broke away. I cursed at the stick and threw it on the floor. I must have been doing...I guess several hundred miles per hour. The ground was coming up fast. Reaching across with my left hand I pulled the lever to eject the canopy several times.

Nothing. I was trapped and sat back to watch the ground rapidly approaching. I couldn't eject and going that fast you can't pull the canopy open. It's impossible, so I ran out of ideas on how to get out of there. I thought 'This is it.' At around, oh I guess 15,000 feet, as I got into heavier air, the plane came out of the dive, and with an ungodly amount of g's. I think I pulled at least 12 g's because my head went down between my knees and my oxygen mask came away from my face. It was really terrible and I blacked out. And when I came to again I was on top of the pullout and I was rolling into another dive. I could see the ground coming right up at me again. As I threw my head back naturally, I noticed a hole in the canopy where the shell had hit, and that's all it took. I just unbuckled my safety belt, stood up on the seat, and my helmet got pulled right off of my head and out of that hole. As my head neared the hole, the terrific suction created by the plane moving that fast popped me from the plane like a champagne cork. I obviously enlarged the hole as I went. The plane must have been going oh... four, five, six hundred miles an hour. I hit the horizontal tail going out, and if I was going 500 miles an hour, I must have hit the tail at 50 miles an hour between the cockpit and the tail. It was a tick...I just felt a tick and I was out in space. The plane hit about 10 seconds after I got out.

"I remember there was a time in WWII where a guy had to bail out of his airplane over England and he was left handed. We found his body with his left hand still trying to reach the ripcord. It was on the left-hand side, and he was scratching his jacket on the right side. He went right into the ground thinking, 'Here's the ripcord.' My right arm was broken and I thought of him. You know when you're in a situation like that you don't panic, but I didn't know how close I was to the ground as my eyes were shut due to the wind blast. I pulled my left arm in close, put my thumb through 'D' ring, made a fist and punched forward.

"The opening shock was terrific as I decelerated from say 200 miles per hour. As the chute cracked open I blacked out. When I came to a short time later, I realized my left leg hung useless below me. The little 'tick' as I left the plane had resulted in a compound leg fracture. My leg had two bleeding holes where the bones broke through the skin.

"After the noise and violence of my escape, hanging in the chute was almost a relief. Then I saw that the wind was blowing me sideways at a good clip, and would rush me into a boulder covered hillside on landing. After twisting to land on my good leg, I hit hard, folded up, and rolled across the boulders down the hill. After stopping on my back, I took stock, laid my right arm across my chest and discovered my left foot up under my knee. I struggled to get my escape kit open to get to flares and start a tourniquet."

At this point he was accosted by a bolo-wielding, nasty-looking ragged peasant. After determining that Heller wasn't armed, he called in his comrades. At Ed's gestured suggestion, his wounds were bound and he was carried away in his parachute. He was soon transferred to a truck and painfully transported to a village doctor who splinted his arm and leg. Then began another agonizing, eight hour journey by bouncing truck to Antung in Manchuria. Once there, he was interrogated by eight Chinese Army officers. After giving his name, rank and serial numher, Heller was promptly accused of being a spy, but that line of questioning ceased when he shot back, "Spies don't fly F-86's."

Then, his interrogator asked him, "What do you think we are going to do with you?"

"I don't know," Ed replied, "probably kill me, I guess." When this was translated to the other officers they all laughed, except for Ed, of course.

"We are not going to kill you," the Chinese interpreter told him. We are taking you to a hospital." Then, after a shot of morphine, he was again trucked out, this time to a civilian hospital in Antung. And so began his two and one-half year imprisonment in China.

Some thirty-four hours after being downed, his broken hones were finally set and properly encased in casts. "Then, for the first time I began to believe they were not going to kill me," Ed remembered.

However, he was to stay in virtual hospital solitary confinement for six weeks. His only relief was by interrogator/interpreter Liu, who had questioned him on arrival. He would 'visit' Ed every four days or so.

Thus began the "good guy/bad guy" cycle of mental torture, interrogation and solitary confinement that would continue throughout his 28-month capitivity.

During this time Ed asked to meet the pilot who downed him to congratulate his opponent. This request was never granted and Ed surmised that he had either been killed in a later engagement or that he was a Russian pilot.

He was transferred to a military barracks outside Antung after six weeks where he met his nemesis, Chiang Chun—the "brain-washer." And so started the hundreds of hours of arguing, pressuring and mental duress that characterized Ed's battle to retain his sanity.

"He was a cold-eyed, pimply, sneering little guy who was my personal villain for the rest of my confinement, always pressing for his version of a confession from me," Ed related.

Ed was held in a 15 x 50 foot room with a straw mattress, fed mainly rice, and was not allowed to bathe. A great concern was his leg. The cast on his arm was removed, and though uselessly stiff, the bones had mended. However, at seven weeks his leg cast was removed and the leg sagged badly. The bones had failed to knit.

At this point Chiang started his Communist doctrine brain-washing in earnest, always pressing for a confession. Through all of the next 20 months of mental torture, Ed's leg was never properly repaired. It is still open to question whether these failed surgeries were intentionally left undone as a tool to increase his mental suffering.

After several months he was transferred to a brick guardhouse.

Chaing intensified his efforts to demand a confession from Ed that he had violated the Chinese border. The erosion of his willpower and stamina was in no way helped by his now useless leg.

Finally, in May of 1953, he was brought to a Chinese Army hospital for surgery. This time they fastened a steel plate to the bones, then placed him in a tumbled-down Chinese house. By now, he was quite gaunt, having lost thirty pounds, was bearded and sickly from lack of treatment. In August he was cheered by the news that the war had ended, but this hope was soon deflated by Chaing telling him, "Your case has nothing to do with the Korean War. You violated the border. You will not go home unless your 'case' is settled." This obviously meant when Ed confessed in the manner the Chinese wanted. Chiang continued to pump him with Chinese doctrine and allowed only their propaganda to reach him. No letters or outside contact was permitted.

Ed was grudgingly transferred hack to the heated hospital as temperatures fell in October. Chiang raged unceasingly at him for a confession that he had been ordered by superiors to cross the border. More pressure was brought to bear and it became obvious that unless some sort of "confession" was signed, Heller would never return home.

Ed finally drafted a hogus confession but it did not admit to being ordered across the border. This was not "acceptable."

Chiang disappeared for the last three months of 1953 and around Christmas Heller received the news that his latest leg operation was a failure. In virtual solitary, Ed grew increasingly despondent. "I would lie helpless and hopeless in bed trying to ignore the terrible bedsores, and had only visions of rotting in China for the rest of my life. Then I remembered Coach Crate's words, "Don't give up, Ed, go all the way.

"I took heed and finally wrote a nasty note to my captors saying, 'I don't want to be an experiment anymore, so just leave my leg alone.' Apparently this was not the Communist intent as I was informed that they were not 'Facist Beasts,' and I was transferred to Mukden in early January of '54 for more surgery. The hospital room at Mukden General Hospital was to be my home for the next 13 months. Another painful operation involved removing the steel and chipping away at the bone ends, ultimately failing again.

On a brighter note, Heller's tormentor was gone, replaced by the friendlier Mr. Chu Ping. "He was ordered to improve my attitude!" Treatment became better, and books and limited letters from home were allowed. One letter sadly brought the news that his mother had passed away in early 1954.

Bad news came in groups. "My last leg operation was a failure, too, and I asked them to 'just amputate it and be done with it.'" The doctor in charge, however, had orders to heal him ... or else. "This surgeon finally performed a bone graft from my hip, and this time, it took. It hurt a lot, but by October I could walk again, some 20 months after I was shot down. My left leg was, and still is, an inch shorter than my right."

Through Chu, Heller discovered three other Americans were also being held—Capt. Harold Fischer, Lt. Roland Parks, and Lt. Lyle Cameron. He was soon allowed limited correspondence with the others. After months of virtual solitary confinement and surgery, this made him pine even more for release. As he noted, "At that point I was ready to 'confess' to their garbage just to get home."

After getting clues from the others as to how to come up with a suitable "confession," Ed finally produced a draft the Chinese would accept in April, 1954. Detailing phony violations of the Chinese border on orders from higher up, this farcical document was laboriously fabricated after hundreds of handwritten pages and memorized bogus details the Chinese required. "They were eventually satisfied with my 'confession' but I can't see what good it could have done them."

Ed, front and center, walks from Chinese captivity in April '55 with Hal Fischer, Roland Parks, and Lyle Cameron. (Ed Heller)

Release At Last

In April, 1955, Heller and his fellow prisoners were hauled to Peking to face a "trial" by a military tribunal of the Supreme Court of the People's Republic of China. A verdict was handed down. "The decision of this court is that Heller, Cameron, Fischer and Parks be deported immediately for flagrantly violating Chinese borders with the intent to cause provocative and harassing acts." Years later he said, "Those were the best words I have ever heard in my life."

Traveling through Peking and Canton, the pilots were transported to the Chinese-Hong Kong border and deported on May 31, 1955. Ed had again made it "all the way," this time through mental and physical torment that would have broken a lesser man. On coming home he was awarded a Purple Heart to add to his many other awards.

Returning to civilization, he rested up in Hawaii and pondered his next move. Returning to duty he had a choice of assignments and selected Ent AFB in Colorado. But after awhile he had had enough. I went to my boss and said, "Let me out!" He was a good guy and told me to take a T-Bird and go find myself a job. I found a spot as Deputy Group Commander at Williams Field." It was at Ent AFB that Ed and his wife, Judy, divorced. After a few years at Williams it was time for another move.

In July of '59 he took a new assignment at Wheelus AFB as Deputy Wing Commander under Pete Everest. It was there he was introduced to the effervescent Johanna "Jo" Schlosser and they really hit it off and were soon married in December, 1960. They were based at Cannon AFB in Mexico for the next two years followed by a five month assignment to the Pentagon.

During the Cuban Missile Crisis, he sat on alert in an F-100 at Homestead AFB for six weeks, again ready to give his life for his country. Weisbaden, Germany, Hqs, USAF Europe beckoned next. It was Johanna's hometown. After that he served two years in Turkey in an underground command post. In 1967, a government cost-saving program brought on his retirement, grounding any pilot 45 years old with 22 years of service. With no prospects for flying, Ed promptly retired and bought a yacht in Athens, Greece and spent the next nine years sailing the Greek islands and the greater Mediterranean. The Hellers returned to the U.S. in January 1976 and settled down in Northern California.

Several years ago Ed suffered a crippling stroke which paralyzed him on one side and destroyed his ability to talk. However, even after being confined to a V.A. facility for the care he required, Ed remained a free spirit. On at least one occasion, we are told, he sat in the lobby in his wheelchair and studied the main exit until he developed a plan to take flight from that place.

Watching his opportunity while near the exit, he wheeled himself out close behind someone who was exiting through the self-locking doors and took off down the street for an outing. Of course, they found him later and returned him to the facility.

A few years ago, a Savannah, Georgia businessman, Robert Jepson, bought and restored a P-51 Mustang and honored Heller by adding Ed's WWII markings on his warbird. Once again a bluenosed Mustang named "HELL-ER BUST" would be taking to the skies honoring this courageous pilot who flew combat in two U.S. wars.

And Robert Jepson did even more. He flew his warbird to California and arranged to have Ed and Johanna and some of their Air Force friends come to dedicate his beautiful warbird carrying Ed's colors. It was probably Ed's last outing before he died in 2004.

Clio Moralez, Heller's Korean War Crew Chief and friend, recalls, "From what I saw of Ed Heller as an officer, pilot, Squadron Commander, and leader, he was one of the best. That is not only my opinion, but also the opinion of the ground personnel and other pilots of the 16th Squadron. Even now, over forty years later, the guys still speak highly of Ed at the reunions. I must say that he was the best C.O. I ever had, and consider it an honor and a privilege to have served with him."

There is no finer tribute that a pilot can have than to be eulogized by the officers and airmen who served with him. It is sometimes said there may not be many fighter pilots in Heaven, but you can bet that Ed Heller went "All the Way."

Dolphin D. Overton, Ed Heller's friend and wingman. Dolph had the hottest streak of victories in the F-86. (Dolph Overton)

Bail Out!
By Merle Olmsted

"Who shot me down?" This was a question asked by Helmut Peter Rix, a Luftwaffe pilot. Difficult to answer in 1945, but forty years later Merle Olmsted recounts some classic research work to find the answer, a story that involves two 352nd pilots.

In the spring of 1945, Luftwaffe fighter pilot Helmut Peter Rix bailed out of his burning Fw-190, ending a short combat career. Through a mutual friend, Rix made contact with the author 41 years later, expressing curiosity as to who might have shot him down in those traumatic times so many years ago.

The date was March 2, 1945, and the long years of war in Europe would be over in less than two months. The steady and long term decline of the Luftwaffe was reaching the point of total disaster. The individual combat units, however, still fought on with what they had against a tidal wave of Allied fighters and bombers now ranging at will across Europe.

On this date, elements of Geschwader JG3O1, based at Stendal, scrambled at 1015 hours to intercept eight Air Force heavy bombers and their escort. Flying a new Fw-190D, and carrying the tactical number Red 4, was newly assigned Fhr Helmut Rix. Although he was a fledgling fighter pilot, he had a respectable 260 hours of flying time. (about the average amount of flight training American pilots were given before getting their wings).

Fahnrich Helmut Rix of 8 JG 301 (Peter Rix)

He was first posted to blind flying school in Denmark as he was destined to fly multi-engined aircraft. The war was far away, but they had a sudden reminder of it when one day two wide-ranging RAF Mosquitos roared across their field on the deck, shooting up facilities and disappearing as quickly as they had arrived.

Rix soon soloed the twin Siebel Si 204 and later the big Junkers 88 and was then engaged in crew training. Graduation came in July, 1944 and he was posted back to Germany as a night fighter pilot, but no assignments came through. Suddenly they were advised that night fighter training had been canceled and they were being reassigned to single-engined fighters. Only two days into this training, however, a faulty

Earl L Mundell in a P-51B promo photo by Sheldon Berlow

rudder control caused a bad crash on landing a 190. With the plane upside down and fuel running from the ruptured tanks, Rix was pulled free, unhurt, by the crash crew.

In December he was posted to JG2 to complete his training and was assigned to the 8 Staffel, II Gruppe, of JG3O1. He recalls that the day he reported to the Staffel was not a happy one. Of the ten pilots he met that day, eight did not return from the day's operations. Due to a shortage of aircraft he had still not flown a combat mission by mid-February, when he was sent to ferry a new Fw-190 to the unit.

Rix recalls: "I got a flight order…to pickup an Fw-190-D9 at Niedermendig…so off I went by train with just my parachute under my arm. It was a hectic journey…taking a little detour to visit my hometown of Krefeld enroute. Bombing was everywhere, day and night, and from my home village we could hear artillery at Rourmont and the fighting in Holland, just 20 miles away. After a few days at home I traveled to Niedermendig and there I got stuck. Allied fighter-bombers, mostly Thunderbolts and Lightnings, were masters of the air. The aircraft I was to ferry was damaged while taxiing and had to have a new propeller. In the meantime I watched the goings-on most of the time under cover of a one-man hole!"

On February 27 with the plane now serviceable, Rix took off and landed at Giessen where it was necessary to lay over due to Allied air activity. Since he carried no ammunition, he continued on his way flying at only fifty meters, arriving in Stendel on February 28.

Although he was scheduled to fly a mission on March 1 his aircraft was, at the last minute, turned over to a more experienced pilot and he did not make his first operational sortie until March 2.

Helmut recalls: "Scramble was at 1015 and I was flying as Red 4. We climbed on a southeast course to 24,000 feet when we spotted our target, a formation of B-17s at 27,000 feet dead ahead. We were in line abreast formation and ready for a frontal attack when my formation leader broke away to the left into a dive. Doing so, we got into a line astern formation with me as No. 4. It was a beautiful morning, clear blue sky, but with a cloud cover of 8/10's at about 14,000 feet."

At approximately 1100 hours, Rix's 190 was apparently hit by gunfire from a source he did not see. "The lead aircraft was disappearing into the clouds when my plane shuddered and flames were licking around the front of my engine."

In that moment of terror when his engine exploded in a gush of flame, his training and sense of survival told him it was time to get out. "In the panic that followed, I got rid of my cockpit hood, but forgot to undo my straps first and so fed the flames with the aircraft out of control. I managed to undo the straps, but how I got out I shall never know. My 'chute opened perfectly and with several burns to my head, face and lower arms, I landed in about six feet of snow and was dragged for some distance really blind."

So, who was it that day so long ago that smashed Rix's engine with a burst of .50 calibre bullets and ended his career as a Luftwaffe pilot?

Finding out who shot him down seemed to be a fairly straight forward research project. We had Rix's recall of the event, times, locations, etc. and now only needed to match up the proper 8th Air Force fighter pilot. The 352nd Fighter Group was engaged in the area which would be correct for the encounter with Rix. It was the southernmost unit, escorting B-17s, most of whom bombed secondary targets at Chemnitz. It was in this area that the 352nd tangled with a force of Fw-190s, given as fifteen in most encounter reports. They also agree that the fifteen German fighters were making a front quarter pass at the bombers, and when bounced, went into a Lufbery Circle. The 352nd claimed 5.5 victories over those 190s.

The mission summary for the 352 reports:

RIV with 10, 11, and 12 C/Grps, 1 Division B-17s in good formation, Weisbaden, 25,000, 0934, visually identified as 41A, B and C c/Grps. Escort to target area, Rositz, where 15+ Fw- 190s seen to attack lead box of bombers from above at 1045. E/A engaged by A. Grp, at 25000.

Captain Edwin Heller of the 486th Squadron was leading six Mustangs. His encounter report gives a vivid picture of the clash with fifteen Fw-190s in the target area:

"I headed north from the target as there were more bombers up there. As I approached them, I noticed fifteen plus bandits coming in from the right. Just then they nosed over and went through a box of bombers. I dropped tanks and told my squadron we would attack.

"I tried to bounce the first of them, but the leading flight turned into me, so I in turn turned into them. Then ensued a Lufbery of all planes. The fight was underway only a few minutes when I noticed at least six Fw-190s burning or spinning down. At one time during the scrap, lining up on one E/A, I glanced back to clear my tail and saw a 190 queuing up on me. At that instant he burst into flame as my wingman, Lt. Paulsen, got him in the sights before he could get a shot at me.

"I got some hits on my 190, but as it was a 90 degree deflection shot, I only damaged him. There were more 190s in the sky so I let him go and got on another Hun's tail. He was pulling up behind a box of bombers when I closed on him. I got many hits in the cockpit and a lot of pieces came off. I am sure I killed him as the plane went down out of control. I looked around for another fight, but there weren't any Jerry planes left."

Heller's encounter report is similar to the others involved in this clash, and none of the circumstances match those of Helmut Rix. Heller was not sure if his victim was a "long nose" 190 or not.

Ed Heller works on his steely eye glare in his HELL-ER-BUST in early 1945. Note the red "Iron Crosses" victory markings. (Heller)

A part of the 352nd Group's 486th Squadron, probably White and Yellow flights, left the target area as can be seen by this continuation of the mission summary:

"Four plus Fw-190s sighted south of target, 24,000, 1100. 486 B Group pursued E/A to area west of Prague where 15 single and twin-engine aircraft observed on A/D at 1100, Prague Ruzyne."

On the day in question the 8th Air Force had launched three forces of heavy bombers to attack targets in central Germany escorted by at least six fighter groups of Mustangs, all of whom made victory claims of fighter aircraft. The 78th FG's claims were all Me-109s, so that eliminated them. The 20th FG claimed only two damaged Fw- 190s, The other groups, the 339th, 352nd, 353rd and 357th, all claimed Fw-190s for a total of seventeen of this type destroyed.

Known losses of Luftwaffe fighters of JG300 and JG3O1, which were engaged in this area, included fourteen FWs of which only five were the long-nosed Fw-190D9s and all of these were from JG3O1, two from Rix's 8 Staffel, two from the 6 and 5 Staffels.

Merle Olmstead sitting on wing of Ed Hyman's P-51 with Armorer Whitey and chief Ray Morrison right. (Olmstead photo)

Since each bomber force had its assigned escort, the general location of combat for each group can be plotted. However, none of the P-51s stayed in their assigned area and their encounter reports referred to the mix-up of Mustangs from other units. With some 300 P-51s engaged in a relatively small area, it was impossible to rule out anyone on location alone.

Rix says he was hit, bailed out and landed near the town of Aussig, near the border area of Germany and Czechoslavakia, some seventy kilometers northwest of Prague. Of the four Fw-190s in of the 8 Staffel in which Rix flew, two were D9s, one flown by the leader, Lieutenant Walter Kropp, the other by Rix. The other two were radial engine A9s. All were shot down during this battle with Rix the only survivor.

All of the encounter reports of the seventeen pilots claiming Fw-1 90s have been studied. Rarely did 8th Air Force pilots mention the markings of aircraft they engaged since things happen fast in aerial combat and in most cases the markings were simply not observed.

In an attempt to match a P-51 pilot with Rix's bailout, consideration must be given to time, location, altitude and circumstances of the combat, with time considered first. Rix says he bailed out at about 1100 hours but since he did not enter this in his logbook until he was released from the hospital, it can only be considered approximate, usually the case in most combat claims. The location situation is similar although his location is verified by where he landed in his 'chute. The casualty list also shows him downed near Aussig. However, the locations of victories given by pilots are seldom accurate. The 339th FG victories were "some 20 miles east of Magdeburg, some 300km from where Rix went down." The 353rd fight was a few miles southeast of Magdeburg. The 357th combat area was north of Leipzig, some 200km from Rix's crash. Of the seventeen claims for 190s, only one encounter report mentions the victim as a "long-nose 190," and this was John Sublett of the 357th.

While over the field, one Fw-190 was destroyed at 15,000 feet and one Fw-189 destroyed during takeoff, the victim of Lt. Col Willie O. Jackson, his seventh and last victory of the war.

The final part of the summary is highly interesting as it closely matches matches our search criteria. The number of 190s pursued by the 486th, the altitude and the time all match the Rix circumstances. The location is close, as Rix's bailout point was only about seventy kilometers north of the Prague-Ruzyne airfield.

From the casualty report of JG300 and JG3OI we know that two other 190s, one a long nose D model, were lost in this area. Flying the D-9 was Uffz G. Schulz, White 14 of the 5 Staffel, and the location is given as eight kilomeres west of Schian Prague. The other loss was Fhr Heinrich Lief, in an Fw 190A.

There was only one known claim in the Prague area, however. This was a joint claim by

Lee Kilgo (Berlow)

Captain Lee Kilgo and Lt Earl Mundell, both of the 486 Squadron of the 352nd FG. Kilgo's encounter report states:

"I was flying No. 4 in Yellow flight when the flight engaged one Fw-190 and began a Lufbery. The E/A turned away, placing me in position on his tail. At long range I fired several bursts and he began smoking and went into a turning dive directly over a town. I broke away—the flight leader saw him crash in flames.

Kilgo was flying P-51 44-15390, PZ-K and Mundell was flying 44-14091 PZ-M. Mundell's report follows:

"My flight was higher, thereby giving me a better chance to close on him. I was making good progress when two P-51s came in from the side. One ship got several hits. The 190 threw out some smoke and broke right and down to an altitude of 15,000 feet. I tacked on, fired a long burst, some parts fell off and the pilot bailed out near 8,000 feet. The ship hit and exploded not far from a small town."

Rix was, of course, No. 4 in a trail formation of four aircraft. The 352nd report does indeed mention four 190s initially, but mentions only the one at the time of the combat. It is possible that the other three had pulled away unobserved by the 352nd pilots involved.

Who, then, was the victim of Kilgo and Mundell—Helmut Rix, Schultz or Lieb in a 190A? Neither Kilgo nor Mundell's encounter reports match Rix's circumstances as closely as would be desired, but it matches closely in time, location, altitude and some other circumstances. We know that both Lieb and Schultz were killed in combat, and possibly significant is the supporting statement in both the Kilgo and Mundell reports by Lt. Cecil Freeman: "While I was flying White 3, I saw the canopy come off the 190 which Kilgo and Mundell were attacking. The pilot bailed out."

Could Red 4 have been hit by gunfire from their intended target, the B-17s? Rix discounted this possibility as he does not believe their four 190s approached any closer than about three miles before turning away.

What happened in the heat of combat over six decades ago in the skies over Sudetenland cannot be positively determined, and we are unable to say who it was that shot down Peter Rix. However, since he was the only one of the three known Fw-190 losses in this area to bail out and survive, there is every indication that Rix's Fw-190D, Red 4, was destroyed by the gunfire from Lee Kilgo and Earl Mundell of the 352 Fighter Group.

Rix takes up the story after he landed in the field near Aussig:

"I was picked up by a local farmer and taken to a hospital in Aussig. Wearing a full combination flying suit probably saved my life although I was severely burned about the head, neck and lower arms. After seven weeks and a transfer to a hospital at Comotou, I discharged myself because things were getting rather hot. The Russians were getting close and I wanted to get back to my squadron.

"It was a terrible journey through the burning Aussig and Dresden and on to Berlin where I arrived on April 20, 1945. I had to dodge Russian shells landing nearby as I made my way to Staaken Airfield on the western outskirts of Berlin. I looked a mess, head and arms still bandaged, my leathers burned, and when I got to the control tower the commanding officer thought I was a ghost! He found out where my squadron was and arranged transport on a convoy heading northwest that same night.

"It was a precarious journey through the night and we arrived at Neudstadt-Glewe unscathed and I was reunited with my squadron. They were surprised to see me as I had been posted missing and no one knew what had happened to me.

"I never flew again, however hard I tried to persuade the medics. It broke my heart to see my comrades take off for fighter bomber attacks into Berlin, and quite a few of them not returning. As the Russians advanced further, the whole JG3O1 transferred northwest to Schleswig Holstein and Leek where we were taken over by the RAF 2nd Tactical Air Force at the end of hostilities and interned. The individual squadrons bivouacked on farms until we were disbanded. It was a long journey by Army trucks and I arrived home on August 18, 1945."

After his burns healed, Helmut Rix moved to England in the early 50's where he and his lovely English wife, Joyce, lived in East Anglia in the heart of 8th Air Force country not far from Bodney Airfield, the home base of the 352nd Fighter Group during those momentous days. Helmut Peter Rix worked at Duxford Airfield, an Allied fighter base during the war and current home of the American Air Museum near Cambridge until his death several years ago. His widow still lives in that area.

Since this story was published in *Fly Past Magazine*, it has been learned that Earl Mundell was killed in a P-51 accident in Alaska in the late 1940's. Lee Kilgo and his wife, Evelyn, lived in Tallapoosa, Georgia, near the Alabama border, until his demise a few years ago.

This story, written by Merle Olmsted of the 352nd FG, has been condensed and edited, with his permission, by 352nd Editor/Historian Bob Powell for inclusion in this book.

Lt. Rudolf Wurff 7 JG 301 in front of Fw 190D9 Stendahl Air Base February 1945. (Jerry Crandall)

II JG 301 Fw 190D-9 aircraft at 15 minute readiness at Stendal air base in early March 1945. (Jean-Bernard Frappe via John R Beaman, Jr)

Days of Glory

by Bill Hess and Bob Powell

Thousands of young Americans responded to the attack on Pearl Harbor and joined the Army Air Corps, secretly dreaming of emulating the great pilots of World War I like Capt. Eddie Rickenbacker, Frank Luke and Canada's Billy Bishop to become famous aces by shooting down five or more enemy aircraft.

Although they possessed the skills and courage to do so and flew numerous missions achieving aerial superiority over the Japanese in the Pacific theatre or the Luftwaffe over Europe, most of them never achieved this dream for lack of the opportunity.

Many flew combat tours without ever engaging the enemy in aerial combat, just not being at the right place at the right time. And even those who became aces sometimes flew for long periods without an opportunity to meet the enemy in combat, thus having to build their score one or two victories at a time to achieve the magic number.

Considering the thousands of sorties flown by pilots of 8th Fighter Command over Europe, only 19 pilots became "aces in a day," by scoring multi-victories when suddenly thrust into air battles involving large numbers of aircraft. Four of those who accomplished this feat were 352nd FG pilots. One, Major George Preddy, scored six victories on one mission. Accounts of these multi-victories by these four 352nd pilots follow.

Capt. Bill Whisner

The 487th Squadron, led by Col. John C. Meyer, found more than 50 Fw-190s in the Merseburg area on November 21, 1944. The enemy fighters were waiting for our bombers and cruising far above the 487th. Meyer took his force up to meet them, climbing to 29,000 feet. Meyer told Capt. Bill Whisner to take a straggler in the enemy formation. As Whisner closed and the straggler moved back into formation, Whisner took a crack at him from about 400 yards but got no hits until he closed to about 200 yards, this time knocking large pieces off the aircraft and it started down in a spin and pouring out black smoke.

Whisner quickly tagged onto another 190 and fired just twice before sending this one down in a smoking spin. Then, closing behind three and went for the leader, who nosed over in a dive. Finding himself almost in formation with 190s to his left and right, Bill banked steeply and fired at the 190, getting good hits and sending it down burning. Meantime, his wingman, Karl Waldron, took care of the one on the left. His next victim tried to escape by taking evasive action, but Bill caught him with a deflection shot and the 190 went down smoking.

Having lost considerable altitude in the fight, Bill climbed back up and saw two 190s off to the side of the enemy formation. Attacking one, he saw it snap, spin and go down, but for this one he only claimed a probable. Then he turned to the rear of the enemy group and attacked another, this one going into a dive and exploding.

Spotting a Mustang with a 190 on its tail, Bill went for it, closing to about 50 yards, knocking pieces off of it before it went down with its engine burning, his fifth for the day.

Lt. Carl J. Luksic

It happened on May 8, 1944 and Carl Luksic said that the 52 missions he flew prior to that date were "milk runs" compared to his day of glory as the first American pilot to shoot down five enemy planes over Germany on one mission. It happened in the skies over Brunswick returning from a bomber escort mission.

Carl grew up in Joliet, Illinois and flying was his boyhood ambition but model planes were the closest he got to flying as a boy. He was working as an apprentice millwright when the Japanese hit Pearl Harbor and he enlisted in the Army Air Corps the next day. After months of hectic flight training he got his wings and commission on his 21st birthday at Luke Field, Arizona. He had been married only 17 days when he shipped out for Europe as a fighter pilot and was assigned to the 352nd FG flying P-51 Mustangs. He named his plane "Elly's Lucky Boy" after his wife, Eleanor.

Years later, in an interview, he said that his first 52 missions were "milk runs" compared to his grandest day, but his action started around Easter. On April 9 he shot down three Ju-88s, shared another kill the next day with Col. J. C. Meyer, then on April 20 destroyed four on the ground in strafing attacks, shot down one Me-109 and shared another with Lt. Bielok, becoming an ace even before his big day.

But on May 8, 1944, Lt. Luksic made history. It had been just another day for the Group. Returning home from their escort mission the 352nd was about half-way home when Luftwaffe 190s and 109s started attacking the bomber formations over Brunswick.

"When White Leader called in the bogey attack and started down, Red Flight, in which I was flying the No. 3 position, went down after them. I followed, picking my own target, but lost him in the clouds. My wingman, Lt. O'Nan spotted a 109 and I told him to go for it since I didn't see it and I followed him. In the chase I saw five or six 190s to my left and turned into them as O'Nan went for the 109.

"I put down some flaps and queued up on the 190 and started getting hits on his canopy and fuselage him at about 300 yards range. He pulled up, rolled over and bailed out.

"I had to break away from one shooting at me and found myself on the tail of another. Shooting short bursts I didn't see hits on him but the pilot spun out, crashed and exploded. I finally joined up with Frank Cutler and his wingman from our Group but pulled away from them when we started attracting heavy ground fire.

"Approaching was another plane I thought was a P-51 but up close I saw it wasn't. I was closing so fast I found myself just off his wing. The pilot looked at me, released his canopy and bailed out.

"I was joined by my wingman, O'Nan and Captain Clayton Davis and we started moving back toward the bombers when we saw twenty or more enemy aircraft in formation off to our left. We turned and went down on them in an attack and I found myself on the tail of a 109 and I got in a good burst on him with hits on his canopy, fuselage and wings. We were at about 800 feet and the plane caught on fire and went straight into the ground.

"Kicking rudder I moved onto the tail of a 190 and started getting good hits on his engine, wing and canopy. He went into a tight spiral and crashed. There were lots of enemy aircraft around and I had to take evasive action by ducking into the clouds and since I was all alone at this time I headed home and used up the rest of my ammo on some oil tankers near the Dutch coast and returned to base where we really celebrated.

"I later learned that I was the first American pilot to get credit for five victories in one mission, but certainly not the last. Then, on May 12th and 19th I got two more."

But on May 24, his World War II days ended. The Group was returning from another escort mission over Germany and Carl was not flying "Elly's Lucky Boy" that day. Another pilot had been shot down using Carl's plane the day before. As they headed for home Carl decided to strafe some planes he spotted on a German airfield below and went down low to destroy them. "It was a setup," he reported later. "They were dummy planes. It was a trap."

The enemy's ground fire tore up his wing and hit his gas tank, setting his plane on fire. He was almost too low to bail out but his parachute opened just seconds before he hit the ground. He suffered some minor burns and was pretty badly bruised up but the Germans captured him and made him their guest for the next 11 months. He was liberated by American ground forces in April, 1945 and came home to a hero's welcome in the city of Joliet.

Carl remained in the Air Force flying F-106 jets and later flew C-130 cargo planes in Vietnam, another three-war 352nd pilot. He retired in 1969 and moved to Panama City, Florida, where he lives today, but is not in good health. "The Air Force was mighty good to me," he said, "and those were grand old days back in 1944 with both good and bad times."

Carl's military awards include the Distinguished Service Cross as well as the Distinguished Flying Cross with one Oak Leaf Cluster, the Air Medal with three OLC's and the Purple Heart.

Capt. Donald S. Bryan

On a mission to Merseburg on November 2, 1944, Capt. Donald S. Bryan, flying his Mustang, "Little One," sighted at least fifty Me-109s attacking our bombers and immediately went for the enemy force to breakup their attack.

At about 300 yards the 328th Squadron ace started scoring hits as he flew through the enemy attackers. With one about to get on his tail, Don took evasive action and lost him. He found himself alone but riding behind several 109s which were flying in string. Using the new K-14 gyro gunsight for the first time he closed on a 109 and and lined up on set it afire. With so many targets he quickly picked another and shot it down with one burst from dead astern, flaming this one as well.

Sighting another 109 bobbing in and out of the clouds attacking a P-51, he dogged the 109 as it dodged in and around the clouds, shooting at him everytime he came into the clear. Finally, he got a good burst into the 109 and saw it dive into the ground.

He climbed back up to about 10,000 feet accompanied by the other P-51 and spotted two 109s flying just above the clouds several thousand feet below them. Diving down on the e/a his fire started raking the wings and fuselage of his adversary as he continued shooting bursts from about 500 to 300 yards. This one went into a violet spin, caught on fire and crashed. He then went for the other 109 but only had one gun firing and did little damage until he closed to within 100 yards and started firing again. This time he saw black smoke pouring out of the 109 as it caught fire and wentdown, thus scoring his fifth victory of the day.

Major George E. Preddy

Suffering from a big party and only an hour or so of sleep, George Preddy was awakened to learn he was to lead a mission which had not been anticipated because of the bad weather over the Continent on August 6, 1944. Noting his condition, Col. John C. Meyer offered to lead the mission but George insisted he was up to it. Shortly after takeoff, however, Preddy got sick in the cockpit but he did not let this deter him, and flew on.

At 27,000 feet escorting the lead boxes of B-17s, he spotted thirty-plus 109s southeast of Hamburg and took his flight into the stern of the enemy gaggle, selecting one of the rear of the formation and started firing at 300 yards. The Me-109 rolled over and nosed down in flames.

He and his wingman, Lt. Sheldon Heyer, continued their attack with George dispatching another 109 from six o'clock. This enemy plane began to burn and spin and the pilot bailed out as it fell to earth. Meanwhile, Heyer blasted another from the sky.

For some unknown reason the enemy gaggle continued to fly together, so Preddy and Heyer closed again on the rear of the formation with Preddy diving in on the tail of a third 109 with a solid burst at close range causing it to go down with parts flying off and smoking.

Four other Mustangs joined in the fray as George closed in on his fourth victim and shattered it with a withering burst that sent it spiraling down and blazing with fire. As the enemy formation began to lose altitude, Preddy nailed his fifth victim, sending it spiraling down in a spin to the ground.

One of the 109s pulled off to the left to try to attack Preddy, who was now alone, but Preddy hit him with a short burst that started him smoking and George pulled up into a tight left turn above his enemy. The Luftwaffe pilot tried to get enough deflection to open fire but as his speed fell off Preddy dropped back down on him. The German pilot jettisoned his canopy at about the time Preddy began to fire and George pulled up and over him and saw him bail out, capping his greatest day.

The author is indebted to Bill Hess and his book, ***Aces in a Day****, from which these accounts were edited and reprinted with his permission; and to Mrs. Carl Luksic for providing the material on Carl Luksic.*

Stalag Luft I

By Major General Luther H. Richmond as told to Marc Hamel

"The rest of us peered through cracks in the blackout shutters and saw the head-digger begin to emerge. About that time, the guard in the tower, 50 feet from the barracks began playing his searchlight up and down the fence line. As he reversed it, the beam swept across our man just as he stood up. The light wavered past him, then came back and fully caught him..."

On April 15th, 1944, Lt. Colonel Luther H. Richmond made a jarring switch from Squadron Leader to Prisoner of War.

Commanding the 486th Fighter Squadron, Luther led his men in a carefully orchestrated strafing attack of the German airfield at Vechta that day. Though the squadron destroyed eight of the German planes caught in the air, it was at the cost of their greatly respected leader who dueled with a flak position and had to bail out of his flaming Mustang PZ-R.

After interrogation and processing, Luther was delivered to Stalag-Luft I, the Luftwaffe POW camp located at Barth, on the Baltic seacoast of north Germany. As a relatively high-ranking officer in the camp, Colonel Richmond was afforded a view of the proceedings denied many of the other 9,000 inhabitants of the camp. Here he shares with us two narratives of his thirteen-month stay as a POW.

Excellent aerial view of Stalag Luft I from over the fire pond. Gustav Lundquist, Luther Richmond, Wally Starck and other 352nd officers were held in this camp. (Linda Lesniewski)

THE GREAT ESCAPE (ALMOST)

"In July of 1944, two of my roommates and I were asked to join in a tunnel digging project. It seems that the Germans had been seen digging up a seismic-type sensor at the fence-line outside of Barracks 6 of the West Compound and hauling it off for repair. Such sensors were buried all along the perimeter fences and wired into the German "Abwehr" or Security Office in the Vorlager. Thus the Germans were aware of tunnel attempts almost as soon as they were begun. They would then find them at will with their long steel probes. Their procedure was to let a tunnel proceed to near-completion, and then "find" it and collapse it with a fire hose and water. The missing sensor provided a blind spot in their security, and a tunnel in that area might just prove successful. For that reason, we three roommates decided to join in the dig and promptly thought up some appropriate nicknames for ourselves: "Mole" McKenzie, "Weasel" Wangeman, and "Ratsy" Richmond.

Luther's POW photo which he liberated from the Commandant's office when the Russians arrived. (Luther Richmond)

Guard tower and fence. (Linda Lesniewski)

"There were 27 of us on this project, and we decided to innovate and devise new tunneling techniques in the interest of speed and efficiency. We were, of course, anxious to complete the job before the Germans could get the sensor repaired and back in place. In that area the soil was all sand with the water table a little over six feet down. Therefore, our tunnels (unlike the fancy ones you have seen in the movies) were about 2 feet in diameter with no bracing or shoring of any kind. We crawled through them using our elbows and toes, and it wasn't hard at all to become claustrophobic; in fact, it was quite easy.

"Our security was outstanding. None of the 27 tunnelers lived in Barracks 6, and we wanted to keep our activities from all the other 'Kriegies' (prisoners) as well as the Germans. Our first task was to dig a 'well' under the end room closest to the perimeter fence. There were three or four dog runs under the barracks so the German police dogs could run through at night to see if

Sawing firewood, which would make nice bracing for tunneling attempts. (Linda Lesniewski)

anything suspicious was going on. Our plan was to use the closest run to gain access under this end room. With three or four of our own sentries lolling about, several men would enter the dog run. The sides were banked right up to the barracks floor, so about midway through the run, we would remove a portion of the banking and dig out a crawl space to get under the end room. The man remaining in the run would repair the banking and leave. Using this method, the well was completed in about a week, fitted with a wooden cover, and covered over with sand against any snooping Germans.

At the bottom of the well, a bellows-type air pump (made from a British barracks bag) was installed. This was used to pump air to the tunnel digger via a nested tin can pipeline buried along one side of the tunnel floor. One man was our head digger, and once the well was finished, he proceeded towards the fence line with his Kreigie lamp and small compass. Air lines were extended as needed with help from the men in the well. One man pumped air to the digger, while another pulled the pan of sand back when full. This was emptied into other containers for disposal later. The pan had a rope attached to each side so it could be pulled back and forth for filling and emptying.

"Everything went quite well until we reached the fence line, where we ran into a concrete wall under the fence. Attempts to penetrate it with our primitive homemade tools proved fruitless. So, we decided to make a right turn along the wall, sloping downward, to try to get under the wall (hopefully before hitting the water table). With the right angle now built into the tunnel, things became more complicated. Our simple rope and pan system would no longer work, so we improvised. Now we used a half-dozen smaller pans which nested compactly and were easy to handle. This required that men lie in the tunnel face down, head to foot, and pass the smaller pans back and forth underneath them. Using this method, we finally came to the bottom of the wall, and the water table as well. Another 90-degree turn left, a gradual slope upward, and we were ready to resume the original course towards the woods.

"I was in the middle of the pan brigade one day and began to feel very faint. The men behind me were also feeling faint so we all backed out of the tunnel to get some air and regroup. We decided we must abolish the pan brigade approach since there was obviously not enough air to go around. There was enough air for the digger and one or two men behind him, but very little for the rest of us. If a man passed out, those ahead of him would never get out, so we decided to use a rope and pan idea. To do this, we had to station a man at each of the turns (including the well) to facilitate getting the pans around the corners. This system worked out fine, and in a few weeks we had the tunnel about 75 feet outside the fence. At this point it was into the bushes and small trees just this side of the woods.

"We began sloping quite sharply upward until we estimated that we were only about

Loading coke. (Linda Lesniewski)

Stalag Luft I guards and prisoners at the gate. (Linda Lesniewski)

eight inches from the top. The tunnel end was then widened a good deal to accommodate the additional material that would have to be pulled in and packed when the tunnel end was opened. Work stopped and the air pump was removed from the well in order to give us some more room there. The air line noted above was left in place, to be used as a speaking tube so that we could pass signals back and forth during the escape. Still, no one in the barracks was aware that a tunnel had been dug from their quarters. Now we waited for the right weather for an escape attempt. Our maps and emergency food were ready, and we even had some chocolate D-Bars that had been hoarded from Red Cross parcels.

"We three roommates planned to stick together once outside. Our plan was to steal a fishing boat off the beach (about a mile away) and head for Sweden (about 100 miles north). There were other groups of two or three who also planned to stick together, so we drew lots to determine the sequence of the groups entering the tunnel. The head digger planned to go alone so we elected him to the number one position, since the tunnel still had to be opened up. Next came a two-some, and then it would be our turn. Finally the right day arrived, cool and breezy. We met and decided to try it that night. With the wind blowing, the fence and the guard towers would creak and groan a bit, and the trees would be rustling, hopefully drowning out any sounds we might make.

"At around 1830 hours (6:30 pm) we casually entered the barracks and were locked in for the night at around 1900 hours by the guard. The barracks was full so most of us stood around in the center hallway that ran the length of it. Our leader went to the barracks commander to advise him of our planned escape. It was getting dark and the blackout shutters were closed. We carefully pried up the floorboards in the end room to get to the tunnel and the head digger went down to open up the end. Since we felt it could be done quickly, we made no provision for air for the digger. We became increasingly concerned as the time wore on and he had not returned. Finally another man went down to check and found him weak and faint from exertion and lack of air. The tunnel was still much deeper than we had earlier thought. However, after a rest in the barracks, the head digger went back and soon opened up a hole large enough to get some air in but too small to get through. Finally, he got the tunnel end open and said he was going out.

"The first man in the twosome now started through while his partner in the well listened for his signal. The rest of us peered through cracks in the blackout shutters and saw the head digger begin to emerge. About that time, the guard in the tower some 50 feet from the barracks began playing his searchlight up and down the fence line. As he reversed it, the beam swept across our man just as he stood up. The light wavered past him, then came back and fully caught him. Our man held his hands high in the air as extra guards came running up shouting. We knew the game was up. All of the lights in the compound came on as we desperately tried to get rid of our escape equipment by hiding it or burning it in the stoves. We had forgotten about the twosome in the tunnel until we heard a pitiful muffled scream saying, 'They got me, they got me!' The man in the tunnel reached the far end, saw the lights, and began backing out as fast as he could. His partner lying in the well was listening to the speaking tube, not knowing what was happening. When his partner's G.I. shoes came up in his face, he grabbed the ankles to avoid being kicked in the face. His partner screamed in panic, thinking the Germans were holding him so they could collapse the tunnel and smother him.

"Soon the Lager officers and guards arrived with the ID card file for the barracks. After sorting out who belonged there, they marched the rest of us off with our hands clasped behind our heads.

It was 0230 hours, and had indeed been a long night! The next morning we were brought before the Commandant for sentencing. My sentence was 10 days in solitary confinement for being in the wrong barrack after lockup. I had to wait about a month or so to serve it, because the jail was full."

THE COMING OF THE RUSSIANS

"In April, 1945, we awakened every morning to a tremendous, though distant, artillery barrage to the east and southeast of Barth. We knew that Marshall Rokassofsy's First Ukrainian Army was attacking across Northern Germany and was getting closer by the day. Our spirits rose at the prospect of being liberated soon.

"The German guards were increasingly nervous, and a bit more friendly than they had been. One day the Camp Commandant (a pretty fair and decent officer) had our senior officers in for a conference. It was obvious that the Russians would be arriving soon and he wanted to march us (along with the German contingent) west to Allied lines. The Allies were some 100 plus miles west, and we felt that 9,000 unidentified males marching down the road would be a very attractive target to roaming Allied pilots (who might not suspect that they were our POWs). Our senior people objected to the plan on those grounds and urged the Commandant to leave with his people. We would stay, take over the camp, and await the Russians' arrival. He finally agreed, feeling he didn't have much choice at this point with Germany's demise so near. It was arranged that the Germans would leave that night after midnight. Prior to leaving, the German guards would unlock certain barracks and let our key people out to run the camp. I was the only one to be let out of Barracks 4, and was waiting at the door at 0100 hours when the guard opened it. Everyone else in the barracks was asleep.

"I felt sorry for the old guard. As he let me out, he said, 'Alles ist Weg. Alles ist weg; Der Krieg ist beendet.' I said a few words of condolence and hope for the future. He said, 'Now we must fight together against the Russians,' and I gave him some non-committal answer.

"By pre-arrangement, all of the chosen few met in the 'Vorlager' and were assigned offices and briefed on our duties. My first act was to find the keys to the jail, remembering that 'Russ' Spicer was in there under sentence of death for inciting a mutiny. Russ had no idea of what was happening, having been in solitary confinement and incommunicado for many months. He was still asleep when I unlocked his cell and shook him by the shoulder saying, 'Wake up Russ, the Germans are gone and the Russians are coming.' He roused a little, looked at his watch, and replied, 'Tell you what Richie, this is my six-month anniversary in here. Just leave the door unlocked and I'll see you in the morning." He turned over and went back to sleep!

"My job was Wing A-4 and I was responsible for food and transportation. Food was the most important issue since we had a population of 9,000 and virtually no food left.

"We had decided to send out a patrol to the east every night in an effort to link up with the

Roll call as seen from the Guard towers. (Linda Lesniewski)

Russian forces to make them aware that there were 9,000 Allied POW's not far to the west. On the third night, our patrol (Russian speaking prisoners driving a charcoal burning truck) encountered a Russian advance patrol also driving a charcoal burning truck. After a tricky period of identification the Russians were very friendly. The two patrols came west together. Upon spying a large German farmhouse, the Russians kicked the door in, roused the inhabitants, and demanded wine, which they received in some quantity. These patrols arrived at the camp about dawn and continued to celebrate for a while before the Russians headed back east to check in with their unit.

The Russian Commander arrives at Stalag Luft I (Linda Lesniewski)

"At about 1100 hours the first Russian troops showed up—five or six soldiers on horseback armed to the teeth. We greeted them and invited them into our senior officer's office (it had been the German Commandant's office). A picture of Hitler was still hanging on the wall and when the Russians saw it they went berserk and took turns smashing it with rifle butts. Next they came up to the main gate, behind which were prisoners jammed-up in an effort to see what was going on. Our people were, of course, manning the gate and keeping the POW's in the camp for their own safety. Outside, it had been a virtual no-man's-land for three days. The obviously drunk Russian leader said through our interpreter, "The Americans are my friends. Why are they behind the fence?" It was explained that this was for their own good, but he would have no part of that. He pulled his pistol and ordered our guard to open the gate, which he did. About 700 dumb Americans (with no identification) streamed out of the gate and disappeared into the countryside. We learned later that a number of them did not survive this adventure. Since the Germans had left behind only old people and kids, any able-bodied individuals appeared to be German soldiers.

"The Russian spearhead proper began arriving the next day. It consisted of horses and wagons (nose to tail), about half a dozen soldiers on each wagon, plus a couple of young girls. They were all riding atop a mound of loot they had collected including silver, linens, china, and other items. Most were drunk, heavily armed, and looked like Mongols. I was able to make contact with a Russian Major in the supply section of the Division that occupied the area. Through him, I was able to arrange a meeting at the Division Headquarters to submit our request for food and other supplies. The headquarters was about 80 miles south of the camp, about halfway to Berlin. British Major Tag Pritchard and I drove down in a commandeered Volkswagen. It was an eerie trip with no sign of life on the road. There were dead horses, burned-out trucks, and knocked-out tanks here and there. We finally found the Division Headquarters in a huge old farmhouse surrounded by a sea of mud (it had been raining quite a bit).

"A female Russian MP motioned us to park close to the house in a fairly dry spot. We were met

The Commandant of Stalag I, Barth. Richmond recalled him being a fair man. (Linda Lesniewski)

by my Russian Major friend, who escorted us into a side room of the house. It was furnished with a beautiful round cherry table and chairs, which Tag, myself, and about six Russians sat around for our conference. I presented our needs from my little notebook: so many tons of meat, potatoes, vegetables, and flour. We also asked for about 2,500 liters of benzine for our vehicles. The Russians approved of most everything except the benzine, but threw in a couple barrels of herring, which we accepted. The meeting had only taken about 20 minutes during which one Russian dressed as a civilian sat there listening while carving on the beautiful cherry tabletop with a large knife! I assumed he was one of the secret police. The Russian Colonel now invited us to lunch and we were happy to accept. He escorted us into the spacious dining room to a large table set with linens, candles, and flowers. There was a water glass at each place and a number of bottles on the table along with scanty helpings of food. A basket held little squares of white bread (I hadn't seen any of that since we got to Germany), and there was a little dish of cut up herring as well. The Colonel stood up and we all rose to join in his toast to Franklin Roosevelt, to the American/Russian alliance, and finally to our victory over the Nazis. I took a sip from my glass, and realized that, whatever it was, it was mighty strong stuff. The Russians all drained their glasses and those on each side of me insisted that I do the same. I finally did, with reluctance. Somebody filled the glasses again and I had enough presence of mind to propose a toast to Joseph Stalin, the great Red Army, and particularly to the unit that liberated us. Again, we went through the same drinking drill. I was beginning to see shooting stars when the Colonel said, "I am sorry we don't have any more German Schnapps, we only have Russian Schnapps," and the glasses were filled again. My memory fails here and I was not aware of anything else until someone pulled me from under the table by my feet. There I was... face down on the floor....

"Everything was in a hazy slow motion after they got me up. A Russian had me on each arm and they walked me out into the muddy farmyard while I "tossed" everything I had eaten. I was aware of baby pigs running around. Tag Pritchard had disappeared and the Russians assured me they would take me home. They finally put me in a Mercedes touring car with red leather upholstery and the top down. Someone crammed a greasy old helmet on my head and a Russian driver drove me back to the prison camp. The fresh air helped a bit. My roommates greeted me with unnecessarily crude remarks as I staggered in and collapsed on my bunk.

"Next morning, I was in my office bright and early (probably trying to atone for the previous day's debacle) when a huge Russian Captain showed up. He was about six feet five inches tall, weighed about 280 pounds, and had no fat on him whatsoever. He was there to inform me that he had the meat ready to deliver but needed some manpower assistance. I later learned that this meant he had 80 head of cattle staked out in several sites, and needed six or seven men to help drive them into camp. Our meat, when we got any, had always been delivered in the form of dressed carcasses on a wagon. I assumed that this delivery would follow the same procedure. I hastily adjusted my thinking and gave the captain seven men to be drovers. Meanwhile, I tapped our resources regarding cowboys who knew how to care for cattle, and ran down some sources of feed for the animals. There wasn't much natural grazing where we were, just salt marshes. The cattle arrived that night and we drove them through the front gate and right out the back gate onto the peninsula in the Baltic Sea. Nine thousand Americans represent a wealth of talent, and we had no trouble rounding up a dozen butchers. The cattle lasted about a week or ten days. In the next few days, the flour, potatoes, and vegetables arrived and we began to feel we would make it to our evacuation.

"It turned out that about the time we received the cattle, our senior officer was not up to date on what was going on foodwise. He spent much of his time partying with the Russian Divisional commander and his staff. Apparently he made a remark that we didn't have any meat, so the Russian general said, 'I will give you 1000 pigs.' An aide made a note, and the next morning the big captain was back to see me with a paper that said he was to deliver 1000 pigs to us. He said it would take him a few days to find this many pigs, but that he had a couple hundred already located. I told him that we didn't need the pigs because we now had the cows, but he repeated, 'The General has ordered me to get you the pigs.' I had a number of German stamps in the office as well as a typewriter, so I took his paper and typed: 'Received, 1000 pigs from Captain so-and-so,' stamped it several times, and signed it. This only partially alleviated his concern, so I suggested that maybe we could take just a few pigs. Finally it was agreed that I would provide two trucks, seven men, and would take as many pigs as they could get into two trucks.

"They left about dawn the following day, and it was well after dark when they returned. I was concerned about them and had stayed in my office awaiting their return. Finally, here they came with two trucks and about 40 huge pigs. They all trooped into my office (*smelling* like pigs) and obviously were very weary. I asked them what had happened, and got the following story.

"They entered a field containing a couple hundred pigs and backed a truck up into one corner. They built a wooden ramp with cleats leading up to the truck bed. They tried to corral

Prisoners and the bread wagon. (Linda Lesniewski)

one pig at a time into the corner and herd it up the ramp. Each pig eventually got away. Next they removed the ramp and tried corralling one pig into the corner so everyone could get their hands under it and lift it into the truck. Each pig wiggled his way out. They finally settled on herding a pig into the corner where all of the Americans would immobilize it. The huge Russian captain would then squat down, get both arms under the pig, and lift it into the truck. He was obviously tired and smelled like a pig.

"While these events were unfolding, we sent a team of around 200 men to the local airport, six miles on the other side of the town of Barth. These men had special training in booby traps, demolitions and the like. Their mission was to clear the airfield of hazards so that aircraft could land and evacuate us. I made several visits to the field to check the progress by riding a commandeered motorcycle. They seemed to be doing well and among other things had the control tower radio back in operation on our frequency. A German Ju-88 bomber had been abandoned on the line. It had an engine fire on the port side that pretty much ruined the wiring harness. Our ingenious guys fixed it and had the engine running. We asked the Russians if we could fly it, and they answered 'Nyet.'

"The Russian Divisional Commander showed up and told our senior officer that he wanted to inspect the airdrome. We all accompanied him and our boss to the airport. The first thing he noticed was the Ju-88 and asked about it.

We told him the status and he replied, 'Let's look at it.' We all trooped over and climbed up on the port wing while he got into the cockpit and sat in the pilot's seat. He asked if it had been cleared of booby traps, and we assured him that it had. He peered around the instrument panel and finally found a wire attached to the landing gear retract lever. It led down behind the panel and a fairing on the port side. He called for some pliers, disconnected the wire, and removed the fairing. He then pulled out a box of nitro-glycerin all fused up and ready to explode if the gear was retracted. He detached the box and we all climbed off the wing a few shades whiter than we had been. On the ground he said, 'I know this fuse. Let me show you. Come here and look.' We turned a bit paler still while he disconnected the fuse and tossed it about 100 feet. He checked his watch and said, 'Two minutes.' Nothing happened for a while and we began to breathe easier…until the fuse detonated. It sounded like a bomb blast. We began shaking again as he began pointing to the bushes over on the perimeter saying, 'What are those?'

"The Germans had strung fused 1000-kilo bombs in threes, placing them across the runway to make a barrier to landing aircraft. We had dragged the bombs over to the bushes on the perimeter and had not attempted to defuse them. He insisted that we all go look at these. He knew this fuse too, and called for more tools while we fidgeted. He unscrewed a fuse and detonated it. Then he pointed to a huge hangar and asked what was in there. We didn't know because all we wanted was to make the airport safe for evacuations. He wanted to see inside so we all approached the electric fence surrounding the building. He touched it with his finger, and said 'See, it is okay. Bring me some wire cutters.' After he cut an opening, we all filed through and went up to the pedestrian door, which was locked. He drew his pistol and fired about five shots into the lock assembly and opened the door. To our amazement, this hangar was filled with partially complete Me-262 aircraft—their first operational jet fighter. There were about 150 of them in various stages of completion along with beautiful machine tools. I am sure the Russians dismantled the whole thing and took it all back to Russia. The Russian general was quite amazing considering he was an Army general. We did not expect him to know that much about Luftwaffe items.

"We learned that the workers in the Me-262 factory had been foreigners kept in a nearby concentration camp. There were only a few of them left, all skin and bones. Despite our efforts to minister to and feed them, they all died in the next few days due to starvation. My wife and I revisited the site in 1985 and found the Russians and East Germans had erected a very impressive memorial to those who had died there. I participated in a wreath-laying ceremony at the memorial.

"Finally arrangements were completed for our evacuation. First came a fleet of C-47's to take out the sick and seriously wounded. I ran into a nurse from one of the aircraft who was looking for her husband (an officer from our Group who was shot down after I was). She had no knowledge of whether he had survived or not, and I sadly had to say he was not in our camp. Next morning the B-17's began arriving and we were ready for them. We had drawn up a very simple plan of evacuation. We would have groups of 35 POW's lined up in blocks at the airport ramp so that a B-17 could stop with the engines running while the people leapt aboard. The plane would then continue to taxi and take off. The aircraft landing interval was about three minutes and they only spent about ten minutes on the ground. Nobody had parachutes, but nobody cared. They stood in bomb bays, crouched in waist gun positions, and sat in every nook and cranny of the aircraft.

"The camp senior staff was assigned the last aircraft, so in late afternoon we 20 or so officers were all that were left. The Russian division commander and his staff were there to see us off. The last aircraft landed, taxied up, and our senior officer signaled the pilot to cut the engines. We all groaned inwardly. Then, in the midst of our farewells, the colonel asked the Russians if they would like to take a ride in a B-17. They accepted.

"The rest of us were on pins and needles until they returned in about 45 minutes with the aircraft still in good shape and ready to take us out. I will never forget the feeling of climbing aboard with my little wooden box that doubled as a suitcase. It contained not much of anything except a few mementos including a couple of pistols, a German bayonet and my POW Log.

"The preceding aircraft had all gone to Rheims, France. We landed in Paris—a ragged looking lot with our little wooden boxes, no money, and not even a wallet. We trooped into the assigned hotel where, to my amazement and great joy, my younger brother, William, arose from a couch in the lobby to greet me. I had not been aware that he was in Europe, but here he was. He was now a captain, and a fighter pilot, and was stationed in Stuttgart, Germany. He had come to Paris looking for me. After a delightful week in Paris, I wound up on a captured Italian ship in the last convoy of the war—bound for New York and home!"

Editor's note: Years later, at the 352nd reunion in Savannah, Ga., we toured the 8th Air Force Heritage Museum and General Richmond was particularly intrigued by the POW room at the museum. Shortly after his return home to Florida, he dug into his wartime memorabilia and sent me something that says more about how his fellow prisoners felt about him as a leader than any words ever could. When I unwrapped the small package from him, I found a pair of pilot's wings, but not the normal shiny silver wings we wore. These wings were dull gray, crudely molded wings. A note from Luther was attached. It read:

> "When I bailed out of my burning plane, my chute straps evidently ripped my wings off my jacket, leaving me wingless when I was captured. I am particularly proud of these wings because they were made for me by the men in my compound and presented to me as their commander. I want you to present these to the 8th Air Force Museum from me."

Using their own wings as a pattern, the POWs had melted some of the cooking pots furnished by the Germans and molded them to replace his original wings. This was, indeed, a stirring tribute to a fine officer and typical of how all those who served under his command felt about him.

General Luther Richmond died in 2003.

Luther Richmond poses with the Nilan Jones painted Vargas-style nude depicting Richmond's wife Jean. "Sweetie" was Richmond's pet name for his wife. (Sheldon Berlow)

Old "Tat"

by Karl Dittmer
Foreword by his son, Kurt

"Fighter pilots give each other tactical call signs for screw-ups, escapades or physical characteristics. His fellow pilots called him "TAT" (for Tired-Assed-Tiger) because he was a veteran of WWII and had a knack for telling war stories. As the first editor of Tactical Air Command's safety magazine, Karl drew "Old TAT" as a crusty cartoon tiger, bringing humor into dry safety lessons. Unlike some ego-driven types, he never claimed to be perfect—in his mistakes, he saw lessons that he could pass on to fellow fighter pilots to keep them from repeating the same errors. It made a difference in my life and it may make a difference in yours.

"Dad has always been my hero . . . he taught my brother and me how to fly—not your schoolhouse, private pilot's course of instruction, but seat-of-the-pants, here's-what-you-need-to-keep-alive lessons. It never dawned on me at the time why he was such a stickler about keeping 'your damn hand on the throttle' . . .The lessons were timeless – the instruction was priceless.

"Karl Dittmer isn't a hero just through an admiring son's eyes. He isn't a hero because he survived three wars. He's a hero because he was a warrior with an uncompromising integrity, a pursuit of excellence and a dedication to the United States Air Force and his family. Although he succeeded through the challenges of three wars, his toughest challenge was his long term battle with Alzheimers which took his life on November 20, 2003 after several years of loving care from his wife, Betty."

— Kurt Dittmer

Dittmer in cockpit of his "Dopey Okie." (Sheldon Berlow)

Ground crew pose by "Dopey Okie."

That Karl Dittmer was a "warrior" is a masterpiece of understatement. He flew a full tour of 30 missions as a B-17 pilot with the 549th Squadron of the 385th Bomb Group from their base at Great Ashfield, near Ipswich, one of the first to complete 30 missions after the Eighth Air Force changed the bomber crew tours from 25 to 30 missions. Because of the security clamp down with D-Day approaching, pilots were not allowed to go home after completing their tours and he was made training officer for a bunch of B-24 pilots who were transitioning into B-17s. Then, he was sent to Ireland to join a ferry squadron delivering aircraft to the various fighter and bomber bases in England.

On one of his flights near Bodney, he scrounged a Jeep and drove to Bodney where he managed to meet Colonel Jim Mayden, then Group Commander of the 352nd Fighter Group. Since he had long harbored a strong desire to fly the Mustang, he persuaded Colonel Mayden to request his transfer into the 352nd FG and soon found himself flying Mustangs in the 487th Squadron commanded by Lt. Col. John C. Meyer. This is Karl Dittmer's story of his combat days with the "Bluenosers."

FROM "FORTRESSES" TO "MUSTANGS"

"One of the officers warned me to report to Colonel Meyer in a military manner. That was somewhat different from the bomb group. We would just slouch into Major Benner's office and start talking. If we were a bit tired, which was most of the time, we'd sit on the edge of his desk and talk.

"Colonel Meyer, or 'J.C.,' as we called him when he wasn't around, was rather strict, so the warning got me off on the right foot. But strict or not, J.C. did well by me. I studied the Dash One, or Pilots' Handbook, for the P-51 and flew the link trainer. Then, Lt. Duerr Schuh, a sharp pilot in the flight I was assigned to, took me up in the Group's AT-6. Two or three days later Schuh took me out to an old P-51 B-model and checked to see if I knew where everything was, then helped me get it started, then jumped off the wing and I was on my own.

"We were stationed at Bodney, a short distance west of Norwich and our field was little more than a cow pasture about 3,500 feet long and almost as wide. It was centered on a low hill fringed with timber and surrounded by clusters of Nissen Huts and revetments floored with pierced steel planking (PSP). The revetments were flanked on three sides by walls built of sandbags, each containing two or three birds.

"The bird was easy to start. I taxied out to the edge of the field, shoved the throttle forward and found myself pressed against the seatback. Wow! In seconds we were in the air. I reached for the gear handle but before I could touch it the

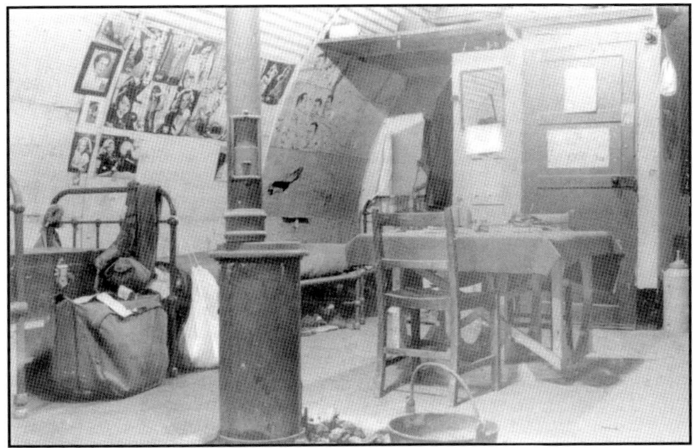

Nice artwork on the interior of the "cozy" Nissen occupied by Bateman, Landrum, Huston, Schuh, and Dittmer at Bodney. (Dittmer)

Dittmer runs up "Dopey Okie" at Y-29. Note the two tone cowling, reflecting a lack of will to repaint the nose art after the noses went to a darker blue. (Dittmer)

bird slowed down. The throttle had come back. I pushed it forward again, only to have it happen again. After a third unsuccessful try I clamped the stick between my knees, reached across with my right hand and twisted the throttle lock full to the right. After that I was able to get the gear up and started to really enjoy the flight. I found there was more difference between the Mustang and the Jug (P-47) than between the Jug and the B-17.

"The Mustang and I soon became friends… I made simulated attacks on a Lancaster, a B-24 and a Mosquito on my second flight in the bird. However, the Mosquito pilot beat me in a simulated dogfight. A few days later I made a simulated attack on a Jug and took him, but in all fairness, the Jug doesn't start to shine until it gets to high altitude.

"On mornings, we could hear the bombers droning around overhead before we had breakfast and briefings. Since I didn't have to get up for briefings yet, I sacked in even later. Occasionally, this resulted in me and my sack ending up inverted on the floor. I always retaliated like setting up a pail of water to dump on the next person to come though the door or booby-trapping the pair who slow-rolled me and my sack by short-sheeting their beds or unjointing the frame and using two loops of string to hold it in position until they sat down on it.

"Duerr Schuh, who sacked near me, received most of this. He was particularly unlucky. He would faithfully pre-flight his bed evening after evening, but the evening he failed to check it would invariably be the one I had chosen to booby-trap. But he got even. One afternoon he unhooked the springs at the pillow end of my cot and reattached each spring with string.

"For some reason I made a flying leap into my bunk, giving the other guys great satisfaction.

"Not all in the same hut were assigned to the same flight, but my flight leader was. He was Captain Clarence Johnson, who had flown a tour in P-38s and was on his second tour. He had eight kills to his credit and was rather aggressive. His voice was a little high pitched and the four of us played quite a bit of rummy evenings. On one such evening, the concrete slab floor shook followed by an explosion and whooshing sound. Johnson squeaked, 'Vee-two,' and calmly selected another card.

"Johnson liked to strafe, and on missions the squadron didn't encounter enemy aircraft, he would generally fly home at low altitude looking for trains, trucks or cars. Just before I started flying combat he was apparently coming back on the deck and the radio chatter indicated they were strafing their way home. They also heard him call to his flight to "watch out for those P-38s." Nobody from that flight returned home. The only air-to-air activity reported that day was from a flight of P-38s that claimed four Me-109s. Years later I read a report by the only survivor of that flight. They were attacking ground targets under a broken cloud deck at about 700 feet when they were attacked by eight to ten Fw-190s. The Krauts shot all four of them down. The number four man was the only survivor, but he was badly wounded and ended up getting excellent care in a small town German hospital.

"Life at Bodney was somewhat more relaxed than at Great Ashfield. On evenings when not writing letters or playing rummy, some of us practiced knife throwing with the bayonets off the carbines we were issued. Our target was a sheet of paper tacked to a wooden door and it wasn't long before we had that door pretty well chewed up. We stopped that practice one night just after someone came through the outer blackout door just as one of the bayonets pierced the inner door, shaking up our visitor when the end of the knife suddenly appeared in front of his face.

"Some of those who bunked in our Nissen Hut were Bill Whisner, Willie Miller, Allen Rigby, Jim Bateman, Dean Huston and Jack 'Moose' Landrum. Moose was somewhat over six feet tall and husky. Like many large men he was easy going, quiet and well liked. He took a hit on a mission that caused one gear to extend and his plane half-rolled and crashed.

"I logged 19 hours in September and flew my first combat mission on the 2nd of October in a P-51C. It had the old bird-cage canopy and I could touch my helmet on the top and both sides of the canopy without moving my head more than a few inches in any direction. The mission was a cramped five hours and I remember having trouble getting my rear end to unbend so I could walk after that one. The next day I flew the same bird on a five-hour, forty-minute haul with no problem.

"I was in another C-model on the fifth when Col. Meyer called out, "Bandits, twelve level!" I punched off my tanks only to have the good sounds abruptly quit. I hastily switched to an internal tank and the Merlin came back to life . . . but by then the fight was over.

"Two days later we launched on another mission, but I didn't get far. I was on Whisner's right wing on takeoff and just after liftoff my bird lurched hard left. I slammed in hard right stick and managed to keep from hitting him. Another pilot in our flight advised that I had lost my right drop tank. I took the bird out to the channel and dumped the remaining tank into the channel. We didn't worry about pollution in those days. On the ninth we had to climb through a heavy overcast. I was on Whisner's right wing again and we were the third flight in the squadron. Whiz had to fly formation off the second flight while the fourth flight leader flew off him. Sometime before I joined the 352nd one of the 487th pilots got too close on this sort of climb out and three pilots died in the pile-up.

"As we climbed the clouds became more dense and all I could see then was a dark smudge where Whiz's bird was. I sweated, but kept position on the smudge. Finally, we popped out of the clouds into beautiful sunshine and my prop was licking eagerly scant inches from his wingtip.

"On October 14th, I was on the mission in another B-model. During climbout all went normally until coolant started spewing all over my windscreen. I called Whiz and told him I had to abort, then headed back to Bodney. Once the bird was headed downhill the coolant quit spewing, so the rest of the flight was without incident. Next day Colonel Meyer told us there was a manual control switch that would override the automatic temperature control. He did not criticize me for aborting, but I got the message.

"We got into bandits again, I think, on the 15th. Just before someone spotted them, I was getting rather uncomfortable and starting corrective action. I had unzipped my flight suit, unbuttoned my OD trousers and was fumbling with my long johns when Colonel Meyer called out "Bandits!"

"This time I got onto an internal tank before shucking my tanks, and even located an Me-109 to shoot at. As I eased over to get in position another bluenoser came from above and behind and cut me out of the game. I will refrain from repeating the names I called him. However, although I never learned just who it was, I have some questions about his mother's sex life.

"After things settled down I pulled power back, reduced rpm from where it was redlined and concentrated on staying in formation during the turn toward home. After some minutes I realized I had not taken care of a certain chore and was in danger of flooding the cockpit. Fortunately, I fumbled my way out of the problem in time.

"I flew missions October 17th and 19th, but don't recall anything of significance. But Bateman made a rather interesting arrival on his return from a mission. He was flying number four position. We always landed in formation and this put him on the far left with the flight leader on the far right. The wind was coming from the right, al-

Ouch! P-51D HO-V "Nancy M 2nd" following the September 26th cross-wind landing attempt by Lt. Bateman. (Sheldon Berlow)

most a direct crosswind. The flight leader lined up with the center of the knoll on our airfield and this put everyone else on the slope of the knoll. The number two man had no trouble. Number three bounced slightly and the wind picked up the right wing a little but he was able to correct. Bateman also bounced slightly but being further down the slope, the wind really caught him. It picked up his right wing enough to cause his left wing tip to dig in, crunching about eighteen inches of said wing. At the same time, it lifted his bird enough to crunch the rudder.

"His bird was being pushed down hill and as it bounced off the left wing, then the tail, the impact rolled enough to catch the right wing. He was still sliding down the hill, curling the right wing from just beyond the outermost machine gun, bending it until the wing tip pointed straight up. Then it dug its nose into the sod and bounded back into a normal attitude. Before you could blink Bateman slammed open the canopy, ran the length of the left wing, and jumped to safety—still wearing his backpack chute and the seat-pack dinghy which was bouncing against his rear. "

I Get My Own Bird

"I don't recall anything about my missions on the 22nd and 24th, but on the 29th I flew my first flight in Dopey Okie. Other pilots who had been in the squadron longer than I had were still waiting for a bird of their own. Of this I was oblivious. The amount of our combat time counted toward such things as who would get their own bird. I was unaware of this and also unaware that Colonel Meyer had ruled that my combat time in B-17s counted, so I got a brand new P-51 D-15, the latest available at that time. After my new bird cleared our group maintenance section and was turned over to S/Sgt Carl Galloway, my crew chief, he helped talk the squadron artist into painting 'Dopey Okie' on the nose panel and my name on the canopy rail.

"On October 29th, I took her up for a test flight. Before takeoff Carl asked me to make a low pass with everything pushed to the wall. Some weeks before, Colonel Mason had said we could get tower clearance for a low pass if we ever wanted to buzz.

"The weather was cruddy. I climbed through the soup and checked all the items they wanted me to verify. Full throttle brought an inch over max allowable manifold pressure, rpm reached a full 3,000 and everything else was in the green. I called for a steer and turned to the heading the tower operator gave me as I started a brisk descent through the weather. At about 2,000 feet, I slowed my descent and broke out under the soup at about 500 feet. After flying long enough to have spotted the field I called for another steer. The heading they gave me was directly aft. I made my turn and saw our cow pasture and went on by, then called for a low pass. The tower approved so I pushed everything forward and came screaming across a wee bit above the brow of the hill, pushed down for an equally low run over our hardstand, pulled up and did a roll.

"I then came in for a landing. As I peeled up onto downwind a gruff voice came over my headset. 'Who's the pilot of the '51 turning on base?'

"'Transport 81, Sir!' (my call sign)

"'No, godammit, I want the pilot's name.'

"'Gulp, Lt. Dittmer, Sir!'

"The next morning Colonel Meyer stopped me in Squadron Ops. 'Dittmer, did you make a low pass over the field yesterday?'

"'Oh, Hell,' I thought. 'I'm in deep stuff now.' Out loud, I said, 'Yessir.'

"'Well, Colonel Mason was right behind you and he said a low roll in that kind of weather was a rather dumb thing to do.' And that was the end of that."

A Wee Bit of Trouble

"My first mission in Dopey Okie was on the last day in October and uneventful. But on the second of November I flew her again and it was darn near the last day of my life. Flying number four position off Sandy Moats' wing, things were uneventful until we were over Holland heading back home. Moats had outstanding eyesight and he spotted an Me-262 crossing below us headed back toward Krautland. So he made a hard turn to the right, then called Whisner to advise him that he had sighted the Kraut jet. By then, we were far enough away that Whiz had no chance of catching us, much less the German, so he said he would see us back home.

"The German pilot was no dummy. We could cut him off on the turns and get closer but as soon as we approached gun range he would roll out of his turn and pull away from us. We played with him for several minutes. I think Sandy was hoping we would run him out of fuel. The Kraut may have been attempting to do the same with us. We had everything forward, of course, and I had my full attention directed toward keeping with Moats and 'checking six' to make certain we weren't becoming the hunted instead of the hunters.

"Finally, Moats gave up and we headed back toward Holland and the channel. I then had time to check the fuel tanks. Whoo, boy! Not so good. I called Sandy and told him how much I had. By then we had the Zuider Zee in sight and were as close to England as we were to the nearest airfields on the Continent. So Sandy called Air/Sea Rescue and had them give us a heading to the nearest airfield and asked them to monitor us as we crossed the channel. I ran two tanks dry and had just switched to the last tank, which was almost empty, when ASR called and advised that our feet were dry. I pulled off power and started letting down and was soon in the clouds.

"After sweating a few moments, I broke clear and that's when my engine sputtered and quit. I looked to my right first, since one can't see much looking forward, and saw an airfield with paved runways a couple of miles or so just off my right wing. I also saw B-24s coming in for landing on the runway that was in line with my wing. I pulled up into the clouds to slow her down, since my airspeed was rather high, then shoved the gear handle down even though my speed was still too high for lowering the gear. I knew I had to get my bird slowed down so I wouldn't get too far from that blessed runway. It slowed quite fast since I had her in a climbing attitude, shoved right rudder and made a hammer-head turn. As I broke out of the clouds again in a dive I saw the runway now ahead and to my left. A brisk turn to the left lined her up with the runway. I was dropping fast and and my airspeed was around 200 mph. We came down in front of a B-24 quite close.

"Another was just about to set down and we were overtaking him. I selected flaps, then realized that we had slowed to 150, which was close to normal approach speed and decided to retract them. As I got the handle up I flared for landing and crossed the runway threshold. Then my bird was thrown violently up and to the right, then nosed down. I found myself looking at grass. I remember pumping the stick back against the left aft stops while pumping the left rudder against its stop, thinking, "Hell, I'm going to mess up my brand new bird." There was no way she could recover, but she did. She righted herself a fraction of an inch above the grass, skidded over to the center of the runway and touched down like a feather. Then my knees started to shake, so badly that when I approached a runway intersection I barely managed to get her to swing onto the other runway. I let her roll to a stop and was still shaking when a GI arrived with a Jeep. He waited while I filled out the Form 1 and managed to calm down enough to function again.

"I had someone call Bodney and, after the fuel truck pulled away, made my way out to my bird. When I tried to fire up, I found the battery dead. I had left everything on. So we had to roundup a battery cart before we could head back home.

"Galloway checked the bird over carefully but found no reason for the excessive fuel consumption. I flew her on another mission on the fourth but don't remember any details except that it was a five-hour haul. We flew again on the fifth and sixth. As I recall, Colonel Meyer had made me Assistant Operations Officer and I had been scheduling myself on every mission that looked interesting. Then on the eighth we went out and, guess what...I was again flying Sandy Moats' wing and again he spotted an Me-262. The details are almost the same as the last time except that I had more fuel and the Air/Sea Rescue controller vectored us to a Spitfire base where I landed with only a few gallons in my tank. We went up to the tower to get them to phone Bodney and the base commander met us there.

"'I say,' he said, 'what time did you chaps takeoff?'

"Sandy checked the inked numbers on the back of his hand and told him our takeoff time. It took him a moment to make the calculation in his head before he said, 'I say, that machine stays up a bloody long time, doesn't it?'

"I flew Moats' wing again on the tenth. But this time, Galloway and the lads at Group Maintenance had found the problem. At high power settings a fair amount of fuel is returned from the carburetor into the left main wing tank. The line was in place, but when they tried to blow air through it, after disconnecting it from the carburetor, they found it was blocked. Someone at the factory had failed to cut a hole through the fuel tank wall.

"On our return from this mission, which was ho hum, Whiz decided he'd buzz the channel. Although I swim pretty well, I get white-knuckled when flying over water and apparently Sandy feels the same. As he started down we moved into close formation—very close formation. Then one of us moved down while the other moved up, reversing our altitude in relation to him. By doing so we had his bird rocking back and forth. He fought to keep his wings level but managed to shake a gloved fist at each of us, switching hands to do so. When he stopped descending we moved out. Whenever he started down again, we repeated our act. That ended the low flying over the channel.

"On the 20th, we ran into sour weather on our return and Colonel Meyer took us into a B-17 base where we spent the night. After supper we went to the Officers' Club where a navigator was doing some nice work on the piano. Someone told us that he had played piano with Gene Krupa's band. Most of the locals were singing to his music and the songs were the usual ribald British Army numbers. As they were singing a particularly dirty number someone called out, 'Red Alert,' and in the two seconds it took for two nurses to walk through the door he had smoothly transitioned into 'Onward Christian Soldiers.'

BAD LUCK FOR ONE KRAUT

"On November 2nd we flew a fairly short mission and I recall nothing of interest. We had a five hour mission on the 25th, and again nothing unusual. I do remember the one we flew on the 27th. I was flying wing on Whisner and we ran into a fair-sized gaggle of Me-109s, attacking them from the rear. The one Whiz attacked rolled inverted and headed straight down. So did we. In a few moments I found my bird porpoising in the dive. I remember using both hands trying to recover from the dive. The next thing I recall is returning to consciousness with my right leg kicking or twitching, a spasm common to complete blackouts. My airspeed indicator was pegged on the dial and the large hand on the altimeter was making a complete revolution every two seconds or so. The small hand indicated we were climbing through 20,000 feet.

"Looking down to the left as I leveled my bird I saw an Me-109 headed toward a cloud. I started to follow him but decided it was foolish. Looking down to my right I saw a bluenosed '51 in a wild flat spin with one wing off. I later learned that it was Smitty, who was leading the element in our flight. I then noticed a couple of bluenosers attacking a truck on the ground and went down to join their strafing, coming in behind Whisner who was making his final pass. I failed to notice which direction he took when he pulled up. Anyhow, I put a burst into the truck and pulled out to the right and I couldn't find Whiz. So I started toward home only to find Miller joining me. Great! We were now a fighting unit. Headed toward home at about 3,000 feet I spotted something on a road but when making an attack discovered it was a manure cart drawn by a pair of oxen. I am ashamed to say that I gave it a burst and then flew on.

"Moments later we ran into the outskirts of Hamburg and I spotted a flak car in what appeared to be a railroad marshalling yard. A German soldier was sitting on the end of the car looking my way. Needless to say I had little love for flak gunners and decided he needed a well-deserved lesson.

"As I aimed my bird at him he swung to his feet and headed for the gun. My first burst kicked up splinters just behind his heels as did my next two bursts. Angry at myself I pulled up and started to make another pass. But Miller called and warned that I was drawing considerable ground fire.

"I headed north to go around the city and spotted a passenger train coming our way on tracks a mile or so to the right. As I started to turn for a pass I noticed an aircraft flying a few hundred feet above the tracks coming up behind the train. For once it was a welcome sight, the hole in the spinner.

"Indicating it was the business end of an Me-109, I bent my bird around and ended up about 1,200 feet behind him. My first burst hit him and he pulled into a hard left turn. I remember caging my K-14 gunsight, pulling my nose ahead of him to get the proper lead. I released the cage button and the pipper settled on his cockpit, and that's where my next burst hit. He continued turning but lost altitude and flew into a high tension line tower, knocking it down, then plowed into a wooded area, knocking down trees like a brick thrown into a patch of dry weeds. The rest of the flight was without incident.

"I scheduled myself for the morning mission but woke up with a nasty pain in my chest. The Flight Surgeon checked me out and drove me to the nearest U.S. hospital. After more thumping and listening and a chest X-ray they came up with a verdict. I had a collapsed lung. I spent the next month in the hospital.

"Shortly after I returned to Bodney, most of the Group went to Belgium leaving only a few of us behind. Being grounded I had no duties and didn't know why they weren't sending me home. However, orders came through promoting me to captain, thanks to Colonel Meyer. I had completed 20 missions with the Bluenosers.

"We had unusually bitter weather during January with several inches of snow. The temperature was so cold that our toilets froze. Not being busy I made a toboggan from a side panel of a belly tank crate and a piece of sheet steel and four of us gave it a go behind a weapons carrier. I only made one error. I didn't tie the tow rope to the vehicle. Instead, I looped it over the bumper for the lead rider to hold. Fortunately, I elected to ride on the tail end. Things went smoothly until the lead rider saw we were headed for a patch of bare pavement and let go of the rope. We came to a sudden stop and I ended up on top of everyone. Our lead rider had to get patched up at the infirmary."

Shortly thereafter, Karl was transferred back to the States, making the long, slow trip back across the angry Atlantic aboard the SS Argentina in convoy with several Liberty ships. He arrived back in New York City and then to Reno, Oklahoma.

While on R&R back home, he attended a youth group meeting at his church and discovered and started dating the girl he was to marry—Betty Booth. They were married after the war and he separated from the service—on Fathers' Day, 1946.

STILL A 'TIGER'

But that did not end his military career. He stayed active in the Air Force Reserve and in March, 1951 his unit was called up for service.

On Thanksgiving Day, he found himself on the way to Korea where he was assigned to the 18th Fighter-Bomber Group near Pusan on the southern tip of Korea. They had an advanced airstrip at Wonju near the 39th parallel where they slept in tents heated by coke stoves. But he said they did have a good latrine—a ten-holer over a metal trough.

When he reported, they told him they had plenty of pilots but were short on aircraft, and he had his choice of the three squadrons. He chose the 39th and again found himself flying Mustangs in combat.

A change of command in the 39th brought him an "it's a small world" surprise when he walked into the mess hall one evening. Two Captains were sitting with a bird Colonel who saw Karl and said, "Hey, Ditt, come here." It was their new C.O., Colonel William T. Halton, who, as a captain, had replaced Col. J. C. Meyer as 487th C.O. just before Karl left England.

The new C.O. was outlining an idea he had for a new attack procedure to make rail cuts with the Mustang. The other two Captains scrunched down in their seats without comment and Halton said, "What do you think, Ditt?"

Karl recalled a rocket pass he had made on a straight stretch of track similar to his plan and said, "Sir, if you make that kind of pass they'll bust your ass." Unfortunately, on one of Halton's first missions with the 18th, he did, and the gooks proved Dittmer was right. Halton was killed in action.

After 29 missions with the 18th he was given the opportunity to transfer into F-86 Sabre Jets and

Dittmer in F-86 "Little Butch" of the 4th FW. (Dittmer)

he joined the 4th Fighter Command under Col. Bud Mahurin, one of the top aces of the 56th FG during WWII at Kimpo. His transition to the Sabres was little more than most of us had when the 352nd switched from the P-47s to the P-51 in April, 1944. Karl had six hours in this completely new type aircraft before flying it on his first combat mission. He completed his tour in the Sabre with missions that included the destruction of two MiGs.

Karl Dittmer stayed in the service and found himself in another war—this time flying O-1 Birddogs as a Forward Air Controller (FAC) and flying another 150 missions in Viet Nam, frequently returning with a bird damaged by groundfire flying "low and slow" marking targets for the fighters.

Few pilots folded their wings with having flown combat in three wars for a total of some 300 combat missions. Karl Dittmer was certainly a "warrior" in the finest sense of the word and earned his "Tired Ass Tiger" moniker. His dedication above and beyond in the service of his country over Europe, Korea and Vietnam proved beyond a doubt that he continued to honor his days as a "Bluenoser." He was a certainly "Second to None."

And, his legacy continued. His two sons, Kurt and Karl, Jr., took up where their Dad left off—both serving in the United States Air Force as F-16 pilots.

(Upper Left) Korea-era photo of Dittmer working on the artwork on "Beep/Val." (Dittmer)
(Upper Right) Dittmer paints "Chuck and Randy." (Dittmer)
(Bottom) Dittmer with Crew Chief of "Betty Boots" in Korea. (Dittmer)

Future Thunderbird Downed
By Bill Pattillo

The loss of the Navy's Sullivan brothers early in WWII resulted in Congress passing a law prohibiting brothers serving in the same unit. However, the 352nd Fighter Group included the Pattillo brothers, Bill and Buck, who distinguished themselves further as pioneer members of the "Thunderbirds," the USAF's famous aerobatic team. When flying with the 352nd Fighter Group, they were assigned to different squadrons but flew on some of the same missions. Both would go on to achieve the rank of general.

April 16, 1945

"We were flying a bomber escort mission to Regensberg. After the bombers had completed their bomb run, our section of the 352nd Fighter Group broke off from the to look for 'targets of opportunity' on the return to home base.

"Captain Ray Littge was leading White flight which included Lieutenants Wittiklend, Creamer, and myself. The second flight, Red Flight, consisted of Lieutenants Waldron, Sykes, Padden and Cole. As we proceeded to let down from escort altitude, we came upon an airfield which had about 50 single engine and twin-engine aircraft dispersed around the perimeter and in grass revetments surrounding the field.

PR Pose—Lt. C.A. Patillo

"Captain Littge was directed to take both flights to knock out the flak positions firing at us as we circled the field. The Group leader had previously made a high speed run across the air field at minimum altitude to test for ground fire. I don't remember who was Group lead that day, either Col. Jim Mayden or Lt. Col. Bill Halton.

"We made about three passes each on the gun positions, either knocking them out or at least temporarily silencing them. After the flak busting, both flights then joined the other aircraft in strafing the parked air craft. Despite the flak and all the small arms fire, it was a field day! I remember the P-51's going in all directions and exploding and setting on fire many different types of aircraft parked around the field. The following is from my Encounter Report, submitted after I returned to the Group after being shot down by ground fire on this mission:

"I was flying White Four and made two passes at flak positions after which we then hit the deck to strafe the enemy aircraft on the airfield at Ganacker.

On my first two passes made on the northeast corner of the field, I got good hits on two dispersed Me-109s but couldn't get them to burn. I then made six more passes on the east side of the field getting six aircraft, three Fw-190's, 2 Me-109's, and 1 Ju-88, hitting one aircraft on each successive pass. I saw each of these planes either explode or burn fiercely.

"On my ninth pass on the southeast corner of the field, I attacked another Fw-190 and got hits but it would not burn either. On this pass I was hit in the engine and oil/coolant lines by enemy ground fire and was forced to crash land my burning plane in enemy territory. I was taken prisoner by the Germans.

I had started my last pass when the Group leader called for rejoin. Since I had already committed myself, I continued the pass at treetop level. As I pulled off the target in a left turn I noticed my engine was running rough so I checked all the gauges and was startled to see that the coolant gauge was pegged in the red and I knew I was in deep trouble. I stayed low and attempted to join the other aircraft which were now at about 3000 feet in a wide left turn. Then I noticed the blue paint on the nose of the airplane starting to bubble and smoke. My engine really started running rough then and I was losing power. When it started smoking badly, I saw flames coming from around the exhaust stacks and I knew I wouldn't be making it back to Bodney.

I had gone about 10 miles northwest of the airfield we strafed and I tried to gain some altitude so I could bail out. I called on the R/T that I was hit and would bail out, but, as confirmed later, no one heard the call. I jettisoned the canopy, unstrapped, and stood up to jump, but realized I was too low for a successful parachute jump. All this time I was over dense woods, and since I could not gain altitude, I knew I would have to crash land into the trees. My engine had now stopped completely. I guess the good Lord was with me at this instant because just as I was reaching stall speed

Bill Pattillo with his mount "Sweet and Lovely," HO-Y.

on top of the trees, a very small plowed field appeared. I lowered the nose and belly landed. With my wheels up, the aircraft didn't cartwheel and I braced myself the best I could with my left arm against the windscreen area as I flew the plane into the ground. I remember the aircraft was slowly turning in a circle and kicking up dirt all over the place. How quiet and peaceful it all seemed when the plane finally stopped and I jumped out of the burning plane on the run! As I looked up and saw the rest of the Group reforming it was about the loneliest feeling I ever had.

"I had gone about 100 yards from the aircraft when German soldiers came out of the woods from all directions with their guns aimed at me, I put up my arms and surrendered. They searched me, took my .45 pistol and hustled me off into the woods to a small command post. There an officer, (who had only one arm and spoke broken English started screaming at me. Not knowing what he was saying, I tried to tell him I wanted to be turned over to the Luftwaffe. We had been briefed by our intelligence officers that if we were shot down to stay clear of civilians, the Hitler Youth, and the German ground forces and, if captured, to make every effort to get with German Air Force people.

"After a couple hours they took me to a farm house nearby and locked me inside a small barn. About dark the door opened, and several German Air Force people appeared. They didn't hurt me, but they did push me around a bit, and put a pistol to my head as if to shoot me. I was then placed in an open vehicle with the gun still at my head. After about 30 minutes, we stopped at a German airfield which I thought was the same airfield we had been strafing. I later found out that this airfield was at Straubing.

"They took my boots and flight jacket and locked me in a very small room in the Luftwaffe stockade on the base. The room was about 5 x 8 ft. and the only furniture was a wooden bunk with straw pillows for a mattress. It had a very narrow window near the ceiling. The door was solid and heavy with a small circular peephole where I often saw an eye-ball looking at me. I was not provided any food that night. The next morning I was given a piece of black bread and a cup of weak ersatz coffee.

"Then they took me to a headquarters building a short distance away where I was interrogated by a middle-aged German officer who spoke very good English. He was polite and firm in his attitude and actions, but he did not attempt to intimidate me when I refused to give any info other than my name, rank and serial number. I guess the Germans knew the war was over and lost to them because none of them at this airfield, though certainly not friendly, mistreated me in any way. I guess the only real complaint I had at the time was not having sufficient and proper food to eat. I was fed a small portion of watery soup (which I swore was made from grass and weeds) along with a piece of hard bread once a day. It seemed I was always hungry but I don't believe the Germans themselves were eating much better at this stage of the war.

"While confined to this small cubicle in the stockade, I witnessed some good air shows and firepower demonstrations. About twice a day the Ninth Air Force P-51s would come in and really work the airfield over. I could stand on the wooden bunk and look out the small window and observe the dive-bombing and strafing. Of course there was a lot of noise and smoke, not only from the strafing fighters but also from the anti-aircraft and small arms firing back at the Americans.

"During these raids I was quite concerned that one of them would hit the building where I was incarcerated. All the German guards and other personnel would run into their bomb shelters or into the surrounding woods when these attacks occurred. I was concerned they would set the building on fire and there would be no one to unlock my cell to let me out. At late evening, just before dark, there were quiet jet operations conducted at the airfield. Apparently, they were conducting jet training in the Me-262s.

"I stayed at Straubing about five days. One morning an older guard—who spoke no English—brought me my flight jacket and motioned me out. I later learned my destination was to be Moosberg, the POW camp near Munich.

"This guard and I walked the back roads and slept in barns along the way. The only food I had was some hard bread and cheese. A couple of days later we met up with a long column of Russian POW's also heading toward Munich, and we joined with them. They had been POW's for quite a long time and were very good thieves and scroungers from the farms they passed, since they seemed to have a good supply of hard salami, eggs, cheese, and bread. They readily shared with me what they had.

"Despite some of this food being old, moldy and greasy, it tasted like a gourmet dinner. While walking on these back roads I experienced some episodes which I shall never forget. Along with the long column of POW's there were all sorts of civilians as well as German soldiers traveling the back roads, apparently wishing to be captured by the Americans rather than the Russians. A few were driving autos using wood-burning petrol generators. Some soldiers were riding motorcycles or bicycles, but most of the civilians were walking, all of them loaded down with what few possessions they could carry or push in a cart. A few of the German officers and their families were riding wagons designed to be pulled by horses but one of these was being pulled by humans!

"The men pulling the wagons were dressed in grey and black-striped clothes and were being beaten with a whip, just as if they were animals. We now know that these people in the black-striped clothes were displaced persons from the German concentration camps. I witnessed one incident where a Russian POW broke from the column to pick up something off the road (it could have been a cigarette butt) and this caused a German SS officer riding a motorcycle to run into him and crash off the motocycle. The German officer was furious and began shouting at the Russian. He then jerked a gun from one of the guards and shot the POW several times and kicked his body into the ditch alongside the road. He then got back on his motorcycle and sped away. It goes without saying that after witnessing this incident, I made it a point to stay very close to my guard.

We never made it to Moosberg. For some reason, I was dropped off at a "tent city" POW camp along the Isar Rriver at Landshut. This camp was established as a mixed Allied-enlisted POW camp since there were Russians, French, British, and a few American GIs there. I would guess there were about 1,500 people in this camp.

"We were liberated on 5 May by tanks of Patton's 3rd Army. And I went out to meet the lead tank to inquire as to how and where to get through the battle lines without becoming a casualty from the Germans as well as the Americans. I didn't intend to hang around that place and get all involved with the bureaucracy and red tape which I knew would follow when the appropriate American units arrived. I proceeded through the lines by hitchhiking with American GIs to Nuremberg where I hitched a ride on a C-47 which was transporting wounded Army personnel to England. After arriving near London, I borrowed several English pounds from the pilot of the C-47, caught a train, and made my way back to Bodney."

B-17s of 305th BG 365th BG at altitude on way to target.

Control Tower Log

Dick Press, one of our Bodney Control Tower officers, maintained a brief, handwritten personal logbook with information obtained from conversations with pilots after their return from missions. His log entries record losses sustained on 352nd missions. Here are selected notations from his notebook:

Control Tower Officer Dick Press

30th November 1943
Lts. Brown, Kramer and Babbitt were bounced on way in to pick up "big friends." When Kramer bailed out he said, "So long boys, see you Christmas." All from 486th Squadron.

22nd December 1943
Lt. Grow shot down in combat by Fw-190s over Holland. Believed killed.

30th December 1943 (41st Mission)
Capts. Button, Dilling and McIntyre all went down in France either due to flak or e/a. All were flight leaders in the 487th.

28th January 1944 (48th mission)
Lt. McPherson was shot down over France by Fw-190 while engaged in combat.
Capt. Preddy, 487th was badly hit by flak—bailed out over the Channel and picked up by Air-Sea-Rescue patrol.

28th January 1944
Major Hyland flying local in a B-24. It caught fire, he bailed out but forgot to hook his leg harness. Impact of chute opening ripped his arms off and he shot out into space to his death. Also with him were two EMs (from his Medical staff) killed in the same crash. He was our Flight Surgeon.

3 February 1944 (51st mission)
Lt. Cornick, 328th, shot down by enemy a/c over France. F/O Sweeney, 487th ran out of gas and believed to have crashed in the Channel.

8th February 1944 (55th Mission)
Lts. Walker, Meagher and Nussman all bounced while escorting crippled B-17. They didn't have a Chinaman's chance. Meagher went down in flames—other two shot down, fate unknown. Lts. Fieg and Phillips who were with above boys managed to bring their riddled ships home. All 328th.

8th March 1944 (69th mission)
Lts. Bond, Miles, McKibben and Miklajcyk were starting off on mission. As they went up through the heavy overcast, one of the planes got caught in another's propwash. All four collided in mid-air. Lt. Bond's a/c exploded and he was killed. Lt. Miles crash landed. The other two bailed out.

11th March 1944 (72nd mission)
Entire Group went out today on the deck. Lt. Riley (487th) hit by ground fire and blew up over France. Lt. Schwenke (328th) b/o over channel at 900 feet, hit water and did not come up. 12 ships returned with battle damage.

8th April 1944 (88th mission)
Entire Group shot up airfields, trains, flak towers, etc. in Germany. Capt. Meroney, only double ace in Group to date, failed to return—believed to have been shot down by small arms fire. (Note: Two squadrons flying P51s).

9th April 1944 (89th mission)
Lt. Yochim shot down over Germany at 300 feet; was seen to bailout. I came over with him and we sailed, just a year ago today.
Lt. Williamson, new pilot in 487th flying relay, started up through clouds over Warwick—plane fell in three (3) pieces. Cause undetermined.

11th April 1944 (91st & 92nd missions)
Lt. A. L. Marshall was apparently hit by flak while returning from mission to Germany. B/O over Channel.

14th April 1944
Lt. Col. Richmond's plane was hit in wing tank while strafing aerodrome. He crashed from 100 feet and was killed in action. *(Note: As shown by story "Stalag Luft I," this Entry was in error; Colonel Richmond bailed out and became a POW—author)*

Lt. Ross is MIA on the same deal—no one saw him go down.

Lt. Fremont Miller B/O over the Channel. Fog set in and it was days before he was found. He was OK but his feet were frozen.

22nd April 1944 (98th mission)
Lt. Long, 328th, developed engine trouble over Holland and had to bailout.

24th April 1944 (100th mission)
Capt. McKean, while strafing, hit by ground fire. Tried to B/O, couldn't and plane crashed and blew up. He was killed somewhere in France. 486th destroyed 29 a/c with 25 probables all on the ground using 9 a/c.

28th April 1944 (106th mission)
Lt. Laing hit by flak over France. B/O.

8th May 1944 (113th mission)
Lt. Kopecky (328th) missing while flying escort to Brunswick, Germany. Group shot down 29 a/c in air today. Luksic 5, Davis 4, Thornell 3, Meyer 3, Cutler 3. (others not listed)

8th May 1944 (non-operational flight)
Lt. Worcester killed while practicing dive-bombing at Mersea Flats

9th May 1944 (non-operational mission)
Capt McMahan killed while practicing dive-bombing at Mersea Flats. Lt. Conard's plane crashed, too, but he bailed out.

12th May 1944 (115th mission)
Lt. Howard MIA while flying escort over Germany.

13th May 1944 (116th mission)
Capt. Cutler MIA while on escort deep in Germany.

19th May 1944 (118th mission)
Lt. Carlone MIA over Germany. No one saw him go down.

21st May 1944 (120th mission)
Lt. Col Clark hit by flak while strafing in Germany. Bellied his plane in and is presumed to have been made a POW.

24th May 1944 (124th mission)
Lt. Luksic B/O and Lt. Hannon crash-landed in Germany.

28th May 1944 (127th mission)
Capt. Anderson attacked by 3 Me-109s—believed to have bailed out over Germany.

5th June 1944 (134th mission)
Lt. Furr hit by flak over France. B/O

6th June 1944 Invasion Day (135th mission)
Lt. Frascotti on pre-dawn T/O, with flight of 5 a/c, flew into new control tower under construction. Plane blew up instantly killing pilot. Lt. R.C. O'Nan hit by flak and forced to crash-land behind enemy lines. Lt. Butler hit by flak and B/O over enemy lines. Today the Group flew 116 sorties from 0330 to 2315 hours. Met no opposition and really had a field day.

7th June 1944 (136th mission)
Lt. Mulkey crashed shortly after T/O carrying bombs—plane completely destroyed—pilot killed. Major Gignac hit by flak over enemy territory and plane exploded at about 2,000 feet Lt. Garney spun in over England on return and was killed. Lt. R.L. Hall MIA over beachhead.

8th June 1944 (137th mission)
Lt. Fahrenwald hit by flak B/O behind enemy lines. Lt. Miller lost in overcast over France MIA. Lt. Campbell crash-landed in France; killed in action while trying to shoot up 6 German tanks.

9th June 1944 (138th mission)
Lt. Markel crashed into some trees while strafing behind enemy lines. Lt. Phillips B/O near LeHarve after hitting ground with ship while ground strafing. Lt. Bundy MIA. No one saw him go down.

19th June 1944 (non-operational mission)
Lt. Daugherty while on training flight over field flew into Lt. McIntire. McIntire B/O. Daugherty was killed.

20 June 1944 (149th mission)
Lt. Raub collided with an Me 109 over Germany and was killed.

25 June 1944 (153rd mission)
Lt. Drisko missing from mission over France.

27th June 1944 (155th mission)
Lt. Berkshire bounced by 3 Me 109s over France. Believed KIA.

28th June 1944 (156th mission)
Lt. Gielok shot down over France. MIA.

29th June 1944
(Russia to Italy shuttle mission)
Lt. Howell went down to engage some 190s and is MIA.

7th July 1944 (Shuttle mission)
Major Andrews hit by flak over Budapest. Crash-landed and managed to get away from plane.

17 July 1944 (170th mission)
Lt. Conard MIA over Germany. Engine cut out on way home.

18 July 1944 (171st mission)
Both Lts. Burr and Galiga last seen entering a cloud somewhere over Germany. Group got 21 victories on this mission.

July 1944 (172nd mission)
Lt. Wilson believed to have gone down in the Channel.

21st July 1944
Lt. Zellner forced to B/O over Ghent, Belgium. He only had two hours to go to finish his tour. Lt.Godfrey spun in somewhere over Belgium. Believed KIA.

5th August 1944 (184th mission)
Lt. French and Wilcox both MIA over Germany.

6th August 1944 (185th mission)
Lt. Gehrke, while in a dive over 500 mph with guns firing. Over Berlin one wing tore off. KIA. (On this day Major Preddy got six 109s).

7th August 1944 (Training flight)
Capt. Quinn augured in cause unknown. KIA.

12th August 1944 (192nd mission)
Lt. Graham, while strafing in France, hit by ground fire and augured in, but got out of the plane OK.

13th August 1944 (193rd mission)
Lt. Allain MIA from mission. No report on him. Lt. Thurman forced to B/O over France. Last seen going in overcast at 500 feet, chute unopened. Lt. Harnes crash-landed in France, got out of it OK. All were on D/B and strafing mission.

18th August 1944
Capt. Davis was heard to say he was bailing out over enemy territory, but not seen. Lt. Reese—crash-landing in enemy occupied territory.

28 August 1944 (202nd mission)
Lt. Fuenstueck missing in action over Berlin. Lt. Ferris B/O in channel, spotted, but later lost and not found.

10 September 1944 (210th mission)
Capt. Donahue hit by flak, forced to B/O over Germany.

11th September 1944 (211th mission)
Lt. Rayborn spun out over the continent. MIA Lt. Combs also spun in over Continent.

12th September 1944 (212th mission)
Lt. Broadwater bounced and shot down over Germany.

23rd September 1944 (219th mission)
Major Blanchard killed while strafing a flak tower.

25th September 1944 (220th mission)
Lt. E. J. Evans engine cutout over Germany. Had to B/O.

27th September 1944 (222nd mission)
Capt Johnson, Lts Waters, Ayers and Clark, whole flight, just disappeared, fate unknown.

5th October 1944 (227th mission)
Capt. Coleman engine cutout over Holland. Had to B/O.

24th October 1944 (236th mission)
Lt. Landrum KIA while strafing in Ruhr district. Hit by ground fire and blew up. Lt. Kyle hit by ground fire had to B/O over target in Germany.

2nd November 1944 (246th mission)
Capt. Miklajcyk and Lt. Clark B/O over Germany. Group went wild today. New high for ETO with 38-1-4. 328th Sqdn high with 28-0-4.

9th November 1944
Lt. Rogers B/O over France; b-out; hasn't been accounted for

27th November 1944
Capt. Starck shot down over Germany after getting two himself. B/O. Capt. Smith's wing came off while chasing a Jerry. Lt. Hudson believed to have been hit by ground fire. Group got 11 victories today.

English-built "Old Tower" watchtower at Bodney airfield's west edge, seen from under the bluenosed Norseman.

THE AIR FORCE'S FIRST JET AEROBATIC TEAM?

by Tom Ivie

Over the years, numerous articles have been written about the USAF's four-plane or larger jet aerobatic teams and various opinions have been expressed as to which of these should be credited as daddy of them all. Some sources suggest that the Aerojets were the first. Not so, said Lawrence E. "Mac" McCarthy, who flew as a member of the 4th Fighter Group team.

McCarthy's records indicate that the 4th Fighter Interceptor Wing's acrobatic team came into existence during the summer of 1948 with World War II ace Captain Vermont Garrison as the team leader. At this time the 4th was based at Andrews AFB, Maryland, and was equipped with the Lockheed F-80 Shooting Star. The team was composed of three members of the 334th FS, Captain Garrison, Lt. Erwin A. Hess, and Lt. Beriger A. Anderson and Lt. Lawrence E. McCarthy of the 336th FS.

When he began his association with the 4th FIW's acrobatic team in 1948, Mac McCarthy was a career officer in the USAF and a very experienced fighter pilot. During WWII he flew both the Republic P-47 Thunderbolt and the North American P-51 Mustang in combat as a member of the famed 352nd Fighter Group during which time he was credited with 2.5 aerial victories while flying 97 missions over Europe. Mac was one of the original pilots assigned to the 328th Fighter Squadron, and gained his combat experience during some of the most intense early air battles of the war.

Mac McCarthy and his crew chief with "Pattie III." (Mac McCarthy)

Lawrence E. McCarthy was born and raised in Detroit, Michigan and graduated from the University of Detroit High School in 1940. He completed one year at the University of Detroit before entering the Army Air Force in early 1942, winning wings and commission as a 2nd Lieutenant on 30 October 1942. From there he was assigned to the 328th Fighter Squadron, 352nd Fighter Group, which was being activated at Trumbell Field, Connecticut. The next six or seven months were quite eventful for the new lieutenant; he became very proficient as a P-47 pilot, married his sweetheart Pattie Phelps and was promoted to 1st Lieutenant.

By June 1943, the 352nd was considered combat ready and shipped out to England, beginning combat operations from its base at Bodney. The Group's first mission was flown on 9 September 1943, and Mac took part in that historic event. A little over five months later, on 20 February 1944, he flew the most successful mission of his long combat tour. On that date, the 352nd FG's mission was to provide withdrawal support as the bombers lumbered over Belgium. No sooner had the 352nd FG made rendezvous with the Forts than the battle began.

The enemy fighters struck just as the Group was positioning itself around the bomber formation and the ensuing battle lasted for about 30 minutes. While the 486th FS engaged a portion of the E/A in the rendezvous area, the 328th FS fought a running battle across the entire withdrawal route, downing seven enemy fighters and damaging several others. Lt. McCarthy showed the way with a devastating assault against a formation of Bf-110s, during which he destroyed two of the twin-engined fighters and damaged two others. For this action, Lt. McCarthy was awarded the Silver Star. His final aerial victory of the war took place on 1 June 1944, when he shared a Bf-109 with Lt. Jack Thornell.

Upon completion of his combat tour in August 1944, Lt. McCarthy returned to the U.S. After a well-earned 30-day leave he trained as a P-47 instructor at Providence, Rhode Island and was sent to Seymour-Johnson AFB, North Carolina in November 1944 as an instructor. Carolina, where he served as a P-47 instructor pilot until early 1946. At that time he was reassigned to Selfridge Field, Michigan, and flew Mustangs until being sent to Chanute Field for a Maintenance Officers school.

The transfer turned out to be of short duration though, as Col. Ernest Beverly, CO of the 4th Fighter Wing, called him back to Selfridge to prepare for a move to Andrews AFB, Maryland, assigned to the 336th Fighter Squadron, 4th FIW.

In March 1947 the brightly marked F-80s of the 4th began filling the skies over Maryland and Virginia. During the next 15 months McCarthy's flying consisted primarily of standard Wing flight operations. In July 1948, however, things began to change, and Mac found himself flying as wingman to Capt. H.M. "Mack" Lane in a two-man acrobatic team. His records indicate that he flew

Original team (left to right) Sandy Hesse, Vermont Garrisin, LE McCarthy, Beriger Anderson.

his first practice missions on 13 July 1948, and that three flights were made that day. Three days later, these two pilots flew a practice flight, and then flew a practice demonstration over Andrews in preparation for their first scheduled air show, scheduled for 17 July in Chattanooga.

The show thoroughly delighted those in attendance and provided Mac a "needed stepping stone to bigger and better things." Upon his return to Andrews AFB, Lt. McCarthy learned that he was to become the fourth member of a new four-plane team under the leadership of Captain Vermont Garrison.

The team began practicing in earnest and, by 10 August 1948 was ready to fly its first official performance. This premier show was viewed by a military audience. Shortly afterwards, the team was notified that it would perform at the National Air Races to be held in Cleveland, Ohio, during September. This would be the first opportunity for the 4th's team to put on a show for the American public and the pilots worked hard during late August to ensure that the aerial maneuvers were perfected. By today's standards, the time allotted to the 4th's acrobatic team was miserly—just five minutes! But, in 1948 when the public's attention was generally directed toward buying a new home and other more domestic issues, the pilots considered it to be a real opportunity and were determined to make the most of it.

Mac, who performed as the solo pilot, describes what the team prepared for the audience at Cleveland and for other air shows which were to follow. "We started with a loop. At the completion of the loop I would break out of the formation and, while the three-ship 'Vie' team was getting in position for the next maneuver, I would fly low and upside down past the stands. The Vie would become a three-ship trail formation and do a trail roll. While they were turning around again. I would do a four-point roll after which they would do another roll. I would then rejoin the formation during the turn and we would do a four-ship roll in a diamond formation, sometimes followed by an inverted Cuban-Eight. The inverted Cuban-Eight was our most hairy maneuver, and to the best of my knowledge, no other team has performed it!

"The inverted Cuban-Eight maneuver was flown in this manner: The team passed in front of the stands at high speed and started a 60-degree climb and, at the same time, was offset about 30 degrees to the right. At about 4500 feet, Garrison would half roll the team to the left, climbing all the while, and then start the split-S recovery. The half roll would put us back on track and we would sail past the stands at maximum speed and do the same thing but in the opposite direction, thus completing the inverted Cuban-Eight...As you can imagine, the hairy part was the split-S recovery. Your eyes arc heavy from the G-forces, you're only 18 inches apart, and the ground is coming up at an alarming rate.

"The show was usually completed with a four-ship loop. At the bottom of the loop I would throttle back a little to clear Garrison's tail, and then I'd square turn the F-80 to a vertical climbing roll as high as I could. In the meantime, Garrison would land the team in V-formation. At that time most runways weren't wide enough to land four planes in formation.

"Incidentally, the square turns got me in hot water with Ben Preston, my squadron commander. He found that I pulled so many Gs on the breakaway from the team that the G-meter needle would bend on the limit peg. I would also pull fillets on the bottom of the aircraft right through the rivet heads."

The team performed its maneuvers in Cleveland in September and proved to be a very popular part of the 1948 National Air Races. Their work drew high praise from Air Force brass and the team earned numerous letters of commendation. One of them was from the Air Force Chief of Staff. General Hoyt S. Vandenburg to the commanding officer of the 4th FIW stating in part:

"I would like to express my appreciation to you and the members of your command for the outstanding exhibition of acrobatic flying of jet propelled aircraft in connection with the Air Force's participation in the Cleveland National Air Races 4-6 September 1948. The impressive maneuvers which you performed received favorable mention from the press and radio.

Mac McCarthy posing by his P-80 at Andrews AFB (Mac McCarthy)

"The personnel in your organization are to be commended for their unselfish devotion to duty, the spirit of friendly cooperation they exhibited, and the industry and enthusiasm which characterized their performance at the races. Their actions in support of this extensive project reflect credit upon themselves and the United States Air Force."

From mid-to-late September, the 4th's jet aerobatic team flew several more air shows, the highlight of which were those flown over Andrews and Boiling Fields on 18 September. McCarthy recalls those shows, "The flights were for, I think, Air Force Day. As I remember, 1948 was the last time the services had individual days; after that we celebrated 'Armed Forces Day.' That morning we put on a show over Andrews, our home base, but the big show was at Bolling AFB. About 100,000 people were there and we had a special ending cooked up for them. After we finished our part of the show I landed at Bolling and the rest of the team went back to Andrews.

"After I landed, Captain Ron Shirlaw and his 334th Engineering crew hooked up two JATO bottles to my F-80, one on each side. The plan was for me to start my takeoff roll and reach takeoff speed just as the Air Force Band reached the 'Off we go into the wild blue yonder' part of the Air Force song. The timing was great, but just as I yanked the aircraft into the air and fired the JATO bottles, Murphy's law made one of its frequent appearances. Only one JATO bottle fired and there was no way to control the beautiful slow roll I did at about a 75-degree climbing attitude. When the bottle quit, I quickly nosed over to avoid a stall. The guys on the ground said it was spectacular. Spectacular or not, I'm glad it never happened again!"

In October 1948, the 336th Fighter Squadron was sent to Alaska to test the F-80 under cold weather conditions. During the Alaska trip, McCarthy's aerobatic flying took a back seat to the mission although he did do a solo performance for the personnel at Ladd AFB on 15 October. The high light of the stay in Alaska, for Mac, was the opportunity to meet Colonel Charles Lindbergh who was there to observe the gunnery missions. As squadron gunnery officer, Lt. McCarthy got the honor of checking Lindbergh out in the F-80 and flying several missions with him. Regarding Lindbergh, Mac noted that "he was one fine gentleman and a super pilot."

During the early months of 1949, Mac was again deeply involved with the 4th's aerobatic team and in March the team performed for the Armed Forces Subcommittee of Congress, the full Congress, and for President Harry S. Truman. There was more than just entertainment value in these flights. Appropriations were hard to come by in these early post-war years, and the Air Force utilized every opportunity to attract the lawmakers' attention.

In June 1949, the 4th FIW moved to Langley Field, Virginia. This move was followed by the an announcement that the Wing would soon be converting to the North American F-86A Sabre. Early in the month, the newly- promoted Captain McCarthy traveled to March Field to check out in the Sabre and then he and Don Stuck brought back the first two F-86s to Langley.

During the June-July period, the aerobatic team continued to fly its shows in the F-80, but everyone knew that the old Shooting Star's days were numbered. The F-80 partially surrendered to the new swept-wing F-86 on 14 and 15 July when Mac used the Sabre for his solo acrobatic demonstration. By the end of August the remainder of the team was checked out in F-86s. In September 1949, the team was invited back to the Cleveland Air Races, but since Vermont Garrison was unable to fly at that time, only a two-man team made the journey.

It was composed of Captain McCarthy and Lt. J.O. Roberts, a recent addition to the team. The two-man team put their Sabres through a number of maneuvers and delighted the large crowd gathered for the races.

By late 1949 or early 1950, the 4th's acrobatic team was renamed as the "Silver Sabres" and was now a five-plane team. The pilots were Captain Vermont Garrison, Captain Lawrence McCarthy, Lt. Colonel Benjamin O. Preston, Lt. J.O. Roberts, and Lt. W.B. Badger, with McCarthy flying the solo routine. The team performed a twelve-minute routine consisting of Cuban-Eights, barrel-rolls, inverted Cuban-Eights, four-point rolls, and Immelmanns. The show ended with McCarthy diving his Sabre toward the field and then pulling straight up in a climb during which he completed eight vertical rolls.

McCarthy's records indicate that the Silver Sabres performed their last show on 23 June 1950. The Korean War started two days later, and the Air Force immediately disbanded all its aerobatic teams until the war ended.

Following his 4th FIW, Mac was sent to France as an exchange pilot and spent three years flying F-84Gs with the French Air Force. After returning to the United States, he was assigned to Fort Monmouth, New Jersey, to teach Army personnel about the mission of the USAF. McCarthy was reassigned to Ent AFB, Colorado. in 1959, and remained there until retiring as a Lt. Colonel in 1965.

Mac's military awards include the Silver Star, the Distinguished Flying Cross with two Oak Leaf Clusters, and the Air Medal with five OLCs.

EDITOR'S NOTE: Ironically, Mac's younger brother, Lt. Francis McCarthy, who was assigned to the 328th Squadron of the 352nd FG after Mac finished his combat tour and returned to the States, was killed in action on the 30th of March, 1945, only a few weeks before the air war over Europe ended.

The author would like to thank the following individuals for their assistance: Lt. Colonel Lawrence E. McCarthy. Robert H. Powell, Jr., David Menard, and J. Griffin Murphey, III.

Lockheed F-80 Shooting Star

Pure Thunder

by Michael Fox

This is the story of the first USAF acrobatic demonstration team, predecessor to today's famous "Thunderbirds." Two of the pilots in this pioneer team first distinguished themselves flying combat with the 352nd Fighter Group in World War II. These were the Pattillo Brothers, Charles C. Pattillo and Cuthbert A. Pattillo, known to the "Bluenosers" as "Buck" and "Bill," both of whom remained in the Air Force and retired as USAF Generals.

Author's note: Any serious aviation history enthusiast should be able to relate to the tremendous impact that the jet airplane has had on aviation, especially since its incorporation into military service during World War II. These airplanes were "pure thunder" in the context of their time. Loud and tremendously fast, they exceeded the speed of piston-engined aircraft by more than 150 mph, a dramatic increase in performance over conventional flying. Gutsy fighter pilots of the day jumped at the chance to fly them, but in the field of aerobatics, only a handful of names come to light. Vince Gordon, the man most responsible for the USAF's jet aerobatic team in the years following WWII, was a P-51 pilot who distinguished himself in combat over Europe with four confirmed victories and five probables against the Luftwaffe. Vince picks up the story:

"We were all returning from the war when Col. Tex Hill, the famous "Flying Tiger" pilot with General Chennault, heard that I was assigned to civil engineering duties and told me to get out of "this engineering thing" and get back into fighters, the first jets.

"I checked out in the P-80 in the same fighter outfit that had several aces, Robin Olds and Pappy Herbst, to name a few. These guys were great, and were among the first fighter pilots to get into the jets. To say the least, it was quite an honor for me to be among such a legendary group. Although I was not permanently assigned to the First Fighter Group, I flew actively in the organization and was fortunate enough to gain status as Chief Test Pilot of the major air depot repair facility at March Field, California. During this period. I was privileged to flight-test all military models that we would get in there, especially the jets. It was a responsible job I truly enjoyed.

"One of the highlights of being there was to observe Robin Olds and Pappy Herbst perform a spectacular maneuver whereby they would fly down the runway in their P-80s in the opposite direction of normal landing traffic and pull up into a formation loop right off the deck. After completing the loop, they would continue to climb out in formation, roll over and "split-S" into their final approach to landing. I'm telling you, it was a most impressive maneuver."

However, flying jet fighters can be dangerous work, especially low-level aerobatics. Safety and perfection of the routines had to remain a priority. Vince explains: "During my very active flight test role, I would often search for a cloud layer somewhere in the sky and use the top of it as my test bed for executing various maneuvers, which provided an opportunity to develop an aerial sequence. Then, when I would come down on the deck, I could skillfully and carefully play the g-forces safely. Down low, there is no room for error. I was perfecting my routine so that from the ground it looked really exciting."

Due to Captain Gordon's success, he was selected to perform at a few airshows in the western United States, including a major event at Lowry Field, Colorado in June of 1946. The local press covered his story and expounded on his background. In the manner of a war hero, Vince was known to make quite a display for his family, friends and other spectators as he arrived for the local air show.

He always obtained authorization to display a brief demonstration before his final approach to the runway in preparation for the following day's formal event.

Critics of the new military jets, including the civilian press, were making waves, however, suggesting that the new aircraft had no future. While this was happening, pilots like Vince were racking up hours in the new airplanes and proving their combat worth. As his career progressed, Gordon soon found a new assignment back in Europe, this time at Furstenfeldbruck, Germany, a former German fighter base outside Munich, which Vince was certain he had strafed during the war.

"In Germany again, I quickly became deeply involved in fighter group flying activity. The quality of pilots in this organization was outstanding and my role, in addition to the primary combat mission, was to represent USAFE at solo jet flying exhibitions and air shows throughout Europe."

This activity continued throughout the summer of 1949 and, during this period, the 36th Fighter Group (22nd, 23rd and 53rd Squadrons) was involved in frequent combat-ready exercises with the RAF as well as gunnery meets on the island of Malta. Among NATO pilots, it was a highly competitive and invigorating atmosphere and Vince recalls an eight g-force tight double loop right off the deck at the gunnery meet at Malta. It seems that this playful activity led to the beginning of precision acrobatic flying, and an acrobatic flight eventually was approved within the 36th Fighter Group.

"Flight Commanders Captains Fogle, Brink and Evans, Lieutenants Buck and Bill Pattillo (both of whom flew combat in WWII with the 352nd Fighter Group) and I participated in various routine squadron formations on return from our gunnery missions. Returning to our home base in Germany, enthusiasm grew regarding the possibility of a four-ship formation team, and Harry Evans and I spent a lot of time discussing it."

Meanwhile, Captain Gordon continued to fly several solo acrobatic demonstrations throughout the summer. During the RAF's Occupation Day Display at RAF Station Gutersloh, Germany in August, 1949, Lt. Col. Douglas led twelve F-80s across the field in a tight formation, high-speed pass. Then, the 36th's acrobatic flight of four aircraft peeled-off to execute a Cuban-Eight in tight formation followed by a neat solo aerobatic sequence piloted by Captain Gordon.

Although still unofficial, Gordon, Harry Evans and other flight commanders, along with the Pattillo twins and Jim Brink, continued practicing a three-ship formation. Vince often took up the slot position when not performing air shows that summer. Then, as to be expected, serious conversation followed about forming a four-ship acrobatic team.

However, they realized they had to follow three major steps: first, to obtain higher command approval; second, to select individual team members who could qualify, and assign them positions in the formation; and, third, to prepare a showcase aerial routine, requiring practice, practice and more practice, keeping safety in mind at all times.

"Initially, it was discussed that Evans would lead, with the Pattillo twins on the wing positions, and I would take the slot. And, of course, I was still scheduled to perform solo demonstrations,"

This was near the end of the 1949 air show season. With winter approaching, the 22nd Fighter Squadron became heavily involved in combat effectiveness mainly to keep the Russians in check, Plans for the formation of a team continued through the winter of 1949-50. The Pattillos returned to the States for a short time to participate in the March, 1950 gunnery meet at Lcs Vegas, Nevada, and in late March, Captain Harry Evans became ill and was hospitalized, resulting in his being removed from flying status for a time.

Further plans for an official acrobatic team were now on the shoulders of Captain Gordon. "The Wing Commander approached me and said, `Hey, you know you are scheduled to put on a solo aerial demonstration in an F-80 at the upcoming International Air Show at Orly Field, Paris. Do you think we can prepare a team to compete with the other teams there?' Naturally, I said with much enthusiasm, `Yes, Sir!' I was the temporary Squadron Commander at the time.

"Much discussion and planning followed. Since I was now leading the team, the first order of business was to select a slot man. Recommended to me by the group was Lt. Larry Damewood. I then scheduled a flight-check with Damewood. a newcomer, and found him highly proficient. He had the job as far as I was concerned, and our team began to take shape."

Sometime in late 1949 or early 1950, while on one of his solo acrobatic missions in Great Britain, Vince witnessed the RAF Meteor team (most likely 263 Squadron, Horsham St. Faith) in action. This team impressed him with a formation "string-loop" to begin their routine and a well-executed "bomb-burst" during the show. "Introducing the string-loop entry and the bomb-burst to my team, we practiced first in a two-ship formation and arranged effective signals. Practicing as a full unit followed, with flight safety being the foremost consideration for the next several weeks. Training became more and more rigorous on each mission, and while this was going on, we had to maintain our regular squadron duties.

"We still did not have a nickname for the team, and several days were spent tossing around various ideas. 'Sky-something' was a must and, as we were clustered together kneeling on the parking ramp after a training mission, we agreed then and there to be known as the 'Skyblazers.'

"We kept practicing our routine and at that time expanded the British 'Orange-Blossom' to include a low-level crossover. This additional maneuver, after the bomb-burst, consisted of our four aircraft split-Sing and closing in at maximum speed from the four cardinal points of the compass, so as to converge on a central point in the middle of the airfield, right before the audience at low-level, followed by a join-up for landing.

"We would blend in barrel-rolls, reverse-backs, or whiffer dills, and other tight maneuvers almost at eye level. As our show opener, we would be in trail or string formation, rising up from the deck into a loop with a diamond formation join-up at the top. Detailed briefing on safety procedures was always paramount before each mission. Putting it all together produced roughly a seventeen minute spectacular. The detailed sequence of maneuvers executed by the Skyblazers have been employed by almost all of the succeeding teams, including the Sabre Knights, Thunderbirds, Blue Angels, and many of the foreign teams. This is truly where and when it all began with the jets."

With their new name and official recognition, the Skyblazers were issued standard F-80s with the team name emblazoned across both sides of the nose. The pilots' helmets were painted a bright crimson with white lightning flashes and each pilot's name was hand-painted just above the visor by the squadron artist. They also received bright red jump suits to finish off the overall image. With USAFE approval, the 36th Fighter Group's Skyblazers became the Official Aerobatic Demonstration Team for United States Air Forces in Europe.

A warm-up show was scheduled for RAF West Raynam, in Great Britain. On June 5th and 6th, 1950, the Skyblazers made their official debut before the British public. They were a huge success and the news media began to report on the new American team. It was now hyped for the big event at Orly Field, Paris, held during June 1950. Vince recalls, "We were ready for Orly, and needless to say, we were a success. I remember the article that appeared in the newspapers: SKYBLAZERS STEAL THE SHOW! People loved our performances. I will never forget some spectators saying that after a very low pass the exhaust from our tail pipes sent the Soviet flag fluttering as it was displayed among the flags of all the participating countries."

The cornerstone of the Skyblazers was now firmly set. The team became very busy in the European air show circuit. Shows were flown in England, Germany, Denmark, the Netherlands, France, Belgium and North Africa. Their performances demonstrated precision flying and pilot skill, thrilling enormous crowds, including military and civilian dignitaries.

"I will always marvel at one particular performance," Vince remembers. "When show time arrived, the weather was almost too marginal for aerial demonstration. The large crowd was obviously disappointed. However, we started engines and taxied out with the intention of just executing a low fly-by. Enjoying complete confidence in the skill of my teammates, upon becoming airborne and on a low-level approach before the crowd, I announced to my team, "Here goes a loop." Upward we disappeared into the overcast, half of the loop in the clouds, and on the backside of the loop, we broke through the overcast, reappearing before the crowd on the cardinal heading of the runway. The audience was overjoyed!

"Looking back, I am very proud of the fact that during my tenure, not one loss or accident occurred with either of my acrobatic teams. I flew with wonderful guys, and we are still close friends to this day. I also must elaborate on the truly professional and dedicated maintenance crews, the unsung heroes who spent many laborious hours ensuring that our aircraft were ready to go. We suffered not one aborted mission, and I take my hat off to those fellows. Without them, it could not have been a success story."

One British writer of the day describes Vince's solo performances at an air show at Ypenburg, Netherlands on July 28, 1950: "The solo demonstration by the Skyblazers' Captain Vince Gordon was literally breathtaking. He is probably the only man who can boast of having flown at 600 mph below sea level, for he flashed past 10 feet from the grass, and Ypenburg Airfield lies 15 feet below sea level!"

Gordon continued as the leader of the Skyblazers throughout the summer of 1950. In late August, Captain Harry Evans rejoined the team and became the new leader, and Vince continued as the team's lone solo pilot, thus handing over to the new leader a magnificently well-trained acrobatic team that could compete with anybody in the world.

"Sweet and Lovely"

The sun sets in the spring of 1945 as Lt. Cuthbert A. "Bill" Pattillo banks his P-51D (HO-O 44-13905) and heads for home. (Courtesy Troy White, Stardust Studios)

IN APPRECIATION

This book was born in the cauldron of war with the advent of Pearl Harbor some 66 years ago, remaining dormant until the year 2003 when three men who had realized how little our children and grandchildren know of the sacrifices made for the freedom of America and the world in which they live today. We three – Marc Hamel, Sam Sox and I – combined our limited talents and resources to bring this book into being, not for personal profit, not for acclaim, but as a labor of love.

But it is not solely the product of the three of us. We only recorded it in words and pictures. It came from the contributions of all those who served in the 352nd Fighter Group and the 1st Service Group and from their families; the fingerprints of numerous authors and historians are on it via marvelous stories that would have faded away in time without their contributions. We thank all of those who joined us in perpetuating the history and achievements of this outstanding fighter group.

We are particularly indebted to Joe Noah and his Preddy Memorial Foundation for their support; to aviation artists Troy White and Susan Ward, who painted and designed our cover; to Troy, Nick Trudgian, Robert Bailey, John Doughty, and Charles Taylor, for use of their paintings; to authors Tom Ivie, Bill Hess, Merle Olmsted, Paul McCue, William Neely and others for use of their works; to our 352nd veterans who told us their stories and made their personal photo albums available to us; to Todd and Alicia Gehrke for their many contributions; to Vally Sharpe, Jan Lowe and Rhonda Powell for their professional help in publishing this book; and last and most importantly, to Betty Noah, Sarah Sox, Betty Powell and Elizabeth Lipscomb for their patience with us during the more than three years we were producing this book. To all of the above, we appreciate you more than you will ever know. Without you this book would never have come to light. We thank you and we salute you! May you always enjoy Blue Skies!

L-R: Marc Hamel, "Punchy" Powell, & Sam Sox, Jr.